UNDERSTANDING GEOMETRIC ALGEBRA

Hamilton, Grassmann, and Clifford for Computer Vision and Graphics

Kenichi Kanatani

Okayama University
Japan

CRC Press
Taylor & Francis Group
Boca Raton London New York

CRC Press is an imprint of the
Taylor & Francis Group, an **informa** business
A CHAPMAN & HALL BOOK

CRC Press
Taylor & Francis Group
6000 Broken Sound Parkway NW, Suite 300
Boca Raton, FL 33487-2742

First issued in paperback 2020

Version Date: 20150217

ISBN 13: 978-0-367-57582-3 (pbk)
ISBN 13: 978-1-4822-5950-6 (hbk)

Visit the Taylor & Francis Web site at
http://www.taylorandfrancis.com

and the CRC Press Web site at
http://www.crcpress.com

Contents

List of Figures

List of Tables

Preface

The aim of this book is to introduce *geometric algebra*, which recently has been attracting attention in various domains of physics and engineering. For better understanding, emphasis is on the background mathematics, including Hamilton algebra, Grassmann algebra, and Clifford algebra. These 19th century mathematics were almost forgotten in the 20th century, where the means for describing and computing geometry was mainly replaced by vector and tensor calculus and matrix computation based on linear algebra. In the late 20th century and the early 21st century, however, scientists began to reevaluate these old mathematics in a new light. A central figure among them is David Hestenes, an American physicist, who called his formulation "geometric algebra" and actively disseminated it. This made a big impact, not only on physicists but also on researchers of engineering applications, including robotic arm control, computer graphics, and computer vision. It is expected that this book will serve a wide range of people engaged in 3D modeling for computer graphics and computer vision.

This book is intended mainly for engineering students higher at undergraduate and graduate levels and for general researchers, but Chapter 2 describes the basis of 3D geometry usually taught in first- or second-year undergraduate courses of all science and engineering departments. The description is detailed and fully comprehensive, so that this chapter can serve as useful material for classes and exercises at lower undergraduate levels. In fact, this chapter alone is almost sufficient for 3D computation in most real applications, including computer graphics and computer vision.

The author obtained extensive knowledge of classical geometry in the 1970s when he was a graduate student of the (then) Department of Mathematical Engineering and Instrumentation Physics, the University of Tokyo, Japan, supervised by (late) Professor Nobunori Oshima. The author still treasures to this day the thick volume of *Lecture Notes on Geometric Mathematical Engineering* of Professor Oshima, to whom the author dedicates this book. He thanks Professor Shun-ichi Amari (currently at RIKEN, Japan), from whom the author has received continuing guidance since his student days at the University of Tokyo. The author also thanks many of his colleagues, including Leo Dorst of the University of Amsterdam, the Netherlands; Eduardo Bayro-Corrochano of CINVESTAV, Mexico; Vincent Nozick of Université Paris-Est, France; Wojciech Chojnacki of the University of Adelaide, Australia; and Reinhard Klette of the University of Auckland, New Zealand, for reading the raw manuscript of this book and giving helpful comments and suggestions.

Introduction

This chapter states the purpose and organization of this book and describes various features of the volume.

1.1 PURPOSE OF THIS BOOK

The aim of this book is to introduce *geometric algebra*, which recently has been attracting attention in various domains of physics and engineering. For beginners, however, describing it directly is rather difficult, so this book takes an indirect approach, which is somewhat different from the standard style. There already exist many textbooks on geometric algebra [2, 3, 4, 5, 12, 16], most of which start by stating "what geometric algebra is," defining numerous symbols and terminologies, and listing fundamental identities and relationships among various quantities. This often gives the impression of reading a formula book, so that beginners tend to shy away from going on further. This is attributed to the fact that geometric algebra has been established through a long history of mathematics. In view of this, this book first gives separate descriptions of various algebras that constitute the background and then shows how they are finally combined to define geometric algebra.

Algebra is a study of operations on symbols. It began in ancient times in an attempt to solve algebraic equations. In the 19th century, techniques for describing geometry in algebraic terms developed, and such unique algebras as the Hamilton, the Grassmann, and the Clifford algebras were born. In the 20th century, however, they were almost forgotten, and the means for describing and computing geometry was mainly replaced by vector and tensor calculus and matrix computation based on linear algebra. In the late 20th century and the early 21st century, however, scientists began to reevaluate old algebras in a new light. A central figure among them is David Hestenes, an American physicist, who called his formulation "geometric algebra" and actively disseminated it. This made a big impact, not only on physicists but also on researchers of engineering applications, including robotic arm control, computer graphics, and computer vision. The background of its spread is the development of software tools for automatically executing algebraic operations that are systematic but tedious for humans. Today, various tools are offered for executing geometric algebra. Thanks to them, users only need to input data for doing complicated geometric computations without the knowledge of the mathematical theories behind them. However, knowing the background would surely bring about a deeper insight and interest in the theory. To provide such background knowledge is the main goal of this book.

the Hamilton, the Grassmann, and the Clifford algebras in turn but not necessarily i historical order. Also, the description is not necessarily in their original formats, whic would be difficult to understand for today's readers; all materials are reorganized to refle the current point of view.

In Chapter 2, 3D Euclidean geometry is described as a preliminary to the subseque chapters based on today's vector calculus. However, the description is somewhat differe from the usual textbooks in that *we do not use linear algebra*. Today, a "vector" is identifie with a column of numbers to which a matrix is multiplied from the left to produce vario results. In contrast, a *vector* in this book is a "symbol" to represent a geometric obje equipped with direction and magnitude, so *it cannot be multiplied by a matrix*. The mai reason for adopting this formulation is to emphasize the fundamental principle of algeb of *defining operations on symbols*, which underlies all algebras discussed subsequently, in cluding Hamilton, Grassmann, and Clifford algebras. Another reason is the fact that line algebra was not established in today's form in the 19th century when Hamilton, Grassman and Clifford introduced their algebras.

Chapter 3 discusses how the descriptions of geometry should be altered if we use non-orthogonal, or *oblique*, coordinate system. The use of an orthogonal, or *Cartesia* coordinate system makes geometric description very simple, so all algebras in this book a defined with respect to an orthogonal coordinate system. This may give a wrong impressio that algebraic methods are valid only for orthogonal coordinate systems. To avoid suc a misunderstanding, it is shown in Chapter 3 that geometric description is possible eve for non-orthogonal coordinate systems with the introduction of a quantity called the *metr tensor*. It is also shown that changes of the description between different coordinate system are systematically specified by a simple transformation rule. The principle stated in th chapter is the basis of a 20th century mathematics called *tensor calculus*. However, sinc the orthogonal coordinate system is used throughout the subsequent chapters for the sak of simplicity, those readers who want to know the idea of geometric algebra quickly can ski this chapter.

Chapter 4 describes Hamilton's quaternion algebra. It is a typical algebraic approach o defining operations on symbols, providing a basis of geometric algebra. In particular, th idea of "sandwiching" a vector by an operator and its conjugate from the left and righ unlike multiplying a column vector by a matrix from the left as in today's linear algebr is the core of geometric algebra. To make a distinction from the standard operator to b acted from the left, this sandwiching operator is termed a *versor*.

Chapter 5 describes Grassmann's outer product algebra. This chapter is relatively lon because the outer product is one of the most fundamental operations of geometric algebr In particular, the fundamental idea of *duality* is described in full detail here.

Chapter 6 introduces the Clifford algebra, which underlies the mathematical structu of geometric algebra. There, three operations are used: one is the standard inner product o vectors; another is the outer product of the Grassmann algebra; in addition, a new operatio called *geometric product* (or *Clifford product*) is introduced. It is shown that the inner an outer products are defined in terms of the geometric product. In this sense, the geometri product is regarded as the most fundamental operation. William K. Clifford, an Englis mathematician, combined Hamilton's quaternion algebra and Grassmann's outer produc algebra to define the general Clifford algebra. Meanwhile, Josiah W. Gibbs, an America physicist, simplified the Hamilton and the Grassmann algebras to the minimum necessar components for describing physics, reducing the basic vector operations to the inner, th geometric relations in 3D can be described by it. The description of 3D Euclidean geometry in Chapter 2 follows this formulation. Overshadowed by the success of vector calculus, the sophisticated Clifford algebra has been almost forgotten except among a small number of mathematicians. Today's geometric algebra is a rediscovery of the Clifford algebra.

Chapter 7 describes points and lines in 3D as objects in 4D. This corresponds to what is known as *projective geometry* using *homogeneous coordinates*. However, the formulation here is somewhat different. In standard projective geometry, the homogeneous coordinates are a quadruplet of numbers with the interpretation that they represent a point at infinity when the last coordinate is 0. In this book, a new symbol e_0 is introduced in addition to the basis $\{e_1, e_1, e_3\}$ of the 3D space, and a Grassmann algebra is defined for these four symbols. This formulation was introduced by Grassmann himself, but 20th century mathematicians gave an elegant formulation as the *Grassmann–Cayley algebra* based on linear algebra and tensor calculus, which are also 20th century mathematics. In this book, however, we do not go into the details of this formulation, merely stating basic ideas. The most fundamental components are the *Plücker coordinates* for representing lines and planes and the *duality theorems* concerning the *joins* and *meets* of points, lines, and planes.

Chapter 8 describes *conformal geometry*, which is the main ingredient of what is now called *geometric algebra*. Here, a new symbol e_∞ is introduced to the 4D space defined in Chapter 7. The resulting 5D space is *non-Euclidean* in the sense that it has a non-positive-definite metric that allows the square norm to be negative; 3D Euclidean geometry is realized in that 5D non-Euclidean space. Next, the Grassmann and Clifford algebras are defined in this 5D space, and *conformal transformations* in 3D, which include translations, rotations, reflections, and dilations, are described in terms of *versors* in 5D. Here, circles and spheres are the most fundamental geometric objects; lines and planes are interpreted to be circles and spheres of infinite radius, respectively, and translations are regarded as rotations around axes placed infinitely far away.

While Chapters 2–7 mainly deal with geometry of lines and planes, we consider, in Chapter 9, camera imaging geometry involving circles and spheres. We start with the ordinary perspective projection camera and then describe the imaging geometry of *fisheye lens* cameras and *omnidirectional* (or *catadioptric*) cameras using a parabolic mirror. It is shown that inversion, which is a special conformal transformation, plays an essential role there. We also describe the imaging geometry of omnidirectional cameras using hyperbolic and parabolic mirrors. The mathematical analysis here is not only interesting in its own right but also closely related to computer vision and autonomous robot applications, because fisheye lenses and omnidirectional cameras have come into practical use more and more today as their prices fall. However, not many textbooks describe such cameras yet, so the description in this chapter is expected to be helpful to many readers.

1.3 OTHER FEATURES

Throughout this book, only 3D geometry is considered. The vector calculus established by Gibbs, which corresponds to the formulation in Chapter 2, is based on inner, vector, and scalar triple products, which have meaning only in 3D. Also, Hamilton's quaternion algebra, described in Chapter 4, only deals with geometry in 3D. In contrast, the Grassmann and Clifford algebras and conformal geometry discussed after Chapter 4 can be straightforwardly extended to general nD. However, description in general nD becomes slightly complicated.

The main aim of this book is the introduction of geometric algebra, which may gi an impression that it has little to do with "today's" mathematics, such as linear algeb and tensor calculus. In order to describe the relationships with traditional mathematic the column entitled *Traditional World* is inserted in many places in all chapters. Throug them, readers can also learn various aspects of today's mathematics, including topolog projective geometry, and group representations. The term "traditional" does not mean th it is historically old; it simply means "20th century mathematics." In contrast, geometr algebra is a 19th century mathematics in its origin and a 21st century mathematics in i development.

In this book, different chapters deal with different geometries and algebras, and relat mathematical concepts and terminologies are separately defined in each chapter. Most them are common to all chapters, so at first sight they appear to be redundant. However, should be kept in mind that the concepts and terminologies defined in each chapter basical concern the geometry or algebra in that chapter.

At the end of each chapter is a section called *Supplemental Note*, which describes histo ical developments of the topics in that chapter and related fields of mathematics. Recon mended references are also given there, but original documents of mere historical intere are not listed. Also, a section called *Exercises* that complements the main points of th chapter and omitted derivations are given. The answers are provided at the end of the boo

3D Euclidean Geometry

3D Euclidean geometry is usually described in terms of numerical vectors, i.e., column and row vectors. In this chapter, a vector is a geometric object equipped with direction and magnitude, not an array of numbers. Here, we present an "algebraic" description of 3D Euclidean geometry, representing objects by "symbols" and defining "operations" on them. From this viewpoint, we introduce the inner, the outer, and the scalar triple products of vectors. Then, we list expressions and relationships involving rotations, projections, lines, and planes. This chapter provides a basis of all the subsequent chapters.

2.1 VECTORS

A *vector* is a geometric object that has "direction" and "magnitude"; one can imagine it as an "arrow" in space. Vectors represent the following quantities and properties:

Displacements, velocities, and force

> Vectors specify in which direction, over what distance, and at what velocity things move or what force is acting. We are interested only in their directions and magnitudes; we are not concerned with the location of the starting point. Such vectors are called *free vectors*.

Directions in space

> Vectors indicate orientations of lines and surface normals to planes. Only directions are important; magnitudes are ignored. Such vectors are called *direction vectors*. Since their magnitudes are irrelevant, they are usually multiplied by appropriate numbers to *unit vectors* with unit magnitude.

Positions in space

> We fix a special point O, called the *origin*, and define the positions of points by their displacements from the origin O. Such vectors are called *position vectors*. Vectors whose starting points are specified are said to be *bound*. Positions vectors are bound to the origin O.

The study of using and manipulating "vectors" for such different quantities and relationships is called *vector calculus*, providing the basis of geometry, classical mechanics, and electrodynamics. In later chapters, the above items will be regarded as different geometric objects, but no distinction is made in this chapter.

We define scalar multiplication and addition of vectors as follows. Multiplication of a vector $\boldsymbol{a}$ by a real number α, written as $\alpha\boldsymbol{a}$, indicates a vector with the same direction as $\boldsymbol{a}$

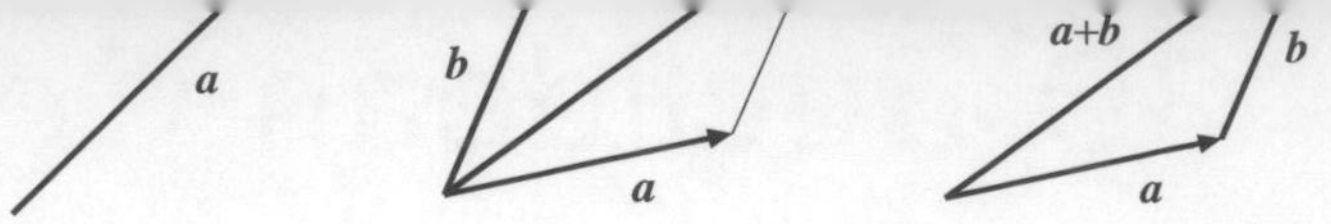

FIGURE 2.1 Scalar multiplication of vector $\boldsymbol{a}$ (a) and the sum of vectors $\boldsymbol{a}$ and $\boldsymbol{b}$ (b), (c).

with a magnitude α times as large (Fig. 2.1(a)). If α is negative, it is interpreted to me the opposite direction with magnitude $|\alpha|$. We simply write $-\boldsymbol{a}$ for $(-1)\boldsymbol{a}$, which is tl vector $\boldsymbol{a}$ with its direction reversed. We call real numbers for multiplying vectors *scala* to distinguish them from vectors. In the following, we use lower case bold face letters $\boldsymbol{a}$, $\boldsymbol{c}$, ... for vectors and Greek letters α, β, γ, ... for scalars. Any vector multiplied by 0 written simply as 0. We also write $\overrightarrow{AB}$ for a bound vector with a specified starting point and an endpoint B.

The *sum* of vectors $\boldsymbol{a}$ and $\boldsymbol{b}$ is the vector constituting the diagonal of the parallelogra defined by $\boldsymbol{a}$ and $\boldsymbol{b}$ after making their starting points coincide (Fig. 2.1(b)). It is als interpreted to be the displacement from the starting point of $\boldsymbol{a}$ to the endpoint of $\boldsymbol{b}$ aft making the endpoint of $\boldsymbol{a}$ and the starting point of $\boldsymbol{b}$ coincide (Fig. 2.1(c)). The *differen* $\boldsymbol{a} - \boldsymbol{b}$ means $\boldsymbol{a} + (-\boldsymbol{b})$. The following rules hold for scalar multiplication and addition vectors as in the the usual arithmetic:

commutativity: $\boldsymbol{a} + \boldsymbol{b} = \boldsymbol{b} + \boldsymbol{a}$.

associativity: $(\boldsymbol{a} + \boldsymbol{b}) + \boldsymbol{c} = \boldsymbol{a} + (\boldsymbol{b} + \boldsymbol{c})$.

distributivity: $(\alpha + \beta)\boldsymbol{a} = \alpha\boldsymbol{a} + \beta\boldsymbol{a}, \quad \alpha(\boldsymbol{a} + \boldsymbol{b}) = \alpha\boldsymbol{a} + \alpha\boldsymbol{b}$.

2.2 BASIS AND COMPONENTS

We fix a point O, called the *origin*, in space and define a Cartesian xyz coordinate systen Let e_1, e_2, and e_3 be the vectors along the x-, y-, and z-axes, respectively, in their positiv directions (Fig. 2.2). They represent directions only; their starting points are irrelevant. Tl triplet $\{e_1, e_2, e_3\}$ is called the *basis* of the coordinate system in the sense that any vect $\boldsymbol{a}$ is expressed as their linear combination in the form

$$\boldsymbol{a} = a_1 e_1 + a_2 e_2 + a_3 e_3. \quad (2.1$$

The numbers a_1, a_2, and a_3 are called the *components* of $\boldsymbol{a}$ with respect to this basis. If w multiply $\boldsymbol{a}$ by a scalar α, the distributivity of scalar multiplication implies

$$\alpha\boldsymbol{a} = \alpha a_1 e_1 + \alpha a_2 e_2 + \alpha a_3 e_3. \quad (2.2$$

If we consider another vector $\boldsymbol{b} = b_1 e_1 + b_2 e_2 + b_3 e_3$, the commutativity and the associativit of vector addition imply

$$\boldsymbol{a} + \boldsymbol{b} = (a_1 + b_1)e_1 + (a_2 + b_2)e_2 + (a_3 + b_3)e_3. \quad (2.3$$

Thus, we obtain

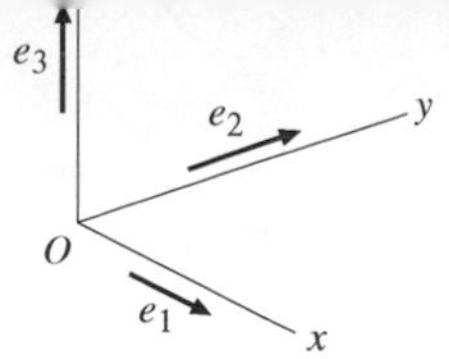

FIGURE 2.2 The basis $\{e_1, e_2, e_3\}$ of a Cartesian xyz coordinate system.

Proposition 2.1 (Components of scalar multiplication and sum) *If vector* $\boldsymbol{a}$ *has components* a_1, a_2, *and* a_3 *and vector* $\boldsymbol{b}$ *components* b_1, b_2, *and* b_3, *scalar multiplication* $\alpha\boldsymbol{a}$ *has components* αa_1, αa_2, *and* αa_3, *and the sum* $\boldsymbol{a}+\boldsymbol{b}$ *has components* a_1+b_1, a_2+b_2, *and* a_3+b_3.

Traditional World 2.1 (Numerical vectors) Traditional treatment of 3D space is called *analytical geometry*: a coordinate system is fixed in space, and vectors are analyzed by means of *linear algebra*, where vectors are vertical arrays of numbers (*column vectors*) or horizontal arrays of numbers (*row vectors*). In this book, however, vectors are not such arrays of numbers. Rather, they are geometric objects equipped with direction and magnitude, as defined in Section 2.1; they can be imagined as "arrows." We follow the convention of writing vectors in boldface, like $\boldsymbol{a}$, but they are merely "symbols" that represent vectors; they are not arrays of numbers.

Traditionally, the array of 0's is written as $\mathbf{0}$ and called the *null vector*, but we use numeral 0 for a vector of magnitude 0. This is for emphasizing the fact that it is not an array of 0's but merely a symbol that denotes "non-existence." Similarly, the basis in traditional treatment is a set of vectors starting from the origin O, written in boldface like $\boldsymbol{e}_1$, $\boldsymbol{e}_2$, and $\boldsymbol{e}_3$, which represent arrays of 1's and 0's; e.g., $\boldsymbol{e}_1$ is an array of 1, 0, and 0. In this book, however, we write e_1, e_2, and e_3 to emphasize the fact that they are merely symbols that represent directions; their starting positions are irrelevant.

These may appear to be insignificant differences. In fact, we would obtain the same results, as far as this chapter is concerned, if vectors are viewed as arrays of numbers. However, the departure from the traditional treatment will be clearer as we proceed to subsequent chapters. Our treatment is *algebraic* in the sense that we define "operations on symbols."

2.3 INNER PRODUCT AND NORM

Assignment of a real value to a pair of vectors $\boldsymbol{a}$ and $\boldsymbol{b}$, written as $\langle \boldsymbol{a}, \boldsymbol{b}\rangle$, is called their *inner product* if the following conditions are satisfied:

Positivity: $\langle \boldsymbol{a}, \boldsymbol{a}\rangle \geq 0$ with equality holding for $\boldsymbol{a} = 0$.

Symmetry: $\langle \boldsymbol{a}, \boldsymbol{b}\rangle = \langle \boldsymbol{b}, \boldsymbol{a}\rangle$.

Linearity: $\langle \boldsymbol{a}, \alpha\boldsymbol{b} + \beta\boldsymbol{c}\rangle = \alpha\langle \boldsymbol{a}, \boldsymbol{b}\rangle + \beta\langle \boldsymbol{a}, \boldsymbol{c}\rangle$.

Since the returned value is a scalar, not a vector, it is also called the *scalar product*. Some authors write $\boldsymbol{a}\cdot\boldsymbol{b}$ and call it the *dot product* of $\boldsymbol{a}$ and $\boldsymbol{b}$.

and use this as the measure of the magnitude of the vector $\boldsymbol{a}$. By the above positivity, the inside of the square root is nonnegative. Hence, $\|\boldsymbol{a}\| = 0$ holds if and only if $\boldsymbol{a} = 0$.

If we express vectors in terms of the basis as in Eq. (2.1), the symmetry and the linearity imply that computation of inner products reduces to the inner products among e_1, e_2, and e_3. If they are mutually orthogonal unit vectors, we say that the basis is *orthonormal*. In that case, their inner products are set as follows:

$$\langle e_1, e_1\rangle = \langle e_2, e_2\rangle = \langle e_3, e_3\rangle = 1, \qquad \langle e_1, e_2\rangle = \langle e_2, e_3\rangle = \langle e_3, e_1\rangle = 0. \tag{2.5}$$

This can be briefly written as

$$\langle e_i, e_j\rangle = \delta_{ij}, \tag{2.6}$$

where the symbol δ_{ij}, called the *Kronecker delta*, takes value 1 for $i = j$ and 0 otherwise.

If vectors $\boldsymbol{a}$ and $\boldsymbol{b}$ are expressed in terms of the basis as in Eq. (2.1), their inner product is computed using the symmetry, the linearity, and Eq. (2.5) as follows:

$$\begin{aligned}
\langle \boldsymbol{a}, \boldsymbol{b}\rangle &= \langle a_1e_1 + a_2e_2 + a_3e_3, b_1e_1 + b_2e_2 + b_3e_3\rangle \\
&= a_1b_1\langle e_1, e_1\rangle + a_1b_2\langle e_1, e_2\rangle + a_1b_3\langle e_1, e_3\rangle + a_2b_1\langle e_2, e_1\rangle + a_2b_2\langle e_2, e_2\rangle \\
&\quad + a_2b_3\langle e_2, e_3\rangle + a_3b_1\langle e_3, e_1\rangle + a_3b_2\langle e_3, e_2\rangle + a_3b_3\langle e_3, e_3\rangle \\
&= a_1b_1 + a_2b_2 + a_3b_3.
\end{aligned} \tag{2.7}$$

In particular, we obtain $\|\boldsymbol{a}\|^2 = a_1^2 + a_2^2 + a_3^2$ by letting $\boldsymbol{b} = \boldsymbol{a}$. Thus, we have

Proposition 2.2 (Inner product and norm of vectors) *If vector $\boldsymbol{a}$ has components a_1, a_2, and a_3 and vector $\boldsymbol{b}$ components b_1, b_2, and b_3, the inner product of $\boldsymbol{a}$ and $\boldsymbol{b}$ is*

$$\langle \boldsymbol{a}, \boldsymbol{b}\rangle = a_1b_1 + a_2b_2 + a_3b_3, \tag{2.8}$$

and the norm of $\boldsymbol{a}$ is

$$\|\boldsymbol{a}\| = \sqrt{a_1^2 + a_2^2 + a_3^2}. \tag{2.9}$$

The following relationships hold for the inner product and the norm ($\hookrightarrow$ Exercises 2.1, 2.2, and 2.3):

Proposition 2.3 (Inner product and angle) *If vectors $\boldsymbol{a}$ and $\boldsymbol{b}$ make an angle θ, then*

$$\langle \boldsymbol{a}, \boldsymbol{b}\rangle = \|\boldsymbol{a}\|\|\boldsymbol{b}\|\cos\theta. \tag{2.10}$$

In particular, vectors $\boldsymbol{a}$ and $\boldsymbol{b}$ are orthogonal if and only if

$$\langle \boldsymbol{a}, \boldsymbol{b}\rangle = 0. \tag{2.11}$$

Proposition 2.4 (Schwartz inequality and triangle inequality) *For vectors $\boldsymbol{a}$ and $\boldsymbol{b}$, the following inequalities hold:*

$$-\|\boldsymbol{a}\|\|\boldsymbol{b}\| \leq \langle \boldsymbol{a}, \boldsymbol{b}\rangle \leq \|\boldsymbol{a}\|\|\boldsymbol{b}\|, \tag{2.12}$$

$$\|\boldsymbol{a} + \boldsymbol{b}\| \leq \|\boldsymbol{a}\| + \|\boldsymbol{b}\|. \tag{2.13}$$

For both, equality holds when $\boldsymbol{a} = \alpha\boldsymbol{b}$ for some scalar α or one of $\boldsymbol{a}$ and $\boldsymbol{b}$ is 0.

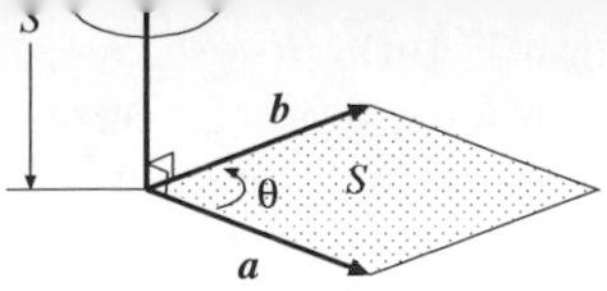

FIGURE 2.3 The vector product $\boldsymbol{a} \times \boldsymbol{b}$ of vectors $\boldsymbol{a}$ and $\boldsymbol{b}$ is orthogonal to both $\boldsymbol{a}$ and $\boldsymbol{b}$ and has the direction a right-handed screw would move if $\boldsymbol{a}$ is rotated toward $\boldsymbol{b}$. Its length equals the area S of the parallelogram defined by $\boldsymbol{a}$ and $\boldsymbol{b}$.

Equations (2.12) and (2.13) are called the *Schwartz inequality* and the *triangle inequality*, respectively.

Traditional World 2.2 (Inner product of numerical vectors) Traditional linear algebra treats vectors as vertical arrays of numbers and defines the inner product by

$$\langle \boldsymbol{a}, \boldsymbol{b} \rangle = \boldsymbol{a}^\top \boldsymbol{b} = \begin{pmatrix} a_1 & a_2 & a_3 \end{pmatrix} \begin{pmatrix} b_1 \\ b_2 \\ a_3 \end{pmatrix} = a_1 b_1 + a_2 b_2 + a_3 b_3, \tag{2.14}$$

where $\top$ denotes the transpose of a column vector into a row vector. In this book, however, "transpose" has no meaning, since vectors are not arrays of numbers. Instead, computation of inner products is reduced to the inner products among the basis vectors e_1, e_2, and e_3. The final results are the same as traditional linear algebra, but we will see the difference in thinking more clearly as we proceed to the subsequent chapters.

2.4 VECTOR PRODUCTS

The *vector product* $\boldsymbol{a} \times \boldsymbol{b}$ of vectors $\boldsymbol{a}$ and $\boldsymbol{b}$ is defined as follows (Fig. 2.3).

- It is orthogonal to both $\boldsymbol{a}$ and $\boldsymbol{b}$ in the direction that a right-handed screw would move if $\boldsymbol{a}$ is rotated toward $\boldsymbol{b}$.
- Its magnitude equals the area of the parallelogram defined by $\boldsymbol{a}$ and $\boldsymbol{b}$ after their starting points are made to coincide.

The term *cross product* is also used for $\boldsymbol{a} \times \boldsymbol{b}$ because of the use of the symbol $\times$. Since $(\alpha \boldsymbol{a}) \times \boldsymbol{b} = \alpha(\boldsymbol{a} \times \boldsymbol{b})$ by definition, parentheses are not necessary. The following holds from the definition:

Proposition 2.5 (Vector product and angle) *If vectors* $\boldsymbol{a}$ *and* $\boldsymbol{b}$ *make an angle* θ, *then*

$$\|\boldsymbol{a} \times \boldsymbol{b}\| = \|\boldsymbol{a}\| \|\boldsymbol{b}\| \sin\theta. \tag{2.15}$$

In particular, vectors $\boldsymbol{a}$ *and* $\boldsymbol{b}$ *are* collinear, *i.e., on the same line when their starting points coincide, if and only if*

$$\boldsymbol{a} \times \boldsymbol{b} = 0. \tag{2.16}$$

The following properties hold:

Antisymmetry: $\boldsymbol{a} \times \boldsymbol{b} = -\boldsymbol{b} \times \boldsymbol{a}$. In particular, $\boldsymbol{a} \times \boldsymbol{a} = 0$.

geometric interpretation (↪ Exercise 2.4).

If vectors are expressed in terms of the basis, computation of vector products reduc using the antisymmetry and the linearity, to the vector products among the basis vecto e_1, e_2, and e_3. From the defining geometric interpretation, we obtain the following rules:

$$
\begin{array}{lll}
e_1 \times e_1 = 0, & e_2 \times e_2 = 0, & e_3 \times e_3 = 0, \\
e_1 \times e_2 = e_3, & e_2 \times e_3 = e_1, & e_3 \times e_1 = e_2, \\
e_2 \times e_1 = -e_3, & e_3 \times e_2 = -e_1, & e_1 \times e_3 = -e_2.
\end{array}
\quad (2.1
$$

If we express vectors $\boldsymbol{a}$ and $\boldsymbol{b}$ as in Eq. (2.1), their vector product is computed, using t antisymmetry, the linearity, and Eq. (2.17) as follows:

$$
\begin{aligned}
\boldsymbol{a} \times \boldsymbol{b} &= (a_1 e_1 + a_2 e_2 + a_3 e_3) \times (b_1 e_1 + b_2 e_2 + b_3 e_3) \\
&= a_1 b_1 e_1 \times e_1 + a_1 b_2 e_1 \times e_2 + a_1 b_3 e_1 \times e_3 + a_2 b_1 e_2 \times e_1 + a_2 b_2 e_2 \times e_2 \\
&\quad + a_2 b_3 e_2 \times e_3 + a_3 b_1 e_3 \times e_1 + a_3 b_2 e_3 \times e_2 + a_3 b_3 e_3 \times e_3 \\
&= (a_2 b_3 - a_3 b_2) e_1 + (a_3 b_1 - a_1 b_3) e_2 + (a_1 b_2 - a_2 b_1) e_3.
\end{aligned}
\quad (2.1
$$

This can be stated in the following form:

Proposition 2.6 (Vector product components) *The vector product of* $\boldsymbol{a} = \sum_{i=1}^{3} a_i$ *and* $\boldsymbol{b} = \sum_{i=1}^{3} b_i e_i$ *is given by*

$$
\boldsymbol{a} \times \boldsymbol{b} = \sum_{i,j,k=1}^{3} \epsilon_{ijk} a_i b_j e_k. \quad (2.1
$$

Here, ϵ_{ijk} is the *permutation signature*, taking value 1 if (i, j, k) is an *even permutatio* of $(1, 2, 3)$ obtained from $(1, 2, 3)$ by swapping two numbers an even number of times, val -1 if (i, j, k) is an *odd permutation* of $(1, 2, 3)$, and value 0 otherwise. The symbol ϵ_{ijk} also called the *Levi–Civita epsilon* or the *Eddington epsilon*.

Associativity does not hold for the vector product, so $(\boldsymbol{a} \times \boldsymbol{b}) \times \boldsymbol{c}$ and $\boldsymbol{a} \times (\boldsymbol{b} \times \boldsymbol{c})$ a not necessarily equal. The product of three vectors in this form is called the *vector trip product*. From Eq. (2.18), we can prove the following identities:

Proposition 2.7 (Vector triple product)

$$
(\boldsymbol{a} \times \boldsymbol{b}) \times \boldsymbol{c} = \langle \boldsymbol{a}, \boldsymbol{c} \rangle \boldsymbol{b} - \langle \boldsymbol{b}, \boldsymbol{c} \rangle \boldsymbol{a}, \qquad \boldsymbol{a} \times (\boldsymbol{b} \times \boldsymbol{c}) = \langle \boldsymbol{a}, \boldsymbol{c} \rangle \boldsymbol{b} - \langle \boldsymbol{a}, \boldsymbol{b} \rangle \boldsymbol{c}. \quad (2.2
$$

Traditional World 2.3 (Tensor calculus) Traditionally, systematic analysis of vect components is done by *tensor calculus*. For instance, the vector products among the bas vectors e_1, e_2, and e_3 in Eq. (2.17) are written as

$$
e_i \times e_j = \sum_{k=1}^{3} \epsilon_{ijk} e_k. \quad (2.21
$$

Let $i = 1$ and $j = 2$, for example. Since ϵ_{12k} is nonzero only for $k = 3$, we obtain $e_1 \times e$

$$\sum_{m=1}^{3} \epsilon_{ijm}\epsilon_{klm} = \delta_{ik}\delta_{jl} - \delta_{il}\delta_{jk}. \tag{2.22}$$

We can confirm this equality from the definition of the Kronecker delta δ_{ij} and the permutation signature ϵ_{ijk} by exhaustively checking all the combinations of the indices. Using this identity, we can obtain Eq. (2.20) of vector triple products. Consider the first formula, for example. Since we can write $\boldsymbol{a} \times \boldsymbol{b} = \sum_{i,j,k=1}^{3} \epsilon_{ijk} a_i b_j e_k$ and $\boldsymbol{c} = \sum_{l=1}^{3} c_l e_l$, we obtain the following result:

$$\begin{aligned}
(\boldsymbol{a} \times \boldsymbol{b}) \times \boldsymbol{c} &= \Big(\sum_{i,j,k=1}^{3} \epsilon_{ijk} a_i b_j e_k \Big) \times \Big(\sum_{l=1}^{3} c_l e_l \Big) = \sum_{i,j,k,l=1}^{3} \epsilon_{ijk} a_i b_j c_l (e_k \times e_l) \\
&= \sum_{i,j,k,l=1}^{3} \epsilon_{ijk} a_i b_j c_l \sum_{m=1}^{3} \epsilon_{klm} e_m = \sum_{i,j,l,m=1}^{3} \Big(\sum_{k=1}^{3} \epsilon_{ijk}\epsilon_{klm} \Big) a_i b_j c_l e_m \\
&= \sum_{i,j,l,m=1}^{3} (\delta_{il}\delta_{jm} - \delta_{im}\delta_{jl}) a_i b_j c_l e_m = \sum_{i,j,l,m=1}^{3} \delta_{il}\delta_{jm} a_i b_j c_l e_m - \sum_{i,j,l,m=1}^{3} \delta_{im}\delta_{jl} a_i b_j c_l e_m \\
&= \sum_{i,j=1}^{3} a_i b_j c_i e_j - \sum_{i,j=1}^{3} a_i b_j c_j e_i = \Big(\sum_{i=1}^{3} a_i c_i \Big) \sum_{j=1}^{3} b_j e_j - \Big(\sum_{j=1}^{3} b_j c_j \Big) \sum_{i=1}^{3} a_i e_i \\
&= \langle \boldsymbol{a}, \boldsymbol{c} \rangle \boldsymbol{b} - \langle \boldsymbol{b}, \boldsymbol{c} \rangle \boldsymbol{a}.
\end{aligned} \tag{2.23}$$

Here, we have noted that the antisymmetry of the indices implies that $\epsilon_{klm} = \epsilon_{lmk}$. The second formula is obtained by rewriting $\boldsymbol{a} \times (\boldsymbol{b} \times \boldsymbol{c})$ as $-(\boldsymbol{b} \times \boldsymbol{c}) \times \boldsymbol{a}$ and applying the above result, but it can also be directly derived using Eqs. (2.21) and (2.22).

2.5 SCALAR TRIPLE PRODUCT

We write $|\boldsymbol{a}, \boldsymbol{b}, \boldsymbol{c}|$ for the volume of the parallelepiped defined by vectors $\boldsymbol{a}$, $\boldsymbol{b}$, and $\boldsymbol{c}$ after making their starting points coincide and call it the *scalar triple product* of $\boldsymbol{a}$, $\boldsymbol{b}$, and $\boldsymbol{c}$ (Fig. 2.4). If $\boldsymbol{c}$ and the direction a right-handed screw would move by rotating $\boldsymbol{a}$ toward $\boldsymbol{b}$ are on the same side of the plane spanned by $\boldsymbol{a}$ and $\boldsymbol{b}$, we say that $\boldsymbol{a}$, $\boldsymbol{b}$, and $\boldsymbol{c}$ are a *right-handed system*. Otherwise, they are a *left-handed system*. We regard the volume $|\boldsymbol{a}, \boldsymbol{b}, \boldsymbol{c}|$ as positive if $\boldsymbol{a}$, $\boldsymbol{b}$, and $\boldsymbol{c}$ are a right-handed system and as negative if they are a left-handed system. The volume $|\boldsymbol{a}, \boldsymbol{b}, \boldsymbol{c}|$ is 0 if $\boldsymbol{a}$, $\boldsymbol{b}$, and $\boldsymbol{c}$ are coplanar. From this definition, we observe the following:

Linearity: $|\boldsymbol{a}, \boldsymbol{b}, \alpha\boldsymbol{c} + \beta\boldsymbol{d}| = \alpha|\boldsymbol{a}, \boldsymbol{b}, \boldsymbol{c}| + \beta|\boldsymbol{a}, \boldsymbol{b}, \boldsymbol{d}|$.

Antisymmetry: $|\boldsymbol{a}, \boldsymbol{b}, \boldsymbol{c}| = -|\boldsymbol{b}, \boldsymbol{a}, \boldsymbol{c}| = -|\boldsymbol{c}, \boldsymbol{b}, \boldsymbol{a}| = -|\boldsymbol{a}, \boldsymbol{c}, \boldsymbol{b}|$.

The above linearity has the geometric meaning of Fig. 2.5(a). The antisymmetry means that the scalar triple product changes its sign if any two vectors are interchanged. Hence, it is 0 for duplicate vectors:

$$|\boldsymbol{a}, \boldsymbol{c}, \boldsymbol{c}| = |\boldsymbol{a}, \boldsymbol{b}, \boldsymbol{a}| = |\boldsymbol{a}, \boldsymbol{a}, \boldsymbol{c}| = 0. \tag{2.24}$$

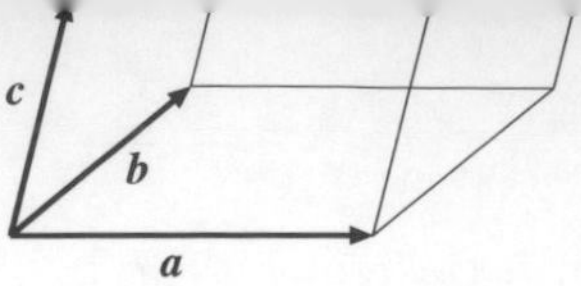

FIGURE 2.4 The scalar triple product $|\boldsymbol{a}, \boldsymbol{b}, \boldsymbol{c}|$ of vectors $\boldsymbol{a}$, $\boldsymbol{b}$, and $\boldsymbol{c}$ is the signed volume of t parallelepiped defined by $\boldsymbol{a}$, $\boldsymbol{b}$, and $\boldsymbol{c}$.

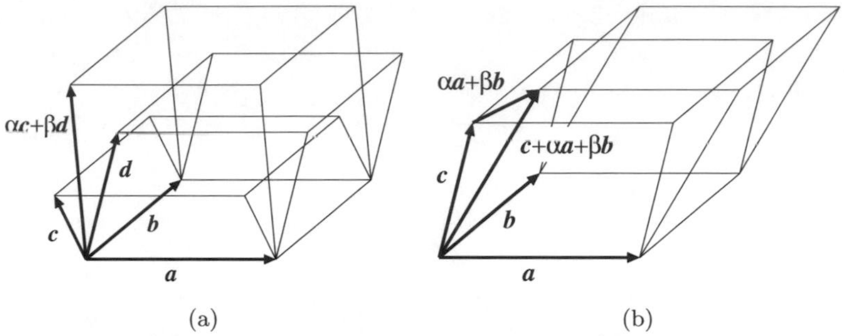

FIGURE 2.5 (a) The volume of the parallelepiped defined by vectors $\boldsymbol{a}$, $\boldsymbol{b}$, and $\alpha\boldsymbol{c} + \beta\boldsymbol{d}$ equals t sum of α times the volume of the parallelepiped defined by $\boldsymbol{a}$, $\boldsymbol{b}$, and $\boldsymbol{c}$ and β times the volume the parallelepiped defined by $\boldsymbol{a}$, $\boldsymbol{b}$, and $\boldsymbol{d}$. (b) The volume of the parallelepiped defined by vecto $\boldsymbol{a}$, $\boldsymbol{b}$, and $\boldsymbol{c}$ equals the volume of the parallelepiped defined by $\boldsymbol{a}$, $\boldsymbol{b}$, and $\boldsymbol{c} + \alpha\boldsymbol{a} + \beta\boldsymbol{b}$.

This implies that the scalar triple product has the same value after multiplying one vect by a scalar and adding it to another:

$$|\boldsymbol{a}, \boldsymbol{b}, \boldsymbol{c} + \alpha\boldsymbol{a} + \beta\boldsymbol{b}| = |\boldsymbol{a}, \boldsymbol{b}, \boldsymbol{c}|. \qquad (2.25)$$

This has the geometric meaning of Fig. 2.5(b). Also, due to the antisymmetry, the scal triple product is invariant to *cyclic permutations*, i.e., the replacement $\boldsymbol{a} \rightarrow \boldsymbol{b} \rightarrow \boldsymbol{c} \rightarrow \boldsymbol{a}$:

$$|\boldsymbol{a}, \boldsymbol{b}, \boldsymbol{c}| = |\boldsymbol{b}, \boldsymbol{c}, \boldsymbol{a}| = |\boldsymbol{c}, \boldsymbol{a}, \boldsymbol{b}|, \qquad |\boldsymbol{c}, \boldsymbol{b}, \boldsymbol{a}| = |\boldsymbol{a}, \boldsymbol{c}, \boldsymbol{b}| = |\boldsymbol{c}, \boldsymbol{b}, \boldsymbol{a}|\,(= -|\boldsymbol{a}, \boldsymbol{b}, \boldsymbol{c}|\,). \qquad (2.26)$$

If vectors are expressed in terms of the basis, computation of scalar triple products re duces, after expansion using the linearity and the antisymmetry, to the scalar triple product of the basis vectors e_1, e_2, and e_3. By definition, we have

$$|e_1, e_2, e_3| = 1. \qquad (2.27)$$

We call $|e_1, e_2, e_3|$ the *volume element* of this coordinate system. Let a_i, b_i, and c_i be th ith components of $\boldsymbol{a}$, $\boldsymbol{b}$, and $\boldsymbol{c}$, respectively. Using the linearity, the antisymmetry, an Eq. (2.27), we can compute their scalar triple product as follows:

$$\begin{aligned} |\boldsymbol{a}, \boldsymbol{b}, \boldsymbol{c}| &= |a_1e_1 + a_2e_2 + a_3e_3, b_1e_1 + b_2e_2 + b_3e_3, c_1e_1 + c_2c_2 + c_3e_3| \\ &= a_1b_2c_3|e_1, e_2, e_3| + a_2b_3c_1|e_2, e_3, e_1| + a_3b_1c_2|e_3, e_1, e_2| \\ &\quad + a_1b_3c_2|e_1, e_3, e_2| + a_2b_1c_3|e_2, e_1, e_3| + a_3b_2c_1|e_3, e_2, e_1| \\ &= (a_1b_2c_3 + a_2b_3c_1 + a_3b_1c_2 - a_1b_3c_2 - a_2b_1c_3 - a_3b_2c_1)|e_1, e_2, e_3| \\ &= a_1b_2c_3 + a_2b_3c_1 + a_3b_1c_2 - a_1b_3c_2 - a_2b_1c_3 - a_3b_2c_1. \end{aligned} \qquad (2.28)$$

Thus, we obtain

$$|\boldsymbol{a}, \boldsymbol{b}, \boldsymbol{c}| = \sum_{i,j,k=1}^{3} \epsilon_{ijk} a_i b_j c_j. \tag{2.29}$$

Note that Eqs. (2.28) and (2.29) are also obtained from Eqs. (2.18) and (2.19), respectively, by replacing e_i with c_i. This implies the following relationship, which is easy to confirm (↪ Exercise 2.9):

Proposition 2.9 (Scalar triple product in terms of inner and vector products) *The scalar triple product can be expressed in terms of the inner and the vector products in the form*

$$|\boldsymbol{a}, \boldsymbol{b}, \boldsymbol{c}| = \langle \boldsymbol{a} \times \boldsymbol{b}, \boldsymbol{c} \rangle. \tag{2.30}$$

From the antisymmetry of the scalar triple product and the symmetry of the inner product, we also see that

$$\begin{aligned} |\boldsymbol{a}, \boldsymbol{b}, \boldsymbol{c}| &= \langle \boldsymbol{a} \times \boldsymbol{b}, \boldsymbol{c} \rangle = \langle \boldsymbol{b} \times \boldsymbol{c}, \boldsymbol{a} \rangle = \langle \boldsymbol{c} \times \boldsymbol{a}, \boldsymbol{b} \rangle \\ &= \langle \boldsymbol{a}, \boldsymbol{b} \times \boldsymbol{c} \rangle = \langle \boldsymbol{b}, \boldsymbol{c} \times \boldsymbol{a} \rangle = \langle \boldsymbol{c}, \boldsymbol{a} \times \boldsymbol{b} \rangle. \end{aligned} \tag{2.31}$$

From the definition of the scalar triple product, we conclude that

Proposition 2.10 (Coplanarity of vectors) *Vectors* $\boldsymbol{a}$, $\boldsymbol{b}$, *and* $\boldsymbol{c}$ *are* coplanar, *i.e., in the same plane if their starting points are made to coincide, if and only if*

$$|\boldsymbol{a}, \boldsymbol{b}, \boldsymbol{c}| = 0. \tag{2.32}$$

Traditional World 2.4 (Determinant) Traditional linear algebra regards vectors $\boldsymbol{a}$, $\boldsymbol{b}$, and $\boldsymbol{c}$ as vertical arrays of numbers. If we write $(\boldsymbol{a}, \boldsymbol{b}, \boldsymbol{c})$ for the matrix that has $\boldsymbol{a}$, $\boldsymbol{b}$, and $\boldsymbol{c}$ as its columns, its *determinant* is

$$|\boldsymbol{a}, \boldsymbol{b}, \boldsymbol{c}| = a_1 b_2 c_3 + b_1 c_2 a_3 + c_1 a_2 b_3 - c_1 b_2 a_3 - b_1 a_2 c_3 - a_1 c_2 b_3, \tag{2.33}$$

where a_i, b_i, and c_i are the ith components of $\boldsymbol{a}$, $\boldsymbol{b}$, and $\boldsymbol{c}$, respectively. Numerically, this equals the scalar triple product $|\boldsymbol{a}, \boldsymbol{b}, \boldsymbol{c}|$. In this book, however, vectors are not arrays of numbers; they are defined by their geometric meaning, and their computation is stipulated by algebraic rules, which reduces the scalar triple product to a multiple of the volume element $|e_1, e_2, e_3|$.

In linear algebra, the determinant of the matrix $\boldsymbol{A}$ whose (ij) element is A_{ij} is formally defined through the permutation signature ϵ_{ijk} by

$$|\boldsymbol{A}| = \begin{vmatrix} a_{11} & a_{12} & a_{13} \\ a_{21} & a_{22} & a_{23} \\ a_{31} & a_{32} & a_{33} \end{vmatrix} = \sum_{i,j,k=1}^{3} \epsilon_{ijk} a_{1i} a_{2j} a_{3k}, \tag{2.34}$$

which is also written as det $\boldsymbol{A}$. In the algebraic treatment of this book, however, we do not use matrices of numerical elements or their determinants.

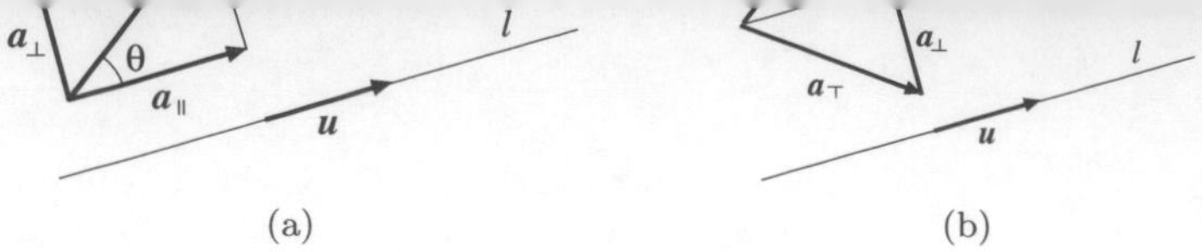

FIGURE 2.6 (a) Projection $\boldsymbol{a}_{\|}$ and rejection $\boldsymbol{a}_{\perp}$ of vector $\boldsymbol{a}$ for line l. (b) Reflection $\boldsymbol{a}_{\top}$ of vect $\boldsymbol{a}$ for line l.

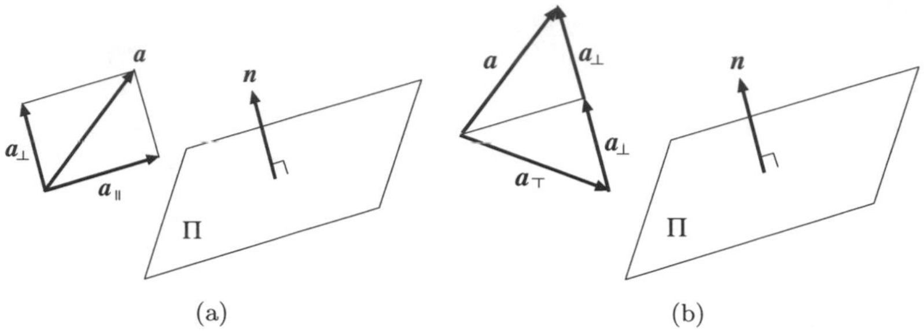

FIGURE 2.7 (a) Projection $\boldsymbol{a}_{\|}$ and rejection $\boldsymbol{a}_{\perp}$ of vector $\boldsymbol{a}$ for plane Π. (b) Reflection $\boldsymbol{a}_{\top}$ of vect $\boldsymbol{a}$ for plane Π.

2.6 PROJECTION, REJECTION, AND REFLECTION

Consider a line l of direction $\boldsymbol{u}$ (unit vector). A vector $\boldsymbol{a}$ is expressed as the sum of a vect $\boldsymbol{a}_{\|}$ parallel to l and a vector $\boldsymbol{a}_{\perp}$ orthogonal to it (Fig. 2.6(a)):

$$\boldsymbol{a} = \boldsymbol{a}_{\|} + \boldsymbol{a}_{\perp}. \tag{2.35}$$

We call $\boldsymbol{a}_{\|}$ the *projection* of $\boldsymbol{a}$ onto line l and $\boldsymbol{a}_{\perp}$ the *rejection* from l. Let θ be the ang made by $\boldsymbol{a}$ and l. Since $\boldsymbol{u}$ is a unit vector, we have $\langle \boldsymbol{a}, \boldsymbol{u} \rangle = \|\boldsymbol{a}\| \cos\theta$ from Eq. (2.10), whic is the (signed) projected length of $\boldsymbol{a}_{\|}$ onto l (positive in the direction of $\boldsymbol{u}$). Hence, $\boldsymbol{a}_{\|}$ $\langle \boldsymbol{a}, \boldsymbol{u} \rangle \boldsymbol{u}$ and $\boldsymbol{a}_{\perp} = \boldsymbol{a} - \boldsymbol{a}_{\|}$. Let us call the vector $\boldsymbol{a}_{\top}$ obtained by reflecting $\boldsymbol{a}$ to the opposi side of line l the (line) *reflection* of $\boldsymbol{a}$ with respect to l (Fig. 2.6(b)). This differs from $\boldsymbol{a}$ t twice the rejection $\boldsymbol{a}_{\perp}$. In summary,

Proposition 2.11 (Projection, rejection, and reflection for a line) *The projectio of vector $\boldsymbol{a}$ onto line l of direction $\boldsymbol{u}$ (unit vector) has length $\langle \boldsymbol{a}, \boldsymbol{u} \rangle$. The projection $\boldsymbol{a}_{\|}$, th rejection $\boldsymbol{a}_{\perp}$, and the reflection $\boldsymbol{a}_{\top}$ of $\boldsymbol{a}$ for line l are given as follows:*

$$\boldsymbol{a}_{\|} = \langle \boldsymbol{a}, \boldsymbol{u} \rangle \boldsymbol{u}, \qquad \boldsymbol{a}_{\perp} = \boldsymbol{a} - \langle \boldsymbol{a}, \boldsymbol{u} \rangle \boldsymbol{u}, \qquad \boldsymbol{a}_{\top} = -\boldsymbol{a} + 2\langle \boldsymbol{a}, \boldsymbol{u} \rangle \boldsymbol{u}. \tag{2.36}$$

Next, consider a plane Π with unit surface normal $\boldsymbol{n}$. A vector $\boldsymbol{a}$ is expressed as th sum of a vector $\boldsymbol{a}_{\|}$ parallel to $\boldsymbol{a}$ and a vector $\boldsymbol{a}_{\perp}$ orthogonal to it in the form of Eq. (2.35 (Fig. 2.6(b)). We call $\boldsymbol{a}_{\|}$ the *projection* of $\boldsymbol{a}$ onto plane Π and $\boldsymbol{a}_{\perp}$ the *rejection* from Π. Sinc the rejection $\boldsymbol{a}_{\perp}$ is the projection of $\boldsymbol{a}$ onto the direction of $\boldsymbol{n}$, we can write $\boldsymbol{a}_{\perp} = \langle \boldsymbol{a}, \boldsymbol{n} \rangle$r Then, the projection $\boldsymbol{a}_{\|}$ is the difference $\boldsymbol{a} - \langle \boldsymbol{a}, \boldsymbol{n} \rangle \boldsymbol{n}$. The vector $\boldsymbol{a}_{\top}$ obtained by reflectin $\boldsymbol{a}$ to the opposite side of plane Π is called the (surface) *reflection*, or *mirror image*, of with respect to Π (Fig. 2.7(b)). This differs from $\boldsymbol{a}$ by twice the rejection $\boldsymbol{a}_{\perp}$. In summar

$a_{\|}$, the rejection $a_{\perp}$, and the reflection $a_{\top}$ of a for plane Π are given as follows.

$$\boldsymbol{a}_{\|} = \boldsymbol{a} - \langle \boldsymbol{a}, \boldsymbol{n} \rangle \boldsymbol{n}, \qquad \boldsymbol{a}_{\perp} = \langle \boldsymbol{a}, \boldsymbol{n} \rangle \boldsymbol{n}, \qquad \boldsymbol{a}_{\top} = \boldsymbol{a} - 2\langle \boldsymbol{a}, \boldsymbol{n} \rangle \boldsymbol{n}. \tag{2.37}$$

Traditional World 2.5 (Matrix representation of linear mappings) A mapping from a vector to a vector is called a *linear mapping* if it *preserves* sums and scalar multiplications, i.e., if the sums or scalar multiplications of vectors are mapped to the sums or scalar multiplications of the mapped vectors. Projection and rejection are typical linear mappings with this property. Traditional linear algebra regards vectors as columns of numbers and expresses linear mappings as multiplication by matrices. In fact, expressing linear mappings by matrices is the essence of linear algebra. In this book, in contrast, vectors are merely symbols, not arrays of numbers, so they cannot be multiplied by matrices.

If we regard vectors as columns of numbers, the inner product is written in the form of Eq. (2.14). Hence, the vector $\boldsymbol{a}_{\|}$ in Eq. (2.36) is written as

$$(\boldsymbol{u}^\top \boldsymbol{a})\boldsymbol{u} = \boldsymbol{u}(\boldsymbol{u}^\top \boldsymbol{a}) = (\boldsymbol{u}\boldsymbol{u}^\top)\boldsymbol{a}, \tag{2.38}$$

so that $\boldsymbol{a}_{\perp}$ and $\boldsymbol{a}_{\top}$ are respectively written as

$$\boldsymbol{a} - (\boldsymbol{u}\boldsymbol{u}^\top)\boldsymbol{a} = (\boldsymbol{I} - \boldsymbol{u}\boldsymbol{u}^\top)\boldsymbol{a}, \qquad -\boldsymbol{a} + 2(\boldsymbol{u}\boldsymbol{u}^\top)\boldsymbol{a} = (2\boldsymbol{u}\boldsymbol{u}^\top - \boldsymbol{I})\boldsymbol{a}, \tag{2.39}$$

where $\boldsymbol{I}$ is the identity matrix. Consequently, the projection $\boldsymbol{a}_{\|}$, the rejection $\boldsymbol{a}_{\perp}$, and the reflection $\boldsymbol{a}_{\top}$ in Eq. (2.36) are expressed as multiplications by the following *projection matrix* $\boldsymbol{P}_{\|}$, *rejection matrix* $\boldsymbol{P}_{\perp}$, and *reflection matrix* $\boldsymbol{P}_{\top}$:

$$\boldsymbol{a}_{\|} = \boldsymbol{P}_{\|}\boldsymbol{a}, \qquad \boldsymbol{P}_{\|} = \boldsymbol{u}\boldsymbol{u}^\top = \begin{pmatrix} u_1^2 & u_1u_2 & u_1u_3 \\ u_2u_1 & u_2^2 & u_2u_3 \\ u_3u_1 & u_3u_2 & u_3^2 \end{pmatrix},$$

$$\boldsymbol{a}_{\perp} = \boldsymbol{P}_{\perp}\boldsymbol{a}, \qquad \boldsymbol{P}_{\perp} = \boldsymbol{I} - \boldsymbol{u}\boldsymbol{u}^\top = \begin{pmatrix} 1-u_1^2 & -u_1u_2 & -u_1u_3 \\ -u_2u_1 & 1-u_2^2 & -u_2u_3 \\ -u_3u_1 & -u_3u_2 & 1-u_3^2 \end{pmatrix},$$

$$\boldsymbol{a}_{\top} = \boldsymbol{P}_{\top}\boldsymbol{a}, \qquad \boldsymbol{P}_{\top} = -\boldsymbol{I} + 2\boldsymbol{u}\boldsymbol{u}^\top = \begin{pmatrix} 2u_1^2-1 & 2u_1u_2 & 2u_1u_3 \\ 2u_2u_1 & 2u_2^2-1 & 2u_2u_3 \\ 2u_3u_1 & 2u_3u_2 & 2u_3^2-1 \end{pmatrix}. \tag{2.40}$$

Similarly, the projection $\boldsymbol{a}_{\|}$, the rejection $\boldsymbol{a}_{\perp}$, and reflection $\boldsymbol{a}_{\top}$ in Eq. (2.37) can be expressed as multiplications by the following *projection matrix* $\boldsymbol{P}_{\|}$, *rejection matrix* $\boldsymbol{P}_{\perp}$, and *reflection matrix* $\boldsymbol{P}_{\top}$:

$$\boldsymbol{a}_{\|} = \boldsymbol{P}_{\|}\boldsymbol{a}, \qquad \boldsymbol{P}_{\|} = \boldsymbol{I} - \boldsymbol{n}\boldsymbol{n}^\top = \begin{pmatrix} 1-n_1^2 & -n_1n_2 & -n_1n_3 \\ -n_2n_1 & 1-n_2^2 & -n_2n_3 \\ -n_3n_1 & -n_3n_2 & 1-n_3^2 \end{pmatrix},$$

$$\boldsymbol{a}_{\perp} = \boldsymbol{P}_{\perp}\boldsymbol{a}, \qquad \boldsymbol{P}_{\perp} = \boldsymbol{n}\boldsymbol{n}^\top = \begin{pmatrix} n_1^2 & n_1n_2 & n_1n_3 \\ n_2n_1 & n_2^2 & n_2n_3 \\ n_3n_1 & n_3n_2 & n_3^2 \end{pmatrix},$$

$$\boldsymbol{a}_{\top} = \boldsymbol{P}_{\top}\boldsymbol{a}, \qquad \boldsymbol{P}_{\top} = \boldsymbol{I} - 2\boldsymbol{n}\boldsymbol{n}^\top = \begin{pmatrix} 1-2n_1^2 & -2n_1n_2 & -2n_1n_3 \\ -2n_2n_1 & 1-2n_2^2 & -2n_2n_3 \\ -2n_3n_1 & -2n_3n_2 & 1-2n_3^2 \end{pmatrix}. \tag{2.41}$$

However, the algebraic treatment of this book does not make use of matrix representation of linear mappings.

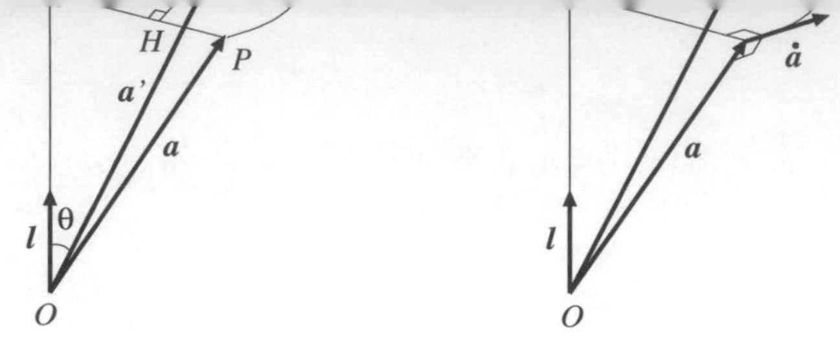

(a) (b)

FIGURE 2.8 (a) Rotation of vector $\boldsymbol{a}$ around axis $\boldsymbol{l}$ by angle Ω (rad). (b) Rotation of vector around axis $\boldsymbol{l}$ with angular velocity ω (rad/sec).

2.7 ROTATION

Suppose a vector $\boldsymbol{a}$ is rotated to $\boldsymbol{a}'$ around an axis l along a unit vector $\boldsymbol{l}$ by angle Ω which we define to be positive in the right-hand screw sense and negative in the opposi sense. The vector $\boldsymbol{a}$ can be anywhere in the space, but for the convenience of analysis w suppose that its starting point is at the origin O. Let P and P' be the endpoints of $\boldsymbol{a}$ an $\boldsymbol{a}'$, respectively. Let Q be the foot of the perpendicular line from P' onto the axis l, and the foot of the perpendicular line from P' onto the line segment OP (Fig. 2.8(a)). We se that

$$\boldsymbol{a}' = \overrightarrow{OQ} + \overrightarrow{QH} + \overrightarrow{HP'}. \qquad (2.4$$

Here, we use the notation $\overrightarrow{AB}$ to denote the vector starting from A and ending at B. Th vector $\overrightarrow{OQ}$ is the projection of $\boldsymbol{a}$ onto l, so from Eq. (2.36) we can write

$$\overrightarrow{OQ} = \langle \boldsymbol{a}, \boldsymbol{l} \rangle \boldsymbol{l}. \qquad (2.4$$

The vector $\overrightarrow{QP}$ is the rejection of $\boldsymbol{a}$ from l, so $\overrightarrow{QP} = \boldsymbol{a} - \langle \boldsymbol{a}, \boldsymbol{l} \rangle \boldsymbol{l}$. Since $\overrightarrow{QH}$ is the projectic of $\overrightarrow{QP'}$ onto the direction of $\overrightarrow{QP}$, we can write

$$\begin{aligned} \overrightarrow{QH} &= \langle \overrightarrow{QP'}, \frac{\overrightarrow{QP}}{\|\overrightarrow{QP}\|} \rangle \frac{\overrightarrow{QP}}{\|\overrightarrow{QP}\|} = \frac{\langle \overrightarrow{QP'}, \overrightarrow{QP} \rangle}{\|\overrightarrow{QP}\|^2} \overrightarrow{QP} = \frac{\|\overrightarrow{QP}\|^2 \cos\Omega}{\|\overrightarrow{QP}\|^2} \overrightarrow{QP} \\ &= (\boldsymbol{a} - \langle \boldsymbol{a}, \boldsymbol{l} \rangle \boldsymbol{l}) \cos\Omega. \end{aligned} \qquad (2.4$$

Suppose $\boldsymbol{l}$ and $\boldsymbol{a}$ make angle θ. From Eq. (2.15), we have $\|\boldsymbol{l} \times \boldsymbol{a}\| = \|\boldsymbol{a}\| \sin\theta = \|\overrightarrow{QP}$ Since $\|\overrightarrow{QP'}\| = \|\overrightarrow{QP}\|$, we have $\|\overrightarrow{HP'}\| = \|\overrightarrow{QP'}\| \sin\Omega = \|\overrightarrow{QP}\| \sin\Omega$. The direction of $\overrightarrow{HP'}$ is equal to the direction of $\boldsymbol{l} \times \boldsymbol{a}$. Hence, $\overrightarrow{HP'}$ is written as

$$\overrightarrow{HP'} = \frac{\boldsymbol{l} \times \boldsymbol{a}}{\|\boldsymbol{l} \times \boldsymbol{a}\|} \|\overrightarrow{QP}\| \sin\Omega = \frac{\boldsymbol{l} \times \boldsymbol{a}}{\|\boldsymbol{l} \times \boldsymbol{a}\|} \|\boldsymbol{l} \times \boldsymbol{a}\| \sin\Omega = \boldsymbol{l} \times \boldsymbol{a} \sin\Omega. \qquad (2.45$$

Substituting Eqs. (2.43), (2.44), and (2.45) into Eq. (2.42), we see that

Proposition 2.13 (Rodrigues formula) *If a vector* $\boldsymbol{a}$ *is rotated around axis* $\boldsymbol{l}$ *(un vector) by angle* Ω*, we obtain*

$$\boldsymbol{a}' = \boldsymbol{a} \cos\Omega + \boldsymbol{l} \times \boldsymbol{a} \sin\Omega + \langle \boldsymbol{a}, \boldsymbol{l} \rangle \boldsymbol{l} (1 - \cos\Omega). \qquad (2.46$$

$\cos\Delta\Omega = 1 + O(\Delta\Omega^2)$ and $\sin\Delta\Omega = \Delta\Omega + O(\Delta\Omega^3)$, Eq. (2.46) reduces to

$$\boldsymbol{a}' = \boldsymbol{a} + \Delta\Omega \boldsymbol{l} \times \boldsymbol{a} + O(\Delta\Omega^2). \tag{2.47}$$

Let us interpret this to be a continuous rotation of $\boldsymbol{a}$ in Δt seconds. Dividing the difference of the left side from the right by Δt and taking the limit of $\Delta t \to 0$, we can express the instantaneous change rate $\dot{\boldsymbol{a}}$ in the form

$$\dot{\boldsymbol{a}} = \omega \boldsymbol{l} \times \boldsymbol{a}, \tag{2.48}$$

where we define the *angular velocity* ω by $\omega = \lim_{\Delta t \to 0} \Delta\Omega/\Delta t$. As we see from Eq. (2.48), $\boldsymbol{a}$ changes in the direction orthogonal to $\boldsymbol{l}$ (Fig. 2.8(b)).

Traditional World 2.6 (Matrix representation of rotation) Since rotation of sums and scalar multiplications of vectors equals the corresponding sums and scalar multiplications of the rotated vectors, rotation is a linear mapping. Hence, if vectors are regarded as vertical arrays of numbers as in the conventional linear algebra, rotation is expressed by multiplication by a matrix. The Rodrigues formula of Eq. (2.46) is written as multiplication by the following *rotation matrix* $\boldsymbol{R}$ ($\hookrightarrow$ Exercises 2.12):

$$\begin{aligned} \boldsymbol{a}' &= \boldsymbol{R}\boldsymbol{a}, \\ \boldsymbol{R} &= \begin{pmatrix} \cos\Omega + l_1^2(1-\cos\Omega) & l_1 l_2(1-\cos\Omega) - l_3\sin\Omega & l_1 l_3(1-\cos\Omega) + l_2\sin\Omega \\ l_2 l_1(1-\cos\Omega) + l_3\sin\Omega & \cos\Omega + l_2^2(1-\cos\Omega) & l_2 l_3(1-\cos\Omega) - l_1\sin\Omega \\ l_3 l_1(1-\cos\Omega) - l_2\sin\Omega & l_3 l_2(1-\cos\Omega) + l_1\sin\Omega & \cos\Omega + l_3^2(1-\cos\Omega) \end{pmatrix}. \end{aligned} \tag{2.49}$$

A mapping from a 3D space to itself is called an *orthogonal transformation* if the norm and the inner product are preserved, i.e., if the norms and inner products of vectors are the same after the mapping. Rotation and reflection are orthogonal transformations. A matrix that defines an orthogonal transformation is said to be *orthogonal*. A matrix $\boldsymbol{A}$ is orthogonal if and only if $\boldsymbol{A}^\top \boldsymbol{A} = \boldsymbol{I}$, which states that the columns (and rows as well) of $\boldsymbol{A}$ are mutually orthogonal unit vectors. It is known that every orthogonal matrix is either a rotation matrix or the product of a rotation matrix and a reflection matrix. In this book, however, we do not represent rotation in matrix form.

2.8 PLANES

We identify the position vector $\boldsymbol{x}$ with the point itself and simply call it "point $\boldsymbol{x}$." By the "equation" of a geometric object, we mean the equation satisfied by points that belong to that object. A plane is one of the most fundamental geometric objects. We specify a plane by its unit surface normal $\boldsymbol{n}$ and the distance h from the origin O (Fig. 2.9(a)), where h is signed, being positive in the direction of $\boldsymbol{n}$ and negative in the opposite direction. Evidently, point $\boldsymbol{x}$ belongs to this plane if and only if its projected length onto the line extending along $\boldsymbol{n}$ is h. Hence, the equation of this plane is written as

$$\langle \boldsymbol{n}, \boldsymbol{x} \rangle = h. \tag{2.50}$$

From this definition, the plane specified by $\boldsymbol{n}$ and h and the plane specified by $-\boldsymbol{n}$ and $-h$ are the same, i.e., one plane is specified in two ways. Since the projected length of a vector $\boldsymbol{x}$ onto the line extending along $\boldsymbol{n}$ is $\langle \boldsymbol{n}, \boldsymbol{x} \rangle$, we obtain the following result (Fig. 2.9(b)):

(a) (b)

FIGURE 2.9 (a) A plane is specified by its unit surface normal $\boldsymbol{n}$ and the distance h from t
origin O. (b) The distance d of point $\boldsymbol{x}$ from plane $\langle \boldsymbol{n}, \boldsymbol{x} \rangle = h$.

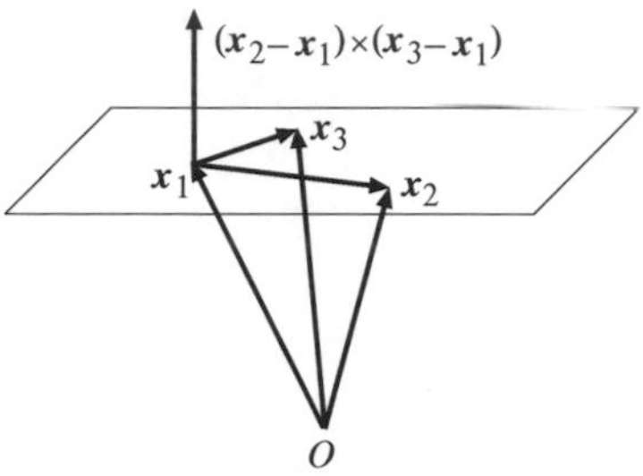

FIGURE 2.10 The plane passing through three points $\boldsymbol{x}_1$, $\boldsymbol{x}_2$, and $\boldsymbol{x}_3$.

Proposition 2.14 (Distance of point from line) *The distance d of point $\boldsymbol{x}$ from pla $\langle \boldsymbol{n}, \boldsymbol{x} \rangle = h$ is given by*

$$d = h - \langle \boldsymbol{n}, \boldsymbol{x} \rangle, \quad (2.51$$

where d is signed and is positive in the direction of $\boldsymbol{n}$.

Evidently, point $\boldsymbol{x}$ is on the plane if and only if the distance d is 0, reducing to the equatio of the plane $\langle \boldsymbol{n}, \boldsymbol{x} \rangle = h$.

Consider a plane passing through three points $\boldsymbol{x}_1$, $\boldsymbol{x}_2$, and $\boldsymbol{x}_3$ (Fig. 2.10). Since vecto $\boldsymbol{x}_2 - \boldsymbol{x}_1$ and $\boldsymbol{x}_3 - \boldsymbol{x}_1$ are on this plane, the vector product $(\boldsymbol{x}_2 - \boldsymbol{x}_1) \times (\boldsymbol{x}_3 - \boldsymbol{x}_1)$ is orthogon to this plane. The unit surface normal $\boldsymbol{n}$ is obtained by normalizing this to unit norm. Th distance h of this plane from the origin O equals the projected length of the vector $\boldsymbol{x}_1$ (c $\boldsymbol{x}_2$ or $\boldsymbol{x}_3$) onto the line extending along the unit vector $\boldsymbol{n}$, so

$$h = \langle \boldsymbol{x}_1, \frac{(\boldsymbol{x}_2 - \boldsymbol{x}_1) \times (\boldsymbol{x}_3 - \boldsymbol{x}_1)}{\|(\boldsymbol{x}_2 - \boldsymbol{x}_1) \times (\boldsymbol{x}_3 - \boldsymbol{x}_1)\|} \rangle. \quad (2.52$$

Noting that $(\boldsymbol{x}_2 - \boldsymbol{x}_1) \times (\boldsymbol{x}_3 - \boldsymbol{x}_1) = \boldsymbol{x}_2 \times \boldsymbol{x}_3 + \boldsymbol{x}_3 \times \boldsymbol{x}_1 + \boldsymbol{x}_1 \times \boldsymbol{x}_2$ and $\langle \boldsymbol{x}_1, (\boldsymbol{x}_2 - \boldsymbol{x}_1) \times (\boldsymbol{x}_3 - \boldsymbol{x}_1)$ $= |\boldsymbol{x}_1, \boldsymbol{x}_2, \boldsymbol{x}_3|$, we obtain the following:

Proposition 2.15 (Plane through three points) *The plane $\langle \boldsymbol{n}, \boldsymbol{x} \rangle = h$ passing throug three points $\boldsymbol{x}_1$, $\boldsymbol{x}_2$, and $\boldsymbol{x}_3$ is given by*

$$\boldsymbol{n} = \frac{\boldsymbol{x}_2 \times \boldsymbol{x}_3 + \boldsymbol{x}_3 \times \boldsymbol{x}_1 + \boldsymbol{x}_1 \times \boldsymbol{x}_2}{\|\boldsymbol{x}_2 \times \boldsymbol{x}_3 + \boldsymbol{x}_3 \times \boldsymbol{x}_1 + \boldsymbol{x}_1 \times \boldsymbol{x}_3\|},$$

$$h = \frac{|\boldsymbol{x}_1, \boldsymbol{x}_2, \boldsymbol{x}_3|}{\|\boldsymbol{x}_2 \times \boldsymbol{x}_3 + \boldsymbol{x}_3 \times \boldsymbol{x}_1 + \boldsymbol{x}_1 \times \boldsymbol{x}_2\|}. \quad (2.53$$

(usually polynomial) satisfied by the coordinates x, y, and z of every point that belongs to that object. In many problems, its derivation reduces to determinant calculation. Every point $\boldsymbol{x}$ on the plane passing through three points $\boldsymbol{x}_1$, $\boldsymbol{x}_2$, and $\boldsymbol{x}_3$ can be expressed in the form

$$\boldsymbol{x} = \lambda_1 \boldsymbol{x}_1 + \lambda_2 \boldsymbol{x}_2 + \lambda_3 \boldsymbol{x}_3, \qquad \lambda_1 + \lambda_2 + \lambda_3 = 1, \tag{2.54}$$

for real numbers λ_1, λ_2, and λ_3. A linear combination with coefficients that sum to 1 is called an *affine combination*. Thus, the affine combination of three points $\boldsymbol{x}_1$, $\boldsymbol{x}_2$, and $\boldsymbol{x}_3$ represents the plane passing through them. If we write $\boldsymbol{x} = \begin{pmatrix} x & y & z \end{pmatrix}^\top$ and $\boldsymbol{x}_i = \begin{pmatrix} x_i & y_i & z_i \end{pmatrix}^\top$, $i = 1, 2, 3$, where $\top$ denotes transpose from a row to a column, Eq. (2.54) is rewritten as a set of linear equations

$$\begin{pmatrix} x_1 & x_2 & x_3 & x \\ y_1 & y_2 & y_3 & y \\ z_1 & z_2 & z_3 & z \\ 1 & 1 & 1 & 1 \end{pmatrix} \begin{pmatrix} \lambda_1 \\ \lambda_2 \\ \lambda_3 \\ -1 \end{pmatrix} = \begin{pmatrix} 0 \\ 0 \\ 0 \\ 0 \end{pmatrix}. \tag{2.55}$$

Linear equations whose constant terms are all zero are said to be *homogeneous*. Homogeneous linear equations always have a trivial solution consisting of 0's. As is well known in linear algebra, homogeneous linear equations have a nontrivial solution if and only if the coefficient matrix is of determinant 0, i.e.,

$$\begin{vmatrix} x_1 & x_2 & x_3 & x \\ y_1 & y_2 & y_3 & y \\ z_1 & z_2 & z_3 & z \\ 1 & 1 & 1 & 1 \end{vmatrix} = 0. \tag{2.56}$$

We can see that this is the equation of the plane by the following reasoning. Cofactor expansion of Eq. (2.56) with respect to the fourth column would lead to a linear equation in x, y, and z. Hence, it should represent a plane. From the properties of the determinant, Eq. (2.56) is identically satisfied for $x = x_1$, $y = y_1$, and $z = z_1$, in which case the first and the fourth columns coincide. Similarly, Eq. (2.56) is satisfied for $x = x_2$, $y = y_2$, and $z = z_2$ and for $x = x_3$, $y = y_3$, and $z = z_3$. Hence, the plane represented by Eq. (2.56) should pass through these three points.

Actually carrying out cofactor expansion, we can rewrite Eq. (2.56) in the form

$$n_1 x + n_2 y + n_3 z = h, \tag{2.57}$$

where

$$\begin{aligned} n_1 &= y_2 z_3 - z_2 y_3 + y_3 z_1 - z_3 y_1 + y_1 z_2 - z_1 y_2, \\ n_2 &= z_2 x_3 - x_2 z_3 + z_3 x_1 - x_3 z_1 + z_1 x_2 - x_1 z_2, \\ n_3 &= x_2 y_3 - y_2 x_3 + x_3 y_1 - y_3 x_1 + x_1 y_2 - y_1 x_2, \end{aligned} \qquad h = \begin{vmatrix} x_1 & x_2 & x_3 \\ y_1 & y_2 & y_3 \\ z_1 & z_2 & z_3 \end{vmatrix}. \tag{2.58}$$

The four number n_1, n_2, n_3, and h are called the *Plücker coordinates* of this plane. Note that Eq. (2.57) holds if n_1, n_2, n_3, and h are multiplied by any nonzero constant. We express this fact by saying that they are *homogeneous coordinates*. If vectors are regarded as vertical arrays of numbers as in the traditional vector analysis, the three equations on the left of Eq. (2.58) are combined into one vector equation

$$\begin{pmatrix} n_1 \\ n_2 \\ n_3 \end{pmatrix} = \begin{pmatrix} x_2 \\ y_2 \\ z_2 \end{pmatrix} \times \begin{pmatrix} x_3 \\ y_3 \\ z_3 \end{pmatrix} + \begin{pmatrix} x_3 \\ y_3 \\ z_3 \end{pmatrix} \times \begin{pmatrix} x_1 \\ y_1 \\ z_1 \end{pmatrix} + \begin{pmatrix} x_1 \\ y_1 \\ z_1 \end{pmatrix} \times \begin{pmatrix} x_2 \\ y_2 \\ z_2 \end{pmatrix}, \tag{2.59}$$

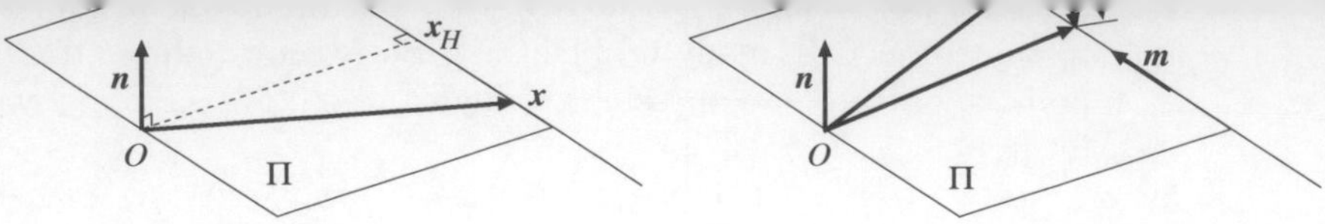

(a) (b)

FIGURE 2.11 (a) Line l is specified by the surface normal $\boldsymbol{n}$ to the supporting plane Π and tl direction vector $\boldsymbol{m}$ of l. (b) The distance d of point $\boldsymbol{x}$ from line $\boldsymbol{x} \times \boldsymbol{m} = \boldsymbol{n}$.

which is equal to the first equation in Eq. (2.53) except for scale normalization. The equa tion on the right of Eq. (2.58) equals the second equation in Eq. (2.53) except for sca normalization.

2.9 LINES

Given a line l in space, we call the plane Π passing through l and the origin O the *supportin plane* of l. If the supporting plane Π is given, the line l on it is specified by its direction an distance from the origin O (Fig. 2.11(a)). Let $\boldsymbol{n}$ be the surface normal to the supportin plane Π, and $\boldsymbol{m}$ the direction vector of l. Since any point $\boldsymbol{x}$ of line l and its direction vect $\boldsymbol{m}$ are on the supporting plane Π, both are orthogonal to the surface normal $\boldsymbol{n}$. Hence, tl vector $\boldsymbol{x} \times \boldsymbol{m}$ is in the direction of $\boldsymbol{n}$. We normalize the magnitudes of $\boldsymbol{m}$ and $\boldsymbol{n}$ so that

$$\boldsymbol{x} \times \boldsymbol{m} = \boldsymbol{n}. \tag{2.6}$$

The sign of $\boldsymbol{m}$ is defined so that movement along $\boldsymbol{m}$ is a positive rotation around $\boldsymbol{n}$. If w let $\boldsymbol{x}_H$ be the foot of the perpendicular line from O onto l, it also satisfies Eq. (2.60). W call this point the *supporting point* of l. Let h be the distance of l from O. Since $\boldsymbol{x}_H$ an $\boldsymbol{m}$ are mutually orthogonal, the norm of $\boldsymbol{x}_H \times \boldsymbol{m}$ is $h\|\boldsymbol{m}\|$ by the definition of the vect product. From Eq. (2.60), this equals $\|\boldsymbol{n}\|$. Hence, the distance h is given by

$$h = \frac{\|\boldsymbol{n}\|}{\|\boldsymbol{m}\|}. \tag{2.61}$$

The vectors $\boldsymbol{m}$ and $\boldsymbol{n}$ are mutually orthogonal from the definition:

$$\langle \boldsymbol{m}, \boldsymbol{n} \rangle = 0. \tag{2.62}$$

The supporting point $\boldsymbol{x}_H$ is in the direction of $\boldsymbol{m} \times \boldsymbol{n}$, whose norm $\|\boldsymbol{m} \times \boldsymbol{n}\|$ equals $\|\boldsymbol{m}\|\|\boldsymbol{n}\|$ since $\boldsymbol{m}$ and $\boldsymbol{n}$ make a right angle. Hence, the supporting point $\boldsymbol{x}_H$ is given by

$$\boldsymbol{x}_H = h\frac{\boldsymbol{m} \times \boldsymbol{n}}{\|\boldsymbol{m}\|\|\boldsymbol{n}\|} = \frac{\boldsymbol{m} \times \boldsymbol{n}}{\|\boldsymbol{m}\|^2}. \tag{2.63}$$

It follows that if we give two vectors $\boldsymbol{m}$ and $\boldsymbol{n}$ such that $\langle \boldsymbol{m}, \boldsymbol{n} \rangle = 0$, we can determine line l having direction $\boldsymbol{m}$ in the distance $\|\boldsymbol{n}\|/\|\boldsymbol{m}\|$ from the origin O on the supportin plane with surface normal $\boldsymbol{n}$. The signs of $\boldsymbol{m}$ and $\boldsymbol{n}$ are determined in such a way that th supporting point $\boldsymbol{x}_H$ and the vectors $\boldsymbol{m}$ and $\boldsymbol{n}$ make a right-hand system in that order. W see from Eq. (2.60) that multiplication of $\boldsymbol{m}$ and $\boldsymbol{n}$ by any nonzero constant defines th same line. So, we normalize their scale so that

$$\|\boldsymbol{m}\|^2 + \|\boldsymbol{n}\|^2 = 1. \tag{2.64}$$

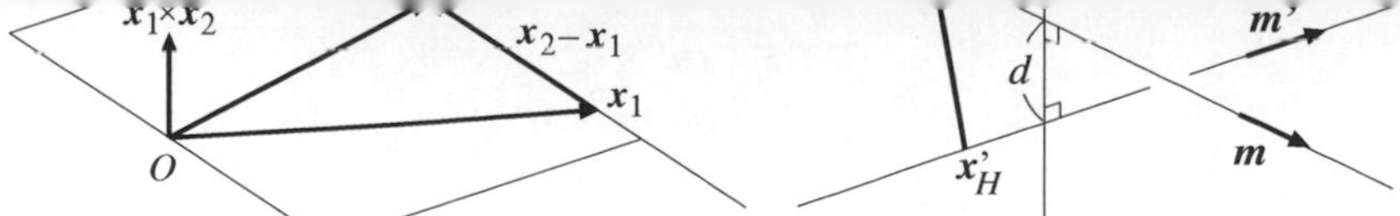

(a) (b)

FIGURE 2.12 (a) The line passing through two points $\boldsymbol{x}_1$ and $\boldsymbol{x}_2$. (b) The distance d between two lines $\boldsymbol{x} \times \boldsymbol{m} = \boldsymbol{n}$ and $\boldsymbol{x} \times \boldsymbol{m}' = \boldsymbol{n}'$ that are in skew position.

From the definition of the surface normal $\boldsymbol{n}$, every point $\boldsymbol{x}$ of l is orthogonal to $\boldsymbol{n}$:

$$\langle \boldsymbol{x}, \boldsymbol{n} \rangle = 0. \tag{2.65}$$

From the above definition, the line specified by $\boldsymbol{m}$ and $\boldsymbol{n}$ and the line specified by $-\boldsymbol{m}$ and $-\boldsymbol{n}$ are the same: one line is specified in two ways.

Let $\boldsymbol{v}$ be the vector starting from point $\boldsymbol{x}$ and perpendicularly intersecting with line l (Fig. 2.11(b)). Since the point $\boldsymbol{x} + \boldsymbol{v}$ is on the line, it satisfies its equation of Eq. (2.60). Since $\boldsymbol{v}$ is orthogonal to $\boldsymbol{m}$, we observe see that

$$\|\boldsymbol{x} \times \boldsymbol{m} - \boldsymbol{n}\| = \|(\boldsymbol{x} + \boldsymbol{v}) \times \boldsymbol{m} - \boldsymbol{n} - \boldsymbol{v} \times \boldsymbol{m}\| = \| - \boldsymbol{v} \times \boldsymbol{m}\| = \|\boldsymbol{v}\|\|\boldsymbol{m}\|. \tag{2.66}$$

The distance of point $\boldsymbol{x}$ from line l is $\|\boldsymbol{v}\|$, so we obtain

Proposition 2.16 (Distance of point from line) *The distance d of point $\boldsymbol{x}$ from line $\boldsymbol{x} \times \boldsymbol{m} = \boldsymbol{n}$ is given by*

$$d = \frac{\|\boldsymbol{x} \times \boldsymbol{m} - \boldsymbol{n}\|}{\|\boldsymbol{m}\|}. \tag{2.67}$$

Evidently, point $\boldsymbol{x}$ is on the line if and only if the distance d is 0, reducing to the equation of the line $\boldsymbol{x} \times \boldsymbol{m} = \boldsymbol{n}$.

Consider a line l passing through two points $\boldsymbol{x}_1$ and $\boldsymbol{x}_2$ (Fig. 2.12(a)). Since the direction vector $\boldsymbol{m}$ is a scalar multiple of $\boldsymbol{x}_2 - \boldsymbol{x}_1$, we can write $\boldsymbol{m} = c(\boldsymbol{x}_2 - \boldsymbol{x}_1)$ for some c. The surface normal $\boldsymbol{n}$ to the supporting plane Π of l is a scalar multiple of $\boldsymbol{x}_1 \times \boldsymbol{x}_2$, so we can write $\boldsymbol{n} = c'\boldsymbol{x}_1 \times \boldsymbol{x}_2$ for some c' It is easily seen that $\boldsymbol{x}_1 \times \boldsymbol{m} = \boldsymbol{n}$ and $\boldsymbol{x}_2 \times \boldsymbol{m} = \boldsymbol{n}$ imply $c = c'$. Applying the normalization of Eq. (2.64), we obtain the following result:

Proposition 2.17 (Line through two points) *The line $\boldsymbol{x} \times \boldsymbol{m} = \boldsymbol{n}$ passing through two points $\boldsymbol{x}_1$ and $\boldsymbol{x}_2$ is given by*

$$\boldsymbol{m} = \frac{\boldsymbol{x}_2 - \boldsymbol{x}_1}{\sqrt{\|\boldsymbol{x}_2 - \boldsymbol{x}_1\|^2 + \|\boldsymbol{x}_1 \times \boldsymbol{x}_2\|^2}}, \qquad \boldsymbol{n} = \frac{\boldsymbol{x}_1 \times \boldsymbol{x}_2}{\sqrt{\|\boldsymbol{x}_2 - \boldsymbol{x}_1\|^2 + \|\boldsymbol{x}_1 \times \boldsymbol{x}_2\|^2}}. \tag{2.68}$$

The condition that two lines $\boldsymbol{x} \times \boldsymbol{m} = \boldsymbol{n}$ and $\boldsymbol{x} \times \boldsymbol{m}' = \boldsymbol{n}'$ are parallel to each other is that their direction vectors $\boldsymbol{m}$ and $\boldsymbol{m}'$ are parallel, i.e., $\boldsymbol{m} \times \boldsymbol{m}' = 0$. If two lines are not parallel, they are said to be in *skew position*. Then, the line segment that connects the two lines in the shortest distance d is orthogonal to the two lines. Hence, it is in the direction of $\boldsymbol{m} \times \boldsymbol{m}'$ (Fig. 2.12(b)). Let $\boldsymbol{x}_H$ and $\boldsymbol{x}'_H$ be the supporting points of the two lines. The

$$
\begin{aligned}
d &= \langle \frac{}{\|\boldsymbol{m}\times\boldsymbol{m}'\|}, \boldsymbol{x}_H - \boldsymbol{x}'_H\rangle = \langle \frac{}{\|\boldsymbol{m}\times\boldsymbol{m}'\|}, \frac{}{\|\boldsymbol{m}\|^2} - \frac{}{\|\boldsymbol{m}'\|^2}\rangle \\
&= \frac{\langle \boldsymbol{m}\times\boldsymbol{m}', \boldsymbol{m}\times\boldsymbol{n}\rangle}{\|\boldsymbol{m}\times\boldsymbol{m}'\|\|\boldsymbol{m}'\|^2} - \frac{\langle \boldsymbol{m}\times\boldsymbol{m}', \boldsymbol{m}'\times\boldsymbol{n}'\rangle}{\|\boldsymbol{m}\times\boldsymbol{m}'\|\|\boldsymbol{m}\|^2} \\
&= \frac{\langle (\boldsymbol{m}\times\boldsymbol{m}')\times\boldsymbol{m}, \boldsymbol{n}\rangle}{\|\boldsymbol{m}\times\boldsymbol{m}'\|\|\boldsymbol{m}\|^2} - \frac{\langle (\boldsymbol{m}\times\boldsymbol{m}')\times\boldsymbol{m}', \boldsymbol{n}'\rangle}{\|\boldsymbol{m}\times\boldsymbol{m}'\|\|\boldsymbol{m}'\|^2} \\
&= \frac{\langle \langle\boldsymbol{m},\boldsymbol{m}\rangle\boldsymbol{m}' - \langle\boldsymbol{m}',\boldsymbol{m}\rangle\boldsymbol{m}, \boldsymbol{n}\rangle}{\|\boldsymbol{m}\times\boldsymbol{m}'\|\|\boldsymbol{m}\|^2} - \frac{\langle \langle\boldsymbol{m},\boldsymbol{m}'\rangle\boldsymbol{m}' - \langle\boldsymbol{m}',\boldsymbol{m}'\rangle\boldsymbol{m}, \boldsymbol{n}'\rangle}{\|\boldsymbol{m}\times\boldsymbol{m}'\|\|\boldsymbol{m}'\|^2} \\
&= \frac{\|\boldsymbol{m}\|^2\langle\boldsymbol{m}',\boldsymbol{n}\rangle}{\|\boldsymbol{m}\times\boldsymbol{m}'\|\|\boldsymbol{m}\|^2} - \frac{-\|\boldsymbol{m}'\|^2\langle\boldsymbol{m},\boldsymbol{n}'\rangle}{\|\boldsymbol{m}\times\boldsymbol{m}'\|\|\boldsymbol{m}'\|^2} = \frac{\langle\boldsymbol{m},\boldsymbol{n}'\rangle + \langle\boldsymbol{m}',\boldsymbol{n}\rangle}{\|\boldsymbol{m}\times\boldsymbol{m}'\|},
\end{aligned} \tag{2.69}
$$

where we have used Eq. (2.63) for the foot of a perpendicular line, Eq. (2.20) for the vecto triple product, Eqs. (2.30) and (2.31) for expressing the scalar triple product in terms the vector product, and the orthogonality relation of Eq. (2.62). The distance d is signe so that it is positive in the direction of $\boldsymbol{m}\times\boldsymbol{m}'$. If the sign is disregarded, we obtain th following result:

Proposition 2.18 (Distance between lines) *For two nonparallel lines* $\boldsymbol{x}\times\boldsymbol{m} = \boldsymbol{n}$ *an* $\boldsymbol{x}\times\boldsymbol{m}' = \boldsymbol{n}'$, *the distance* d *between them is*

$$
d = \frac{|\langle\boldsymbol{m},\boldsymbol{n}'\rangle + \langle\boldsymbol{m}',\boldsymbol{n}\rangle|}{\|\boldsymbol{m}\times\boldsymbol{m}'\|}. \tag{2.70}
$$

In particular, they intersect if and only if

$$
\langle\boldsymbol{m},\boldsymbol{n}'\rangle + \langle\boldsymbol{m}',\boldsymbol{n}\rangle = 0. \tag{2.71}
$$

If the two lines $\boldsymbol{x}\times\boldsymbol{m} = \boldsymbol{n}$ and $\boldsymbol{x}\times\boldsymbol{m}' = \boldsymbol{n}'$ are parallel, there exists some c ($\neq 0$) suc that $\boldsymbol{m} = c\boldsymbol{m}'$. Hence, $\langle\boldsymbol{m},\boldsymbol{n}'\rangle = \langle c\boldsymbol{m}',\boldsymbol{n}\rangle = c\langle\boldsymbol{m}',\boldsymbol{n}'\rangle = 0$ and $\langle\boldsymbol{m}/c,\boldsymbol{n}\rangle = \langle\boldsymbol{m},\boldsymbol{n}\rangle/c =$ 0. Hence, Eq. (2.71) is satisfied (↪ Exercise 2.13). If we interprete two parallel lines to b intersecting at infinity, Eq. (2.71) can be regarded as the general condition for two lines t intersect.

Traditional World 2.8 (Equation of line via determinants) A point $\boldsymbol{x}$ is on the lin that passes through two points $\boldsymbol{x}_1$ and $\boldsymbol{x}_2$ if and only if it is expressed as their affin combination:

$$
\boldsymbol{x} = \lambda_1\boldsymbol{x}_1 + \lambda_2\boldsymbol{x}_2, \qquad \lambda_1 + \lambda_2 = 1. \tag{2.72}
$$

In other words, the affine combination of $\boldsymbol{x}_1$ and $\boldsymbol{x}_2$ represents a line. In traditional analysi using coordinates, Eq. (2.72) is written in the form

$$
\begin{pmatrix} x_1 & x_2 & x \\ y_1 & y_2 & y \\ z_1 & z_2 & z \\ 1 & 1 & 1 \end{pmatrix}
\begin{pmatrix} \lambda_1 \\ \lambda_2 \\ -1 \end{pmatrix} =
\begin{pmatrix} 0 \\ 0 \\ 0 \\ 0 \end{pmatrix}, \tag{2.73}
$$

where $\boldsymbol{x} = \begin{pmatrix} x & y & z \end{pmatrix}^\top$ and $\boldsymbol{x}_i = \begin{pmatrix} x_i & y_i & z_i \end{pmatrix}^\top$, $i = 1, 2$. Equation (2.73) is a set c homogeneous linear equations. As is well known in linear algebra, a nontrivial solutio

$$\begin{vmatrix} y_1 & y_2 & y \\ z_1 & z_2 & z \\ 1 & 1 & 1 \end{vmatrix} = 0, \qquad \begin{vmatrix} x_1 & x_2 & x \\ z_1 & z_2 & z \\ 1 & 1 & 1 \end{vmatrix} = 0,$$

$$\begin{vmatrix} x_1 & x_2 & x \\ y_1 & y_2 & y \\ 1 & 1 & 1 \end{vmatrix} = 0, \qquad \begin{vmatrix} x_1 & x_2 & x \\ y_1 & y_2 & y \\ z_1 & z_2 & z \end{vmatrix} = 0. \tag{2.74}$$

If we carry out cofactor expansion with respect to the third columns and let

$$\begin{aligned} m_1 &= x_2 - x_1 & n_1 &= y_1 z_2 - z_1 y_2 \\ m_2 &= y_2 - y_1 & n_2 &= z_1 x_2 - x_1 z_2 \\ m_3 &= z_2 - z_1 & n_3 &= x_1 y_2 - y_1 x_2, \end{aligned} \tag{2.75}$$

Eq. (2.74) is rewritten as

$$\begin{aligned} y m_3 - z m_2 &= n_1 \\ z m_1 - x m_3 &= n_2 \\ x m_2 - y m_1 &= n_3 \end{aligned} \qquad n_1 x + n_2 y + n_3 z = 0. \tag{2.76}$$

The six numbers m_1, m_2, m_3, n_1, n_2, and n_3 are called the *Plücker coordinates* of this line. Since Eq. (2.76) holds if they are multiplied by any nonzero constant, they are homogeneous coordinates. From Eq. (2.75), we see that

$$m_1 n_1 + m_2 n_2 + m_3 n_3 = 0, \tag{2.77}$$

which is known as the *Plücker condition.* If vectors are regarded as vertical arrays of numbers as in the traditional vector analysis, Eq. (2.75) can be written as

$$\begin{pmatrix} m_1 \\ m_2 \\ m_3 \end{pmatrix} = \begin{pmatrix} x_2 \\ y_2 \\ z_2 \end{pmatrix} - \begin{pmatrix} x_1 \\ y_1 \\ z_1 \end{pmatrix}, \qquad \begin{pmatrix} n_1 \\ n_2 \\ n_3 \end{pmatrix} = \begin{pmatrix} x_1 \\ y_1 \\ z_1 \end{pmatrix} \times \begin{pmatrix} x_2 \\ y_2 \\ z_2 \end{pmatrix}, \tag{2.78}$$

and the three equations on the left of Eq. (2.76) are combined into one vector equation

$$\begin{pmatrix} x \\ y \\ z \end{pmatrix} \times \begin{pmatrix} m_1 \\ m_2 \\ m_3 \end{pmatrix} = \begin{pmatrix} n_1 \\ n_2 \\ n_3 \end{pmatrix}. \tag{2.79}$$

Hence, Eq. (2.76) corresponds to Eqs. (2.60) and (2.65), and Eq. (2.77) corresponds to Eq. (2.62). We can also see that Eq. (2.75) agrees with Eq. (2.68) except for scale normalization.

2.10 PLANES AND LINES

We now show how to compute the plane passing through a given point and a given line, the intersection point of a given plane and a given line, and the intersection line of two given planes.

2.10.1 Plane through a point and a line

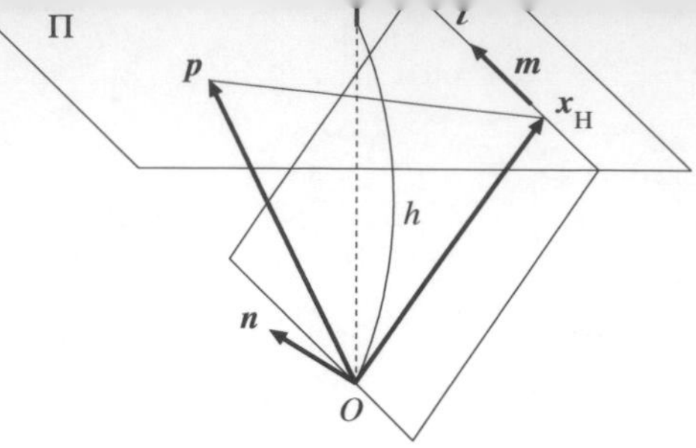

FIGURE 2.13 The plane Π passing through point $\boldsymbol{p}$ and line $\boldsymbol{x} \times \boldsymbol{m} = \boldsymbol{n}$.

Let l be a line with equation $\boldsymbol{x} \times \boldsymbol{m} = \boldsymbol{n}$. Consider the plane Π that passes through l ar a point $\boldsymbol{p}$ (Fig. 2.13). Since line l is on plane Π, the direction vector $\boldsymbol{m}$ and the supportir point $\boldsymbol{x}_H = \boldsymbol{m} \times \boldsymbol{n}/\|\boldsymbol{m}\|^2$ are also on Π. Hence, the surface normal to Π is in the directic of

$$(\boldsymbol{x}_H - \boldsymbol{p}) \times \boldsymbol{m} = \left(\frac{\boldsymbol{m} \times \boldsymbol{n}}{\|\boldsymbol{m}\|^2} - \boldsymbol{p}\right) \times \boldsymbol{m} = \frac{(\boldsymbol{m} \times \boldsymbol{n}) \times \boldsymbol{m}}{\|\boldsymbol{m}\|^2} - \boldsymbol{p} \times \boldsymbol{m} = \boldsymbol{n} - \boldsymbol{p} \times \boldsymbol{m}, \qquad (2.80)$$

where we have used the relationship $\langle \boldsymbol{m}, \boldsymbol{n} \rangle = 0$ and Eq. (2.20) of the vector triple produc After normalization to unit norm, the unit surface normal to the plane Π is given by

$$\boldsymbol{n}_\Pi = \frac{\boldsymbol{n} - \boldsymbol{p} \times \boldsymbol{m}}{\|\boldsymbol{n} - \boldsymbol{p} \times \boldsymbol{m}\|}. \qquad (2.81)$$

The distance h of this plane from the origin O is given by the projected length of tl supporting point $\boldsymbol{x}_H$ onto the line along the unit surface normal $\boldsymbol{n}_\Pi$. Hence,

$$\begin{aligned} h = \langle \boldsymbol{n}_\Pi, \boldsymbol{x}_H \rangle &= \langle \frac{\boldsymbol{n} - \boldsymbol{p} \times \boldsymbol{m}}{\|\boldsymbol{n} - \boldsymbol{p} \times \boldsymbol{m}\|}, \frac{\boldsymbol{m} \times \boldsymbol{n}}{\|\boldsymbol{m}\|^2} \rangle = -\frac{\langle \boldsymbol{p} \times \boldsymbol{m}, \boldsymbol{m} \times \boldsymbol{n} \rangle}{\|\boldsymbol{m}\|^2 \|\boldsymbol{n} - \boldsymbol{p} \times \boldsymbol{m}\|} \\ &= -\frac{|\boldsymbol{p}, \boldsymbol{m}, \boldsymbol{m} \times \boldsymbol{n}|}{\|\boldsymbol{m}\|^2 \|\boldsymbol{n} - \boldsymbol{p} \times \boldsymbol{m}\|} = -\frac{\langle \boldsymbol{p}, \boldsymbol{m} \times (\boldsymbol{m} \times \boldsymbol{n}) \rangle}{\|\boldsymbol{m}\|^2 \|\boldsymbol{p} \times \boldsymbol{m} - \boldsymbol{n}\|} \\ &= -\frac{\langle \boldsymbol{p}, -\|\boldsymbol{m}\|^2 \boldsymbol{n} \rangle}{\|\boldsymbol{m}\|^2 \|\boldsymbol{n} - \boldsymbol{p} \times \boldsymbol{m}\|} = \frac{\langle \boldsymbol{p}, \boldsymbol{n} \rangle}{\|\boldsymbol{n} - \boldsymbol{p} \times \boldsymbol{m}\|}, \end{aligned} \qquad (2.82)$$

where we have used Eq. (2.63) for the foot of a perpendicular line, Eq. (2.20) for the vectc triple product, Eqs. (2.30) and (2.31) for expressing the scalar triple product in terms c the vector product, and the orthogonality relation of Eq. (2.62). From this we obtain

Proposition 2.19 (Plane through a point and a line) *The plane* $\langle \boldsymbol{n}_\Pi, \boldsymbol{x} \rangle = h$ *th passes through point* $\boldsymbol{p}$ *and line* $\boldsymbol{x} \times \boldsymbol{m} = \boldsymbol{n}$ *is given by*

$$\boldsymbol{n}_\Pi = \frac{\boldsymbol{n} - \boldsymbol{p} \times \boldsymbol{m}}{\|\boldsymbol{n} - \boldsymbol{p} \times \boldsymbol{m}\|}, \qquad h = \frac{\langle \boldsymbol{p}, \boldsymbol{n} \rangle}{\|\boldsymbol{n} - \boldsymbol{p} \times \boldsymbol{m}\|}. \qquad (2.83)$$

2.10.2 Intersection of a plane and a line

Consider a plane Π with equation $\langle \boldsymbol{n}_\Pi, \boldsymbol{x} \rangle = h$ and a line l with equation $\boldsymbol{x} \times \boldsymbol{m} = \boldsymbol{n}$. Their intersection point $\boldsymbol{p}$ is given as follows. The supporting plane Π_l of line l has surfac

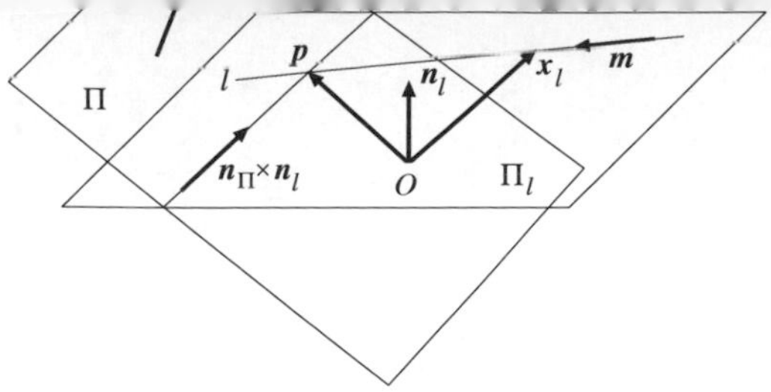

FIGURE 2.14 The intersection $\boldsymbol{p}$ of plane $\langle \boldsymbol{n}_\Pi, \boldsymbol{x} \rangle = h$ and line $\boldsymbol{x} \times \boldsymbol{m} = \boldsymbol{n}_l$.

normal $\boldsymbol{n}_l$, and the plane Π has surface normal $\boldsymbol{n}_\Pi$. Hence, their intersection line is in the direction of $\boldsymbol{n}_\Pi \times \boldsymbol{n}_l$ (Fig. 2.14). This line intersects line l at $\boldsymbol{p}$, the intersection point of plane Π and line l. Choose a point $\boldsymbol{x}_l$ on l in such a way that $\boldsymbol{x}_l$ is parallel to $\boldsymbol{n}_\Pi \times \boldsymbol{n}_l$. This point $\boldsymbol{x}_l$ is given by

$$\boldsymbol{x}_l = c\boldsymbol{n}_\Pi \times \boldsymbol{n}_l \tag{2.84}$$

for some c. Since this point satisfies the equation of line l, we have

$$(c\boldsymbol{n}_\Pi \times \boldsymbol{n}_l) \times \boldsymbol{m} = \boldsymbol{n}_l. \tag{2.85}$$

Using Eq. (2.20) of the vector triple product, the left side reduces to

$$c\Big(\langle \boldsymbol{n}_\Pi, \boldsymbol{m} \rangle \boldsymbol{n}_l - \langle \boldsymbol{n}_l, \boldsymbol{m} \rangle \boldsymbol{n}_\Pi\Big) = c\langle \boldsymbol{n}_\Pi, \boldsymbol{m} \rangle \boldsymbol{n}_l. \tag{2.86}$$

Hence, $c = 1/\langle \boldsymbol{n}_\Pi, \boldsymbol{m} \rangle$, so we can write

$$\boldsymbol{x}_l = \frac{\boldsymbol{n}_\Pi \times \boldsymbol{n}_l}{\langle \boldsymbol{n}_\Pi, \boldsymbol{m} \rangle}. \tag{2.87}$$

The intersection point $\boldsymbol{p}$ is in the direction $\boldsymbol{m}$ from this point, so it is written as

$$\boldsymbol{p} = \boldsymbol{x}_l + C\boldsymbol{m} \tag{2.88}$$

for some C. This point satisfies the equation of plane Π, hence

$$\langle \boldsymbol{n}_\Pi, \boldsymbol{p} \rangle = \langle \boldsymbol{n}_\Pi, \boldsymbol{x}_l \rangle + C\langle \boldsymbol{n}_\Pi, \boldsymbol{m} \rangle = \langle \boldsymbol{n}_\Pi, \frac{\boldsymbol{n}_\Pi \times \boldsymbol{n}_l}{\langle \boldsymbol{n}_\Pi, \boldsymbol{m} \rangle} \rangle + C\langle \boldsymbol{n}_\Pi, \boldsymbol{m} \rangle = C\langle \boldsymbol{n}_\Pi, \boldsymbol{m} \rangle = h. \tag{2.89}$$

Thus, $C = h/\langle \boldsymbol{n}_\Pi, \boldsymbol{m} \rangle$, and we obtain

$$\boldsymbol{p} = \boldsymbol{x}_l + \frac{h\boldsymbol{m}}{\langle \boldsymbol{n}_\Pi, \boldsymbol{m} \rangle} = \frac{\boldsymbol{n}_\Pi \times \boldsymbol{n}_l + h\boldsymbol{m}}{\langle \boldsymbol{n}_\Pi, \boldsymbol{m} \rangle}. \tag{2.90}$$

In summary,

Proposition 2.20 (Intersection of plane and line) *The intersection* $\boldsymbol{p}$ *of plane* $\langle \boldsymbol{n}_\Pi, \boldsymbol{x} \rangle = h$ *and line* $\boldsymbol{x} \times \boldsymbol{m} = \boldsymbol{n}_l$ *is given by*

$$\boldsymbol{p} = \frac{\boldsymbol{n}_\Pi \times \boldsymbol{n}_l + h\boldsymbol{m}}{\langle \boldsymbol{n}_\Pi, \boldsymbol{m} \rangle}. \tag{2.91}$$

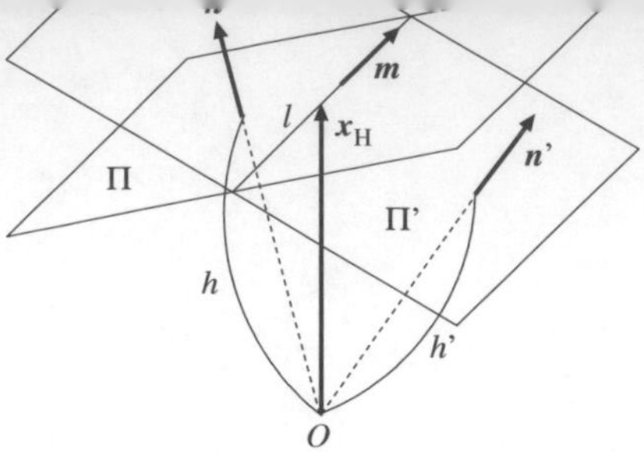

FIGURE 2.15 The intersection l of two planes $\langle \boldsymbol{n}, \boldsymbol{x} \rangle = h$ and $\langle \boldsymbol{n}', \boldsymbol{x} \rangle = h'$.

2.10.3 Intersection of two planes

The intersection of two planes $\langle \boldsymbol{n}, \boldsymbol{x} \rangle = h$ and $\langle \boldsymbol{n}', \boldsymbol{x} \rangle = h'$ is computed as follows. Since the direction $\boldsymbol{m}$ of the intersection line is orthogonal to the surface normals $\boldsymbol{n}$ and $\boldsymbol{n}'$ of the two planes, we can write $\boldsymbol{m} = c\boldsymbol{n} \times \boldsymbol{n}'$ for some c (Fig. 2.15). The supporting point $\boldsymbol{x}_H$ of the intersection line l is on the plane spanned by $\boldsymbol{n}$ and $\boldsymbol{n}'$, so we can write $\boldsymbol{x}_H = a\boldsymbol{n} + b\boldsymbol{n}'$ for some a and b. Since this point is on the two planes, we have

$$\langle \boldsymbol{n}, \boldsymbol{x}_H \rangle = a + b\langle \boldsymbol{n}, \boldsymbol{n}' \rangle = h, \qquad \langle \boldsymbol{n}', \boldsymbol{x}_H \rangle = a\langle \boldsymbol{n}, \boldsymbol{n}' \rangle + b = h'. \tag{2.92}$$

Solving these equations for a and b, we obtain

$$a = \frac{h - h'\langle \boldsymbol{n}, \boldsymbol{n}' \rangle}{1 - \langle \boldsymbol{n}, \boldsymbol{n}' \rangle^2}, \qquad b = \frac{h' - h\langle \boldsymbol{n}, \boldsymbol{n}' \rangle}{1 - \langle \boldsymbol{n}, \boldsymbol{n}' \rangle^2}. \tag{2.93}$$

Hence, the supporting point $\boldsymbol{x}_H$ is given by

$$\boldsymbol{x}_H = \frac{(h - h'\langle \boldsymbol{n}, \boldsymbol{n}' \rangle)\boldsymbol{n} + (h' - h\langle \boldsymbol{n}, \boldsymbol{n}' \rangle)\boldsymbol{n}'}{1 - \langle \boldsymbol{n}, \boldsymbol{n}' \rangle^2}. \tag{2.94}$$

Let $\boldsymbol{m} \times \boldsymbol{x} = \boldsymbol{n}_l$ be the equation of the intersection line l. Since the supporting point $\boldsymbol{x}_H$ is on this line, we have

$$\begin{aligned}
\boldsymbol{n}_l = \boldsymbol{m} \times \boldsymbol{x}_H &= \frac{(h - h'\langle \boldsymbol{n}, \boldsymbol{n}' \rangle)\boldsymbol{m} \times \boldsymbol{n} + (h' - h\langle \boldsymbol{n}, \boldsymbol{n}' \rangle)\boldsymbol{m} \times \boldsymbol{n}'}{1 - \langle \boldsymbol{n}, \boldsymbol{n}' \rangle^2} \\
&= c\frac{(h - h'\langle \boldsymbol{n}, \boldsymbol{n}' \rangle)(\boldsymbol{n} \times \boldsymbol{n}') \times \boldsymbol{n} + (h' - h\langle \boldsymbol{n}, \boldsymbol{n}' \rangle)(\boldsymbol{n} \times \boldsymbol{n}') \times \boldsymbol{n}'}{1 - \langle \boldsymbol{n}, \boldsymbol{n}' \rangle^2} \\
&= c\frac{(h - h'\langle \boldsymbol{n}, \boldsymbol{n}' \rangle)(\boldsymbol{n}' - \langle \boldsymbol{n}, \boldsymbol{n}' \rangle\boldsymbol{n}) - (h' - h\langle \boldsymbol{n}, \boldsymbol{n}' \rangle)(\boldsymbol{n} - \langle \boldsymbol{n}, \boldsymbol{n}' \rangle\boldsymbol{n}')}{1 - \langle \boldsymbol{n}, \boldsymbol{n}' \rangle^2} \\
&= c\frac{(1 - \langle \boldsymbol{n}, \boldsymbol{n}' \rangle^2)h\boldsymbol{n}' - (1 - \langle \boldsymbol{n}, \boldsymbol{n}' \rangle^2)h'\boldsymbol{n}}{1 - \langle \boldsymbol{n}, \boldsymbol{n}' \rangle^2} = c(h\boldsymbol{n}' - h'\boldsymbol{n}),
\end{aligned} \tag{2.95}$$

where we have used Eq. (2.20) for the vector triple product. Applying the scale normalization of Eq. (2.64) to $\boldsymbol{m}$ $(= c\boldsymbol{n} \times \boldsymbol{n}')$ and this $\boldsymbol{n}_l$, we obtain the following:

Proposition 2.21 (Intersection of planes) *The intersection line $\boldsymbol{m} \times \boldsymbol{x} = \boldsymbol{n}_l$ of two planes $\langle \boldsymbol{n}, \boldsymbol{x} \rangle = h$ and $\langle \boldsymbol{n}', \boldsymbol{x} \rangle = h'$ is given by*

$$\boldsymbol{m} = \frac{\boldsymbol{n} \times \boldsymbol{n}'}{\sqrt{\|\boldsymbol{n} \times \boldsymbol{n}'\|^2 + \|h\boldsymbol{n}' - h'\boldsymbol{n}\|^2}}, \qquad \boldsymbol{n}_l = \frac{h\boldsymbol{n}' - h'\boldsymbol{n}}{\sqrt{\|\boldsymbol{n} \times \boldsymbol{n}'\|^2 + \|h\boldsymbol{n}' - h'\boldsymbol{n}\|^2}}. \tag{2.96}$$

a fundamental basis of classical mechanics and electromagnetism. It is also indispensable to shape modeling and rendering for computer graphics. In the subsequent chapters, we introduce new elements and operations to this formulation and describe Hamilton's quaternion algebra, Grassmann's outer product algebra, the Clifford algebra, and the Grassmann–Cayley algebra, in turn. However, this is not the historical order. Historically, Hamilton's and Grassmann's algebras are the oldest. It is *Gibbs* (Josiah Willard Gibbs: 1839–1902), an American physicist, who simplified the algebras of Hamilton and Grassmann to minimum necessary components that are sufficient to describe physics, establishing today's vector calculus.

The "inner product" introduced in Sec. 2.3 is, strictly speaking, called a *Euclidean metric*. A space equipped with such a metric is said to be a *Euclidean space*. If the positivity is removed, the resulting product is called a *non-Euclidean metric*, and a space equipped with such a metric is said to be *non-Euclidean*. We introduce a non-Euclidean space in Chapter 8.

There are some authors who call the "vector product" in Sec. 2.4 the "outer (or exterior) product," but we define in Chapter 5 the outer (or exterior) product of Grassmann, so we do not use this term to mean the vector product, although they are closely related; actually they mean the same thing in a sense. The term "rejection" in Sec. 2.6 was introduced by Hestenes and Sobczyk [12]. The Rodrigues formula of Eq. (2.46) was introduced by *Benjamin Olinde Rodrigues* (1795–1851), a French mathematician.

2.12 EXERCISES

2.1. Show that Eq. (2.10) holds.

2.2. Show that the Schwarz inequality of Eq. (2.12) holds.

2.3. Show that the triangle inequality of Eq. (2.13) holds. Why is this called the "triangle" inequality?

2.4. Show that the linearity relation $\boldsymbol{a} \times (\alpha \boldsymbol{b} + \beta \boldsymbol{c}) = \alpha \boldsymbol{a} \times \boldsymbol{b} + \beta \boldsymbol{a} \times \boldsymbol{c}$ holds for the vector product.

2.5. Show that the parallelogram defined by two vectors $\boldsymbol{a} = a_1 e_1 + a_2 e_2$ and $\boldsymbol{b} = b_1 e_1 + b_2 e_2$ in the xy plane after making their starting points coincide has area

$$S = a_1 b_2 - a_2 b_2,$$

where the rotation of $\boldsymbol{a}$ toward $\boldsymbol{b}$ around the z-axis is in the right-handed screw sense.

2.6. Show that the parallelogram defined by two vectors $\boldsymbol{a} = a_1 e_1 + a_2 e_2 + a_3 e_3$ and $\boldsymbol{b} = b_1 e_1 + b_2 e_2 + b_3 e_3$ after making their starting points coincide has area

$$S = \sqrt{(a_2 b_3 - a_3 b_2)^2 + (a_3 b_1 - a_1 b_3)^2 + (a_1 b_2 - a_2 b_1)^2}.$$

2.7. Let S be the area of the parallelogram defined by vectors $\boldsymbol{a}$ and $\boldsymbol{b}$, and let S_{yz}, S_{zx}, and S_{xy} be the areas of the projections of that parallelogram onto the yz, the zx, and the xy planes, respectively (Fig. 2.16). Show that the following relationship holds:

$$S = \sqrt{S_{yz}^2 + S_{zx}^2 + S_{xy}^2}.$$

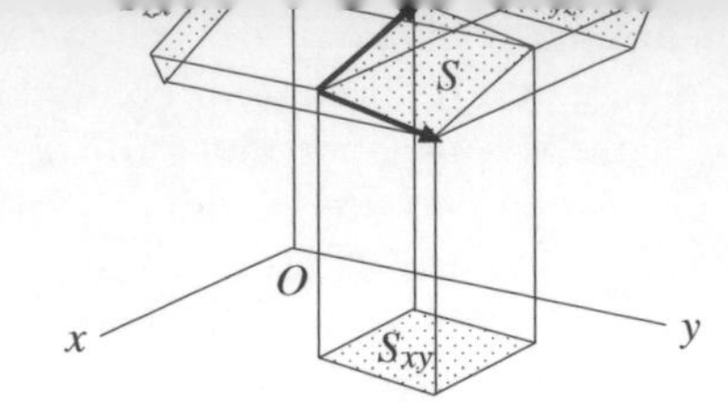

FIGURE 2.16 Projections of a parallelogram onto the xy, yz, and zx planes.

2.8. Confirm that the vector product $\boldsymbol{a} \times \boldsymbol{b}$ in Eq. (2.18) is orthogonal to both $\boldsymbol{a}$ and $\boldsymbol{b}$.

2.9. Show that $\langle \boldsymbol{a} \times \boldsymbol{b}, \boldsymbol{c} \rangle$ equals the signed volume of the parallelepiped defined by vecto $\boldsymbol{a}$, $\boldsymbol{b}$, and $\boldsymbol{c}$ after making their starting points coincide, being positive if $\boldsymbol{a}$, $\boldsymbol{b}$, and are a right-handed system and negative otherwise.

2.10. Show the following identities:

$$(\boldsymbol{a} \times \boldsymbol{b}) \times \boldsymbol{c} + (\boldsymbol{b} \times \boldsymbol{c}) \times \boldsymbol{a} + (\boldsymbol{c} \times \boldsymbol{a}) \times \boldsymbol{b} = 0,$$
$$\boldsymbol{a} \times (\boldsymbol{b} \times \boldsymbol{c}) + \boldsymbol{b} \times (\boldsymbol{c} \times \boldsymbol{a}) + \boldsymbol{c} \times (\boldsymbol{a} \times \boldsymbol{b}) = 0.$$

2.11. Show that the following equality holds:

$$\langle \boldsymbol{x} \times \boldsymbol{y}, \boldsymbol{a} \times \boldsymbol{b} \rangle = \langle \boldsymbol{x}, \boldsymbol{a} \rangle \langle \boldsymbol{y}, \boldsymbol{b} \rangle - \langle \boldsymbol{x}, \boldsymbol{b} \rangle \langle \boldsymbol{y}, \boldsymbol{a} \rangle.$$

2.12. Let $\boldsymbol{a} = a_1 e_1 + a_2 e_2 + a_3 e_3$ and $\boldsymbol{a}' = a_1' e_1 + a_2' e_2 + a_3' e_3$ in the Rodrigues formula i Eq. (2.46), and express a_1', a_2', and a_3' as expressions in a_1, a_2, and a_3.

2.13. If two lines $\boldsymbol{x} \times \boldsymbol{m} = \boldsymbol{n}$ and $\boldsymbol{x} \times \boldsymbol{m}' = \boldsymbol{n}'$ are parallel, show that the distance betwee them is given by

$$d = \left\| \frac{\boldsymbol{n}}{\|\boldsymbol{m}\|} - \frac{\boldsymbol{n}'}{\|\boldsymbol{m}'\|} \right\|.$$

2.14. Let Π be the plane that passes through line l of equation $\boldsymbol{x} \times \boldsymbol{m} = \boldsymbol{n}_l$ and contair the unit direction vector $\boldsymbol{u}$. If we write the equation of the plane Π as $\langle \boldsymbol{n}, \boldsymbol{x} \rangle = h$ show that $\boldsymbol{n}$ and h are given by

$$\boldsymbol{n} = \frac{\boldsymbol{m} \times \boldsymbol{u}}{\|\boldsymbol{m} \times \boldsymbol{u}\|}, \qquad h = \frac{\langle \boldsymbol{n}_l, \boldsymbol{u} \rangle}{\|\boldsymbol{m} \times \boldsymbol{u}\|}.$$

2.15. Let Π be the plane that passes through point $\boldsymbol{p}$ and contains the unit direction vector $\boldsymbol{u}$ and $\boldsymbol{v}$. If we write the equation of the plane Π as $\langle \boldsymbol{n}, \boldsymbol{x} \rangle = h$, show that $\boldsymbol{n}$ and are given by

$$\boldsymbol{n} = \frac{\boldsymbol{u} \times \boldsymbol{v}}{\|\boldsymbol{u} \times \boldsymbol{v}\|}, \qquad h = \frac{|\boldsymbol{p}, \boldsymbol{u}, \boldsymbol{v}|}{\|\boldsymbol{u} \times \boldsymbol{v}\|}.$$

Oblique Coordinate Systems

In the preceding chapter, an orthogonal Cartesian coordinate system was assumed. Here, we consider an "oblique coordinate system" that is not necessarily orthogonal. If the axes are not orthogonal, the basis vectors are not orthogonal, either. Then, vector components are defined in two different ways: a vector can be expressed as a linear combination of the basis vectors, or it can be expressed as a linear combination of another set of vectors, called "reciprocal basis vectors," that are orthogonal to the basis vectors. The inner product, the vector product, and the scalar triple product have different expressions depending on which convention we use. It is shown, however, that these different expressions can be transformed to each other by means of the "metric tensor" that specifies the inner products among the basis vectors. If we use another oblique coordinate system, the same vector has a different expression, but the "coordinate transformation" can be described in a systematic way. For simplicity, the Cartesian coordinate system is used throughout the subsequent chapters, so the readers who want to know the idea of geometric algebra quickly can skip this chapter in the first reading. This chapter requires some knowledge of linear algebra.

3.1 RECIPROCAL BASIS

Consider an xyz coordinate system whose axes are not necessarily orthogonal to each other. Moreover, the unit of scale is not assumed to be the same for each axis. Such a coordinate system is said to be *oblique*. Let e_1, e_2, and e_3 be the vectors parallel to the x-, y-, and z-axes, respectively, with magnitude equal to the unit of length on each axis (Fig. 3.1(a)). If a vector $\boldsymbol{a}$ is expressed as a linear combination of them in the form

$$\boldsymbol{a} = a^1 e_1 + a^2 e_2 + a^2 e_3, \tag{3.1}$$

we call a^1, a^2, and a^3 the *components* of $\boldsymbol{a}$ with respect to that coordinate system. We follow the convention of using upper indices for the components with respect to an oblique coordinate system.

Let e^1 be the vector orthogonal to the basis vectors e_2 and e_3; its length is determined so that $\langle e_1, e^1 \rangle = 1$. Similarly, let e^2 be the vector orthogonal to e_3 and e_1, and e^3 the vector orthogonal to e_1 and e_2; their lengths are determined so that $\langle e_2, e^2 \rangle = \langle e_3, e^3 \rangle = 1$ (Fig. 3.1(b)). In other words, we define e^1, e^2, and e^3 so that

$$\langle e_i, e^j \rangle = \delta_i^j, \tag{3.2}$$

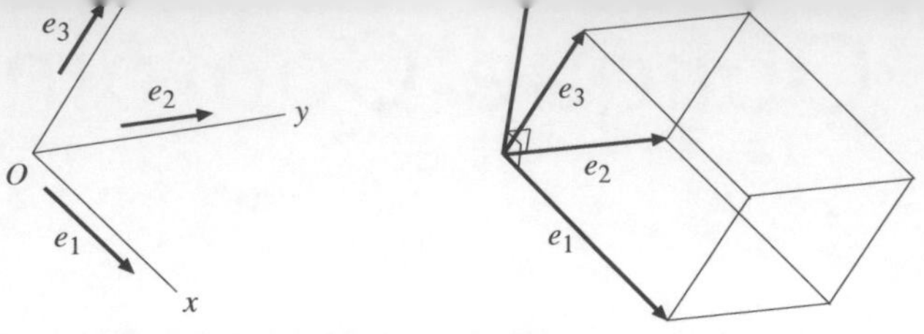

(a) (b)

FIGURE 3.1 (a) The basis $\{e_1, e_2, e_3\}$ of an xyz oblique coordinate system. (b) The reciproc basis vector e^3 is orthogonal to the basis vectors e_1 and e_2.

where the symbol δ_i^j, which we call the *Kronecker delta* just as δ_{ij}, assumes the value 1 fo $i = j$ and 0 otherwise. The set of vectors $\{e^1, e^2, e^3\}$ is called the *reciprocal basis* of th original basis $\{e_1, e_2, e_3\}$.

Since e^1 is orthogonal to e_2 and e_3, we can write $e^1 = ce_2 \times e_3$ for some c. From

$$\langle e_1, e^1 \rangle = \langle e_1, ce_2 \times e_3 \rangle = c|e_1, e_2, e_3| = 1, \quad (3.3$$

we see that $c = 1/|e_1, e_2, e_3|$. The same holds for e^2 and e^3, too. Hence, the reciprocal bas has the following expression:

Proposition 3.1 (Reciprocal basis in terms of basis) *The reciprocal basis* $\{e^1, e^2, e^3$ *of the basis* $\{e_1, e_2, e_3\}$ *of an oblique coordinate system is given by*

$$e^1 = \frac{e_2 \times e_3}{|e_1, e_2, e_3|}, \qquad e^2 = \frac{e_3 \times e_1}{|e_1, e_2, e_3|}, \qquad e^3 = \frac{e_1 \times e_2}{|e_1, e_2, e_3|}. \quad (3.4$$

The scalar triple product $|e_1, e_2, e_3|$ is called the *volume element* of this oblique coord nate system and denoted by the symbol I:

$$I = |e_1, e_2, e_3|. \quad (3.5$$

Computing the inner products of Eq. (3.1) with the reciprocal basis vectors e^1, e^2, and e we see from the orthogonality relation of Eq. (3.2) that

$$\langle e^1, \boldsymbol{a} \rangle = a^1, \qquad \langle e^2, \boldsymbol{a} \rangle = a^2, \qquad \langle e^3, \boldsymbol{a} \rangle = a^3. \quad (3.6$$

Hence,

Proposition 3.2 (Vector components) *The components* a^1, a^2, *and* a^3 *of vector* $\boldsymbol{a}$ *wit respect to an oblique coordinate system are given by*

$$a^i = \langle e^i, \boldsymbol{a} \rangle. \quad (3.7$$

From Eqs. (2.17) and (2.27) in Chapter 2, we observe

Proposition 3.3 (Orthonormal basis) *An orthonormal basis* $\{e_1, e_2, e_3\}$ *is the recip rocal basis of itself:*

$$e^1 = e_1, \qquad e^2 = e_2, \qquad e^3 = e_3. \quad (3.8$$

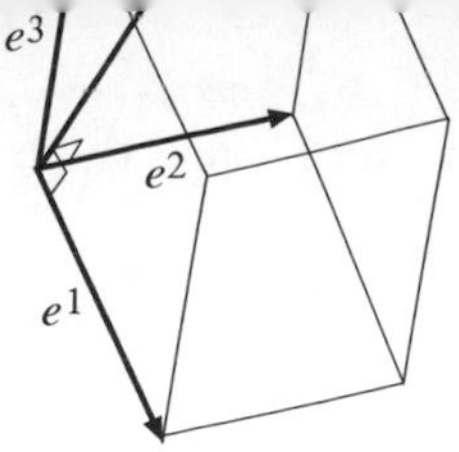

FIGURE 3.2 The basis vector e_3 is orthogonal to the reciprocal basis vectors e^1 and e^2.

3.2 RECIPROCAL COMPONENTS

By definition, the reciprocal basis vectors e^1 and e^2 are both orthogonal to e_3 (Fig. 3.2). Hence, we can write $e_3 = c' e^1 \times e^2$ for some c'. From

$$\langle e_3, e^3 \rangle = \langle c' e^1 \times e^2, e^3 \rangle = c' |e^1, e^2, e^3| = 1, \tag{3.9}$$

we see that $c' = 1/|e^1, e^2, e^3|$. The same holds for e^1 and e^2, too. Hence, we obtain the following expression corresponding to Eq. (3.4):

Proposition 3.4 (Basis in terms of reciprocal basis) *The basis $\{e_1,\ e_2,\ e_3\}$ of an oblique coordinate system is expressed in terms of its reciprocal basis $\{e^1,\ e^2,\ e^3\}$ in the form*

$$e_1 = \frac{e^2 \times e^3}{|e^1, e^2, e^3|}, \qquad e_2 = \frac{e^3 \times e^1}{|e^1, e^2, e^3|}, \qquad e_3 = \frac{e^1 \times e^2}{|e^1, e^2, e^3|}. \tag{3.10}$$

Hence, we observe the following (Fig. 3.2):

Proposition 3.5 (Reciprocal of reciprocal) *The reciprocal of the reciprocal basis $\{e^1,\ e^2,\ e^3\}$ coincides with the original basis $\{e_1,\ e_2,\ e_3\}$.*

From Eq. (3.10), the volume element I has the following expression in terms of the reciprocal basis:

$$\begin{aligned} |e_1, e_2, e_3| &= \left| \frac{e^2 \times e^3}{|e^1, e^2, e^3|}, \frac{e^3 \times e^1}{|e^1, e^2, e^3|}, \frac{e^1 \times e^2}{|e^1, e^2, e^3|} \right| = \frac{|e^2 \times e^3, e^3 \times e^1, e^1 \times e^2|}{|e^1, e^2, e^3|^3} \\ &= \frac{\langle (e^2 \times e^3) \times (e^3 \times e^1), e^1 \times e^2 \rangle}{|e^1, e^2, e^3|^3} = \frac{\langle \langle e^2, e^3 \times e^1 \rangle e^3 - \langle e^3, e^3 \times e^1 \rangle e^2, e^1 \times e^2 \rangle}{|e^1, e^2, e^3|^3} \\ &= \frac{\langle |e^2, e^3, e^1| e^3, e^1 \times e^2 \rangle}{|e^1, e^2, e^3|^3} = \frac{|e^2, e^3, e^1||e^3, e^1, e^2|}{|e^1, e^2, e^3|^3} = \frac{1}{|e^1, e^2, e^3|}. \end{aligned} \tag{3.11}$$

Here, we have used Eqs. (2.20) and (2.31) in Chapter 2 for the vector triple product and the scalar triple product. Thus, we conclude that

Proposition 3.6 (Reciprocal volume element) *The volume element of the basis $\{e_1, e_2, e_3\}$ and the volume element of the reciprocal basis $\{e^1, e^2, e^3\}$ are reciprocal of each other:*

$$|e_1, e_2, e_3||e^1, e^2, e^3| = 1. \tag{3.12}$$

In other words, if we write $|e_1, e_2, e_3| = I$, then $|e^1, e^2, e^3| = I^{-1}$.

$a = a_1 e^1 + a_2 e^2 + a_3 e^3,$ (3.1

the coefficients a_1, a_2, and a_3 are called the *reciprocal components* of $\boldsymbol{a}$. Computing the inn products of Eq. (3.13) with the basis vectors e_1, e_2, and e_3, we see from the orthogonali relation of Eq. (3.2) that

$$\langle e_1, \boldsymbol{a}\rangle = a_1, \qquad \langle e_2, \boldsymbol{a}\rangle = a_2, \qquad \langle e_3, \boldsymbol{a}\rangle = a_3. \qquad (3.1$$

Hence, corresponding to Eq. (3.7),

Proposition 3.7 (Reciprocal components) *The reciprocal components a_1, a_2, and of vector $\boldsymbol{a}$ are given by*

$$a_i = \langle e_i, \boldsymbol{a}\rangle. \qquad (3.1$$

Proposition 3.3 implies that for the Cartesian coordinate system there is no distinctic between the usual and the reciprocal components.

3.3 INNER, VECTOR, AND SCALAR TRIPLE PRODUCTS

In the following, we omit the symbol $\sum_{i=1}^{3}$ in Eqs. (3.1) and (3.13) and simply write

$$\boldsymbol{a} = a^i e_i, \qquad \boldsymbol{a} = a_i e^i. \qquad (3.1$$

To be specific, if the same letter appears in the lower and upper indices, they are understoc to be summed over 1, 2, and 3. This is called *Einstein's summation convention*. Unpaire indices are understood to mean the set with their values running over 1, 2, and 3. F example, a^i is interpreted to represent the set $\{a^1, a^2, a^3\}$.

The inner product of vectors $\boldsymbol{a}$ and $\boldsymbol{b}$ is written from the orthogonality relation Eq. (3.2) as follows ($\hookrightarrow$ Exercise 3.2):

$$\langle \boldsymbol{a}, \boldsymbol{b}\rangle = \langle a^i e_i, b_j e^j\rangle = a^i b_j \langle e_i, e^j\rangle = a^i b_j \delta_i^j = a^i b_i. \qquad (3.1$$

Note that any letter can be used for summation as long as it corresponds between th lower and upper indices. To emphasize this, we often say that the letters for summation a *dummy*. This fact can be utilized to avoid confusion about summation. In Eq. (3.17), fc example, we use different letters for the dummy indices to make clear over which letter th sum is taken. From Eq. (3.17), we observe that

Proposition 3.8 (Inner product and norm) *The inner product of vectors $\boldsymbol{a} = a^i e_i$ an $\boldsymbol{b} = b_i e^i$ is given by*

$$\langle \boldsymbol{a}, \boldsymbol{b}\rangle = a^i b_i. \qquad (3.1$$

In particular, the norm of vector $\boldsymbol{a} = a^i e_i = a_i e^i$ is given by

$$\|\boldsymbol{a}\| = \sqrt{a^i a_i}. \qquad (3.1$$

If vectors $\boldsymbol{a}$ and $\boldsymbol{b}$ are expressed with respect to the basis e_i in the form $\boldsymbol{a} = a^i e_i$ and $= b^i e_i$, the vector product computation of Eq. (2.18) in Chapter 2 becomes as follows:

$$\begin{aligned}
\boldsymbol{a} \times \boldsymbol{b} &= (a^1 e_1 + a^2 e_2 + a^3 e_3) \times (b^1 e_1 + b^2 e_2 + b^3 e_3) \\
&= a^1 b^1 e_1 \times e_1 + a^1 b^2 e_1 \times e_2 + a^1 b^3 e_1 \times e_3 + a^2 b^1 e_2 \times e_1 + a^2 b^2 e_2 \times e_2 \\
&\quad + a^2 b^3 e_2 \times e_3 + a^3 b^1 e_3 \times e_1 + a^3 b^2 e_3 \times e_2 + a^3 b^3 e_3 \times e_3 \\
&= (a^2 b^3 - a^3 b^2) e^1 |e_1, e_2, e_3| + (a^3 b^1 - a^1 b^3) e_2 |e_1, e_2, e_3| + (a^1 b^2 - a^2 b^1) e_3 |e_1, e_2, e_3| \\
&= \Big((a^2 b^3 - a^3 b^2) e^1 + (a^3 b^1 - a^1 b^3) e^2 + (a^1 b^2 - a^2 b^1) e^3 \Big) I. \qquad (3.2
\end{aligned}$$

Namely,

$$a \times b = I\epsilon_{ijk}a^i b^j e^k, \tag{3.21}$$

where I is the volume element of the basis e_i.

As in the computation of Eq. (2.28) in Chapter 2, the scalar triple product has the following expression:

$$\begin{aligned}
|\boldsymbol{a}, \boldsymbol{b}, \boldsymbol{c}| &= |a^1 e_1 + a^2 e_2 + a^3 e_3, b^1 e_1 + b^2 e_2 + b^3 e_3, c^1 e_1 + c^2 e_2 + c^3 e_3| \\
&= a^1 b^2 c^3 |e_1, e_2, e_3| + a^2 b^3 c^1 |e_2, e_3, e_1| + a^3 b^1 c^2 |e_3, e_1, e_2| \\
&\quad + a^1 b^3 c^2 |e_1, e_3, e_2| + a^2 b^1 c^3 |e_2, e_1, e_3| + a^3 b^2 c^1 |e_3, e_2, e_1| \\
&= (a^1 b^2 c^3 + a^2 b^3 c^1 + a^3 b^1 c^2 - a^1 b^3 c^2 - a^2 b^1 c^3 - a^3 b^2 c^1)|e_1, e_2, e_3|.
\end{aligned} \tag{3.22}$$

This can be summarized in the following form:

Proposition 3.10 (Scalar triple product) *The scalar triple product of $\boldsymbol{a} = a^i e_i$, $\boldsymbol{b} = b^i e_i$, and $\boldsymbol{c} = c^i e_i$ is given by*

$$|\boldsymbol{a}, \boldsymbol{b}, \boldsymbol{c}| = I\epsilon_{ijk} a^i b^j c^k, \tag{3.23}$$

where I is the volume element of the basis e_i.

From the above argument, we can see that the inner product, the vector product, and the scalar triple product have simple forms if expressions in terms of the basis e_i and expressions in terms of the reciprocal basis e^i are properly combined. The rule of thumb is that *expressions are so arranged that upper and lower indices are summed.*

3.4 METRIC TENSOR

In the preceding section, we computed the inner product by combining expressions in terms of the basis e_i and expressions in terms of the reciprocal basis e^i so that summation takes place over upper and lower indices. Here, we compute the inner product using only expressions in terms of the basis e_i. Exploiting the symmetry and the linearity of the inner product, we can reduce the computation to the inner products among the basis vectors e_1, e_2, and e_3. For an oblique coordinate system, however, they are not necessarily unit vectors, and the inner product of different vectors may not be 0. So, we let

$$\langle e_i, e_j \rangle = g_{ij}. \tag{3.24}$$

Thus, g_{ij} is a quantity that specifies the lengths of the individual basis vectors e_1, e_2, and e_3 and their pairwise angles. For example, the vector e_1 has length $\sqrt{g_{11}}$, and the vectors e_1 and e_2 make an angle θ_{12} such that $\cos\theta_{12} = g_{12}/\sqrt{g_{11}}\sqrt{g_{22}}$. We call the set of numbers g_{ij} the *metric tensor* or simply the *metric*. The term "tensor" is used for a set of vector components or matrix elements that have some geometric or physical meaning. By definition, g_{ij} is symmetric with respect to the indices:

$$g_{ij} = g_{ji}. \tag{3.25}$$

We express this fact by saying that g_{ij} is a *symmetric tensor*. Equation (2.6) in Chapter 2 means that the metric tensor of the Cartesian coordinate system is δ_{ij}. In terms of the metric tensor g_{ij}, the inner product of vectors $\boldsymbol{a} = a^i e_i$ and $\boldsymbol{b} = b^i e_i$ is expressed as follows:

$$\langle \boldsymbol{a}, \boldsymbol{b} \rangle = \langle a^i e_i, b^j e_j \rangle = a^i b^j \langle e_i, e_j \rangle = a^i b^j g_{ij}. \tag{3.26}$$

we obtain

Proposition 3.11 (Inner product and norm in terms of metric) *The inner produ of vectors* $\boldsymbol{a} = a^i e_i$ *and* $\boldsymbol{b} = b^i e_i$ *is given by*

$$\langle \boldsymbol{a}, \boldsymbol{b} \rangle = g_{ij} a^i b^j. \tag{3.27}$$

In particular, the norm of vector $\boldsymbol{a} = a^i e_i$ *is given by*

$$\|\boldsymbol{a}\| = \sqrt{g_{ij} a^i a^j}. \tag{3.28}$$

The reason that the symbol g_{ij} is called the "metric" is that g_{ij} provides the basis length measurement, as shown in Eq. (3.28).

3.5 RECIPROCITY OF EXPRESSIONS

If vector $\boldsymbol{a}$ has expression $\boldsymbol{a} = a_i e^i$ with respect to the reciprocal basis, its inner produc with vector $\boldsymbol{b} = b^i e_i$ is $\langle \boldsymbol{a}, \boldsymbol{b} \rangle = a_i b^i$ from Eq. (3.18). Comparing this with Eq. (3.27), w obtain $g_{ij} a^i b^j = a_j b^j$. This must hold identically whatever b^j is, so

$$g_{ij} a^i = a_j. \tag{3.29}$$

This can be regarded as a set of simultaneous linear equations in a^i. We write the solutic a^i in the form

$$a^i = g^{ij} a_j, \tag{3.30}$$

where g^{ij} is the (i, j) element of the inverse of the matrix whose (i, j) element is g_{ij} (Exercise 3.3(1)). Since the inverse of a symmetric matrix is also symmetric, Eq. (3.2 implies that g^{ij} is also symmetric:

$$g^{ij} = g^{ji}. \tag{3.31}$$

The product of a matrix and its inverse is an identity, whose (i, j) element is δ_i^j. This fa is written as

$$g_{ik} g^{kj} = \delta_i^j. \tag{3.32}$$

Thus, we obtain

Proposition 3.12 (Reciprocity of components) *The components with respect to* e_i *an the reciprocal components with respect to* e^i *of vector* $\boldsymbol{a}$ *are related by*

$$a_i = g_{ij} a^j, \qquad a^i = g^{ij} a_j. \tag{3.33}$$

Vector $\boldsymbol{a}$ can be expressed in two ways, i.e., as $a^i e_i$ with respect to the basis e_i and a $a_i e^i$ with respect to the reciprocal basis e^i. Hence,

$$a^i e_i = a_i e^i = g_{ij} a^j e^i = a^i (g_{ij} e^j), \tag{3.34}$$

where we have noted the symmetry of the indices of g_{ij} and changed the dummy indice for summation. Equation (3.34) must identically hold whatever a^i is, so

$$g_{ij} e^j = e_i. \tag{3.35}$$

$$e^i = g^{ij} e_j. \tag{3.36}$$

Hence, we obtain

Proposition 3.13 (Reciprocity of basis) *The basis e_i and its reciprocal e^i are related by*

$$e_i = g_{ij} e^j, \qquad e^i = g^{ij} e_j. \tag{3.37}$$

Equations (3.33) and (3.37) imply that g_{ij} can be interpreted as an "operator" to *lower upper indices* and g^{ij} to *raise lower indices* both for vector components and for the basis vectors. From the second equation in Eq. (3.37), we obtain

$$\langle e^i, e^j \rangle = \langle g^{ik} e_k, e^j \rangle = g^{ik} \langle e_k, e^j \rangle = g^{ik} \delta_k^j = g^{ij}, \tag{3.38}$$

which corresponds to Eq. (3.24) for the basis e_i. Using this, we can express the inner product of $\boldsymbol{a} = a_i e^i$ and $\boldsymbol{b} = b_i e^i$ as

$$\langle \boldsymbol{a}, \boldsymbol{b} \rangle = \langle a_i e^i, b_j e^j \rangle = a_i b_j \langle e^i, e^j \rangle = a_i b_j g^{ij}, \tag{3.39}$$

which corresponds to Eq. (3.27). This result is also obtained by applying the second equation in Eq. (3.33) to a^i in Eq. (3.18).

If we apply the second equation in Eq. (3.33) to a^i and b^j in Eq. (3.21), the vector product is expressed in the form

$$\boldsymbol{a} \times \boldsymbol{b} = I \epsilon_{ijk} g^{kl} a^i b^j e^l = I \epsilon_{ijk} g^{il} g^{jm} a_l b_m e^k. \tag{3.40}$$

Similarly, applying the second equation in Eq. (3.33) to a^i, b^j, and c^k in Eq. (3.23), we can express the scalar triple product in the form

$$|\boldsymbol{a}, \boldsymbol{b}, \boldsymbol{c}| = I \epsilon_{ijk} g^{il} g^{jm} g^{nk} a_l b_m c_n. \tag{3.41}$$

Consider the volume element $I^{-1} = |e^1, e^2, e^3|$ of the reciprocal basis. Since the reciprocal basis vectors e^1, e^2, and e^3 can be written as $\delta_i^1 e^i$, $\delta_i^2 e^i$, and $\delta_i^3 e^i$, respectively, we obtain from Eq. (3.41)

$$I^{-1} = |e^1, e^2, e^3| = I \epsilon_{ijk} g^{il} g^{jm} g^{nk} \delta_l^1 \delta_m^2 \delta_n^3 = I \epsilon_{ijk} g^{i1} g^{j2} g^{n3} = I \epsilon_{ijk} g^{1i} g^{2j} g^{3k}. \tag{3.42}$$

We see from Eq. (2.34) in Chapter 2 that $\epsilon_{ijk} g^{1i} g^{2j} g^{3k}$ equals the determinant of the matrix whose (i, j) element is g^{ij}. This determinant is the reciprocal of its inverse, i.e., the matrix whose (i, j) element is g_{ij}. We denote the determinant of g_{ij}, i.e., the determinant of the matrix whose (i, j) element is g_{ij}, by

$$g = \sum_{i,j,k=1}^{3} \epsilon_{ijk} g_{1i} g_{2j} g_{3k}, \tag{3.43}$$

where Einstein's summation convention is not used, since the sum is not over upper and lower indices. From the above argument, we conclude that

$$I^{-1} = I g^{-1}, \tag{3.44}$$

which is rewritten as $g = I^2$. Hence, we obtain

$$I = \pm\sqrt{g}. \tag{3.45}$$

where the sign is positive if e_i is a right-handed basis and negative if it is left-handed.

Traditional World 3.1 (Curvilinear coordinate systems) In physics, *curvilinear c ordinate systems* consisting of coordinate curves are frequently used, typical ones being t $r\phi\theta$ *spherical coordinates system* and the $\rho\phi z$ *cylindrical coordinate system* (↪ Exercise 3. 3.5). Changing r for fixed ϕ and θ in the spherical coordinate system results in radial ra from the origin; changing ϕ alone results in parallel circles around a sphere; changing θ alo results in meridians along a sphere. In the cylindrical coordinate system, changing ρ alo results in planar radial rays orthogonal to the z-axis; changing ϕ alone results in parall circles around a cylinder; changing z alone results in lines parallel to the z-axis. Curviline coordinate systems are suitable for describing physical phenomena when the surroundi space has some symmetry, or something is constant in some directions, or boundaries specified shapes exist. Then, the description is based on a *local coordinate system* who origin is at each point and whose axes are tangent to the coordinate curves there. This is general an oblique coordinate system. For the spherical and cylindrical coordinate system the local coordinate axes are orthogonal to each other, but their coordinate scales do n directly reflect the physical length. Hence, the metric tensor g_{ij} is not necessarily δ_{ij}. F such a coordinate system, the mathematical formulation described in this chapter is sui able, but since the local coordinate system smoothly changes as we move in space, quantiti like the metric tensor g_{ij} and the volume element I are position-dependent smooth field Mathematics that describes such a situation is the traditional *tensor calculus*.

3.6 COORDINATE TRANSFORMATIONS

Since coordinate systems are used merely for the convenience of description, we can use i principle any coordinate system we like. Suppose we use a new $x'y'z'$ coordinate syste different from the current xyz coordinate system. Let $e_{i'}$ be the basis of the $x'y'z'$ coordina system. Suppose this $e_{i'}$ is expressed in terms of the original basis e_i in the form

$$e_{i'} = A^i_{i'} e_i. \tag{3.46}$$

Here, we are following the convention that the primes are put not to the symbols but t their indices. If vector $\boldsymbol{a} = a^i e_i$ is expressed as $\boldsymbol{a} = a^{i'} e_{i'}$ for the new basis, we can wri from Eq. (3.46)

$$\boldsymbol{a} = a^i e_i = a^{i'} e_{i'} = a^{i'}(A^i_{i'} e_i) = (A^i_{i'} a^{i'}) e_i. \tag{3.47}$$

Hence, we obtain

$$A^i_{i'} a^{i'} = a^i. \tag{3.48}$$

This can be regarded as a set of simultaneous linear equations in $a^{i'}$. We write the solutio $a^{i'}$ in the form

$$a^{i'} = A^{i'}_i a^i, \tag{3.49}$$

where $A^{i'}_i$ is the (i', i) element of the inverse of the matrix whose (i, i') element is $A^i_{i'}$ (↪ Exercise 3.6(1)). Since the product of a matrix and its inverse is the identity matrix, w have the following relationships:

$$A^i_{i'} A^{i'}_j = \delta^i_j, \qquad A^{i'}_i A^i_{j'} = \delta^{i'}_{j'}. \tag{3.50}$$

Proposition 3.15 (Transformation of vector components) *If the original basis e_i and the new basis $e_{i'}$ are related by*

$$e_{i'} = A^i_{i'} e_i, \qquad e_i = A^{i'}_i e_{i'}, \tag{3.51}$$

and if vector $\boldsymbol{a}$ is expressed with respect to them as $\boldsymbol{a} = a^{i'} e_{i'} = a^i e_i$, the components $a^{i'}$ and a^i are related by

$$a^{i'} = A^{i'}_i a^i, \qquad a^i = A^i_{i'} a^{i'}. \tag{3.52}$$

Let $e^{i'}$ be the reciprocal basis of the new coordinate system. It is expressed as a linear combination of the reciprocal basis e^i of the original coordinate system. Suppose we have

$$e^{i'} = B^{i'}_i e^i, \qquad e^i = B^i_{i'} e^{i'}, \tag{3.53}$$

where $B^{i'}_i$ is the (i', i) element of the inverse of the matrix whose (i, i') element is $B^i_{i'}$. From the definition of the reciprocal basis, the following equalities hold:

$$\delta^{i'}_{j'} = \langle e^{i'}, e_{j'} \rangle = \langle B^{i'}_i e^i, A^j_{j'} e_j \rangle = B^{i'}_i A^j_{j'} \langle e^i, e_j \rangle = B^{i'}_i A^j_{j'} \delta^i_j = B^{i'}_i A^i_{j'}. \tag{3.54}$$

Comparing this with the second equation in Eq. (3.50), we obtain $B^{i'}_i = A^{i'}_i$ and hence $B^i_{i'} = A^i_{i'}$ as well. Thus, we observe that

Proposition 3.16 (Transformation of reciprocal basis) *The original reciprocal basis e^i and the new reciprocal basis $e^{i'}$ are related by*

$$e^{i'} = A^{i'}_i e^i, \qquad e^i = A^i_{i'} e^{i'}. \tag{3.55}$$

Suppose vector $\boldsymbol{a}$ is expressed in terms of the new reciprocal basis as $\boldsymbol{a} = a_{i'} e^{i'}$ and also expressed as $\boldsymbol{a} = a_i e^i$ with respect to the original reciprocal basis. Then,

$$\boldsymbol{a} = a_i e^i = a_i A^i_{i'} e^{i'} = (A^i_{i'} a_i) e^{i'}. \tag{3.56}$$

This implies that $a_{i'} = A^i_{i'} a_i$. Solving this for a_i, we obtain $a_i = A^{i'}_i a_{i'}$ (↪ Exercise 3.6(4)). In summary,

Proposition 3.17 (Transformation of reciprocal components) *If vector $\boldsymbol{a}$ is expressed in terms of the new and the original reciprocal bases as $\boldsymbol{a} = a_{i'} e^{i'} = a_i e^i$, the components $a_{i'}$ and a_i are related by*

$$a_{i'} = A^i_{i'} a_i, \qquad a_i = A^{i'}_i a_{i'}. \tag{3.57}$$

Equations (3.51), (3.52), (3.55), and (3.57) imply that we can interpret $A^{i'}_i$ and $A^i_{i'}$ as "operators" to *interchange the corresponding indices i and i'* both for the basis and for vector components. It can also be seen that we can move $A^i_{i'}$ from one side to the other by changing it to $A^{i'}_i$ and move $A^{i'}_i$ from one side to the other by changing it to $A^i_{i'}$, just as in linear algebra, where we can move matrix $\boldsymbol{A}$ from one side to the other by changing it to $\boldsymbol{A}^{-1}$.

If g_{ij} is the metric tensor of the original basis e_i, the metric tensor $g_{i'j'}$ of the new basis $e_{i'}$ is given by

$$g_{i'j'} = \langle e_{i'}, e_{j'} \rangle = \langle A^i_{i'} e_i, A^j_{j'} e_j \rangle = A^i_{i'} A^j_{j'} \langle e_i, e_j \rangle = A^i_{i'} A^j_{j'} g_{ij}. \tag{3.58}$$

This can be rewritten as $g_{ij} = A^{i'}_i A^{j'}_j g_{i'j'}$, where $A^{i'}_i$ is the inverse of the transformation $A^i_{i'}$ (↪ Exercise 3.6(5)). In summary, we obtain

related by

$$g_{i'j'} = A^i_{i'} A^j_{j'} g_{ij}, \qquad g_{ij} = A^{i'}_i A^{j'}_j g_{i'j'}. \tag{3.59}$$

If the original coordinate system is Cartesian, in particular, we can write the metric tensor $g_{i'j'}$ in the form

$$g_{i'j'} = A^i_{i'} A^j_{j'} \delta_{ij} = \sum_{i=1}^{3} A^i_{i'} A^i_{j'}, \tag{3.60}$$

where we use the summation symbol $\sum_{i=1}^{3}$ since the sum is not over the upper and lower indices. From Eq. (3.38), we see that the tensor $g^{i'j'}$ for the new coordinate system is related to the tensor g^{ij} for the original coordinate system by

$$g^{i'j'} = \langle e^{i'}, e^{j'} \rangle = \langle A^{i'}_i e^i, A^{j'}_j e^j \rangle = A^{i'}_i A^{j'}_j \langle e^i, e^j \rangle = A^{i'}_i A^{j'}_j g^{ij}, \tag{3.61}$$

which corresponds to Eq. (3.58). Using the inverse $A^i_{i'}$ of $A^{i'}_i$, we can rewrite this as $g^{ij} = A^i_{i'} A^j_{j'} g^{i'j'}$ (↪ Exercise 3.6(6)). In summary, we obtain

Proposition 3.19 (Transformation of g^{ij}) *The tensor g^{ij} for the original coordinate system and the tensor $g^{i'j'}$ for the new coordinate system are related by*

$$g^{i'j'} = A^{i'}_i A^{j'}_j g^{ij}, \qquad g^{ij} = A^i_{i'} A^j_{j'} g^{i'j'}. \tag{3.62}$$

Equations (3.58) and (3.62) imply that we can interpret $A^{i'}_i$ and $A^i_{i'}$ as "operators" to *interchange the corresponding indices i and i'* of tensors in the same way as in the case of the basis and vector components.

Now, the volume element I' of the new coordinate system is defined by

$$I' = |e_{1'}, e_{2'}, e_{3'}| = |A^i_{1'} e_i, A^j_{2'} e_j, A^k_{3'} e_k| = A^i_{1'} A^j_{2'} A^k_{3'} |e_i, e_j, e_k|, \tag{3.63}$$

and the scalar triple product $|e_i, e_j, e_k|$ equals $|e_1, e_2, e_3| = I$ if (i, j, k) is an even permutation of (1,2,3), $-I$ if it is an odd permutation, and 0 otherwise. Hence,

$$|e_i, e_j, e_k| = I\epsilon_{ijk}. \tag{3.64}$$

As pointed out in Eq. (2.34) in Chapter 2, $\epsilon_{ijk} A^i_{1'} A^j_{2'} A^k_{3'}$ is simply the determinant of the matrix whose (i, i') element is $A^i_{i'}$. So, let us write

$$|A| = \epsilon_{ijk} A^i_{1'} A^j_{2'} A^k_{3'}. \tag{3.65}$$

Then, we obtain from Eq. (3.63)

Proposition 3.20 (Transformation of volume element) *The volume element I of the original coordinate system and the volume element I' of the new coordinate system are related by*

$$I' = |A|I. \tag{3.66}$$

of a *linear space*, and their elements are called *vectors*. The set of matrices or the set of functions (over some domain) are vector spaces in this sense, since matrices and functions can be added or multiplied by scalars. The set of arrays of numbers such as a_i and b^i also makes a vector space in this sense. Hence, in tensor calculus, such arrays of numbers (whether their alignment is horizontal or vertical does not matter) are called "vectors," and we say "vector a_i" and "vector b^i." It appears at first sight that the omission of the basis vectors e_i or e^i does not make much difference. However, a crucial issue is involved.

The problem is that while a^i transforms in the form of Eq. (3.52) when the coordinate system is changed, a_i transforms in the form of Eq. (3.57). In the framework of viewing arrays of numbers as vectors, we call a vector a^i that transforms in the form of Eq. (3.52) a *contravariant vector* and a vector a_i that transforms in the form of Eq. (3.57) a *covariant vector*. These terms stem from the fact that, as we see from Eq. (3.51), a_i transforms in the "opposite" way to the basis e_i, while a_i transforms in the "same" way as e_i. We also call Eqs. (3.52) and (3.57) the *rules of coordinate transformation* for contravariant and covariant vectors, respectively. Furthermore, we call a symbol with multiple lower indices, like the metric tensor g_{ij}, that transforms in the form of Eq. (3.58) a *covariant tensor* and a symbol with multiple upper indices, like g^{ij}, that transforms in the form of Eq. (3.63) a *contravariant tensor*. Equations (3.58) and (3.63) are the rules of coordinate transformation for covariant and contravariant tensors (of degree 2), respectively.

Making such distinctions is very convenient in physics, because quantities such as velocity and displacement that depend on positions are usually described as contravariant vectors, while quantities that act on movement, such as force, electric field, and magnetic field, as well as gradients of intensities, such as temperature gradient and pressure gradient, that indicate the directions normal to their equivalue contours are usually described as covariant vectors. This distinction makes it easy to understand the physical meaning. In writing equations of physical laws, contravariant vectors are never added to covariant vectors, and both sides of an equation must be of the same kind. Also, indices over which summation takes place must correspondingly appear in upper and lower positions. Such consistency of indices plays an important role in Einstein's general theory of relativity.

Traditional World 3.3 (Axial and polar vectors) Regarding arrays of numbers as vectors causes some problems. For example, if we define from vectors a^i and b^i a new vector $c_k = \epsilon_{ijk}a^i b^j$, what does it represent? In other words, what magnitude and direction does $\boldsymbol{c} = c_k e^k$ have? From Eq. (3.21), we see that $I\boldsymbol{c}$ equals the vector product $\boldsymbol{a} \times \boldsymbol{b}$ of $\boldsymbol{a} = a^i e_i$ and $\boldsymbol{b} = b^i e_i$. However, the volume element I has different signs depending on whether the coordinate system is right-handed or left-handed. We call vectors, like c_k, that change sign depending on the handedness of the coordinate system *axial vectors* or *pseudovectors*; those that do not change sign are called *polar vectors*. In physics, axial vectors appear in relation to the axis of rotation or the direction of rotational motion; the term "axial" originated from this. Again, the consistency that quantities of different kinds, such as axial and polar, are never added or equated plays an important role in describing physical laws. Note that such characterization was caused by viewing arrays of numbers as vectors. If vectors are regarded as geometric objects equipped with magnitude and direction as we do in this book, we need not make any distinction as to whether "vector $\boldsymbol{a}$" is contravariant or covariant or whether it is axial or polar, because it is defined independently of the coordinate system; the term "coordinate-free" is used to mean this. Hence, the vector $\boldsymbol{a}$ has the same meaning

3.7 SUPPLEMENTAL NOTE

The purpose of this chapter is to show that when the coordinate system is oblique, there a two ways to express vectors as linear combinations of the basis vectors: the direct use of th basis, and the use of the reciprocal basis. Each has advantages and disadvantages, and th inner product, the vector product, and the scalar triple product have different expressio depending on which basis we use. However, the different expressions can be transforme to each other by means of the metric tensor. Another important fact is that when th coordinate system is changed, the resulting changes of expressions are described by simp rules. Hence, we can obtain equivalent descriptions whatever coordinate system we us Einstein's general theory of relativity is based on the principle that "laws of physics mu be equivalently expressed whatever coordinate system we use." This is the main reason th the traditional tensor calculus plays a central role there. Well-known classical textbooks tensor calculus are the books of Schouten [21, 22].

3.8 EXERCISES

3.1. If vectors $\boldsymbol{a}$, $\boldsymbol{b}$, and $\boldsymbol{c}$ are not coplanar, show that an arbitrary vector $\boldsymbol{x}$ can b expressed as their linear combination in the following form:

$$\boldsymbol{x} = \frac{|\boldsymbol{x},\boldsymbol{b},\boldsymbol{c}|}{|\boldsymbol{a},\boldsymbol{b},\boldsymbol{c}|}\boldsymbol{a} + \frac{|\boldsymbol{a},\boldsymbol{x},\boldsymbol{c}|}{|\boldsymbol{a},\boldsymbol{b},\boldsymbol{c}|}\boldsymbol{b} + \frac{|\boldsymbol{a},\boldsymbol{b},\boldsymbol{x}|}{|\boldsymbol{a},\boldsymbol{b},\boldsymbol{c}|}\boldsymbol{c}.$$

3.2. Show the following relationships:

$$\delta_i^j a^i = a^j, \qquad \delta_i^j a_j = a_i.$$

In other words, show that the Kronecker delta δ_i^j can be regarded as an *operator* f replacing index i by index j or replacing index j by index i.

3.3. Using the tensors g_{ij} and g^{ij} that satisfy Eq. (3.32), show the following. Note th summation is always implied over corresponding upper and lower indices, which a dummy.

(1) Equation (3.30) is obtained by multiplying Eq. (3.29) by g^{ij} on both sides, an Eq. (3.29) is obtained by multiplying Eq. (3.30) by g_{ij} on both sides.

(2) Equation (3.36) is obtained by multiplying Eq. (3.35) by g^{ij} on both sides, an Eq. (3.35) is obtained by multiplying Eq. (3.36) by g_{ij} on both sides.

3.4. A position $\boldsymbol{x}$ in 3-D is expressed in terms of the spherical coordinates r, θ, and ϕ a follows (Fig. 3.3):

$$\boldsymbol{x} = e_1 r \sin\theta\cos\phi + e_2 r \sin\theta\sin\phi + e_3 r\cos\theta.$$

(a) Regarding the vectors that represent the differential changes of the coordinate

$$e_r = \lim_{\Delta r\to 0}\frac{\boldsymbol{x}(r+\Delta r,\theta,\phi)-\boldsymbol{x}(r,\theta,\phi)}{\Delta r} = \frac{\partial\boldsymbol{x}}{\partial r},$$

$$e_\theta = \lim_{\Delta \theta\to 0}\frac{\boldsymbol{x}(r,\theta+\Delta\theta,\phi)-\boldsymbol{x}(r,\theta,\phi)}{\Delta \theta} = \frac{\partial\boldsymbol{x}}{\partial \theta},$$

$$e_\phi = \lim_{\Delta \phi\to 0}\frac{\boldsymbol{x}(r,\theta,\phi+\Delta\phi)-\boldsymbol{x}(r,\theta,\phi)}{\Delta \phi} = \frac{\partial\boldsymbol{x}}{\partial \phi},$$

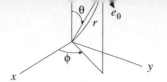

FIGURE 3.3 Spherical coordinate system.

as new basis vectors, compute the corresponding metric tensor g_{ij}, $i, j = r, \theta, \phi$.

(b) Compute the volume element $I_{r\theta\phi} = |e_r, e_\theta, e_\phi|$ for this coordinate system. Using this result, show that a sphere of radius R has volume $4\pi R^3/3$.

3.5. A position $\boldsymbol{x}$ in 3-D is expressed in terms of the cylindrical coordinates r, θ, and z as follows (Fig. 3.4):

$$\boldsymbol{x} = e_1 r \cos\theta + e_2 r \sin\theta + e_3 z.$$

(a) Regarding the vectors that represent the differential changes of the coordinates

$$e_r = \lim_{\Delta r \to 0} \frac{\boldsymbol{x}(r + \Delta r, \theta, z) - \boldsymbol{x}(r, \theta, z)}{\Delta r} = \frac{\partial \boldsymbol{x}}{\partial r},$$

$$e_\theta = \lim_{\Delta \theta \to 0} \frac{\boldsymbol{x}(r, \theta + \Delta\theta, z) - \boldsymbol{x}(r, \theta, z)}{\Delta \theta} = \frac{\partial \boldsymbol{x}}{\partial \theta},$$

$$e_z = \lim_{\Delta z \to 0} \frac{\boldsymbol{x}(r, \theta, z + \Delta z) - \boldsymbol{x}(r, \theta, z)}{\Delta z} = \frac{\partial \boldsymbol{x}}{\partial z},$$

as new basis vectors, compute the corresponding metric tensor g_{ij}, $i, j = r, \theta, z$.

(b) Compute the volume element $I_{r\theta z} = |e_r, e_\theta, e_z|$ of this coordinate system. Using this result, show that a cylinder of height h and radius R has volume $\pi R^2 h$.

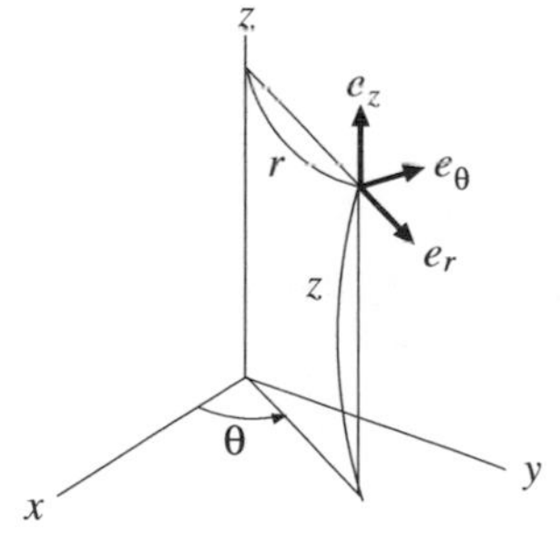

FIGURE 3.4 Cylindrical coordinate system.

3.6. Using $A_i^{i'}$ and $A_{i'}^{i}$ that satisfy Eq. (3.50), show the following. Note that summation is always implied over corresponding upper and lower indices, which are dummy.

(1) The second equation of Eq. (3.51) is obtained by multiplying the first by $A_i^{i'}$ on both sides, and the first equation is obtained by multiplying the second by $A_{i'}^{i}$ on both sides.

on both sides.

(3) The second equation of Eq. (3.55) is obtained by multiplying the first by $A^{i}_{i'}$ o both sides, and the first equation is obtained by multiplying the second by A on both sides.

(4) The second equation of Eq. (3.57) is obtained by multiplying the first by $A^{i'}_{i}$ o both sides, and the first equation is obtained by multiplying the second by A on both sides.

(5) The second equation of Eq. (3.59) is obtained by multiplying the first by $A^{i'}_{i} A$ on both sides, and the first equation is obtained by multiplying the second b $A^{i}_{i'} A^{j}_{j'}$ on both sides.

(6) The second equation of Eq. (3.62) is obtained by multiplying the first by $A^{i}_{i'} A$ on both sides, and the first equation is obtained by multiplying the second b $A^{i'}_{i} A^{j'}_{j}$ on both sides.

Hamilton's Quaternion Algebra

Hamilton's quaternions exhibit a typical algebraic method for defining operations on symbols to describe geometry. A quaternion can be regarded as a combination of a scalar and a vector, and the quaternion product can be viewed as simultaneous computation of the inner product and the vector product. Furthermore, division can be defined for quaternions. This chapter explains the mathematical structure of the set of quaternions and shows how quaternions are suitable for describing 3D rotations. Also, various mathematical facts related to rotations are given.

4.1 QUATERNIONS

The expression

$$q = q_0 + q_1 i + q_2 j + q_3 k \tag{4.1}$$

for four real numbers q_0, q_1,q_2, and q_3 and the three symbols i, j, and k is called a *quaternion*. One may wonder how symbols can be multiplied by numbers or added to other symbols, but additions and scalar multiplications are only formal without having particular meanings. For example, $2i$ only means that the symbol i is counted twice, and the addition "+" merely indicates the set of summands. This is the same as dealing with complex numbers. For example, $2 + 3i$ is nothing but the set of a real number 2 and an imaginary number $3i$, which means that the imaginary unit i is counted three times; it is not that adding a real number and an imaginary number creates something new. Such an addition as a "set operation" is called a *formal sum*. However, the formal sum is not just a mere enumeration of elements. We require that the commutativity, the associativity, and the distributivity that we showed for vectors in Sec. 2.1 of Chapter 2 be satisfied too. In other words, we are allowed to change the order of additions and distribute a real coefficient to each term. Hence, if

$$q' = q_0' + q_1' i + q_2' j + q_3' k \tag{4.2}$$

is another quaternion, we can write

$$2q + 3q' = (2q_0 + 3q_0') + (2q_0 + 3q_0')i + (2q_0 + 3q_0')j + (2q_0 + 3q_0')k. \tag{4.3}$$

Mathematically, we say that the set of all quaternions constitutes a vector space *generated* by the basis $\{1, i, j, k\}$.

Thus, we are free to add quaternions and multiply them with scalars. Next, we define products of the symbols i, j, and k. We define

$$i^2 = -1, \qquad j^2 = -1, \qquad k^2 = -1, \tag{4.}$$

$$jk = i, \qquad ki = j, \qquad ij = k,$$

$$kj = -i, \qquad ik = -j, \qquad ji = -k. \tag{4.}$$

Equation (4.4) implies that each of i, j, and k is the "imaginary unit." Equation (4. suggests that if i, j, and k are identified with the orthonormal basis vectors e_1, e_2, and e_3 Chapter 2, the rule of the vector product of Eq. (2.17) in Chapter 2 applies. From the ru of Eqs. (4.4) and (4.5), the product of Eqs. (4.1) and (4.2) has the following expression:

$$\begin{aligned} qq' &= (q_0 + q_1 i + q_2 j + q_3 k)(q_0' + q_1' i + q_2' j + q_3' k) \\ &= q_0 q_0' + q_1 q_1' i^2 + q_2 q_2' j^2 + q_3 q_3' k^2 + (q_0 q_1' + q_1 q_0')i + (q_0 q_2' + q_2 q_0')j + (q_0 q_3' + q_3 q_0')k \\ &\quad + q_1 q_2' ij + q_1 q_3' ik + q_2 q_1' ji + q_2 q_3' jk + q_3 q_1' ki + q_3 q_2' kj \\ &= q_0 q_0' - q_1 q_1' - q_2 q_2' - q_3 q_3' + (q_0 q_1' + q_1 q_0')i + (q_0 q_2' + q_2 q_0')j + (q_0 q_3' + q_3 q_0')k \\ &\quad + q_1 q_2' k - q_1 q_3' j - q_2 q_1' k + q_2 q_3' i + q_3 q_1' j - q_3 q_2' i \\ &= (q_0 q_0' - q_1 q_1' - q_2 q_2' - q_3 q_3') + (q_0 q_1' + q_1 q_0' + q_2 q_3' - q_3 q_2')i \\ &\quad + (q_0 q_2' + q_2 q_0' + q_3 q_1' - q_1 q_3')j + (q_0 q_3' + q_3 q_0' + q_1 q_2' - q_2 q_1')k. \end{aligned} \tag{4.}$$

4.2 ALGEBRA OF QUATERNIONS

Equation (4.6) states that the product of quaternions is again a quaternion. It can b confirmed from Eq. (4.6) that the associativity

$$(qq')q'' = q(q'q'') \tag{4.}$$

holds. This can also be seen from Eqs (4.4) and (4.5), which imply that the multiplicatio rule of 1, i, j, and k is associative, i.e., $(ij)k = i(jk) = -1$, $(ij)i = i(ji) = j$, etc. Howeve the commutativity $qq' = q'q$ does not necessarily hold, as Eq. (4.5) shows. If associati multiplication is defined in a vector space, and if the space is closed under that multiplica tion, i.e., the product of arbitrary elements also belongs to that space, the vector space said to be an *algebra*. In this sense, the set of all quaternions is an algebra.

If we let $q_1 = q_2 = q_3 = 0$ in Eq. (4.1), then q $(= q_0)$ is a real number. Hence, the se of all real numbers is included in the set of quaternions, i.e., quaternions are an extensio of real numbers. On the other hand, if we identify a quaternion $q = q_1 i + q_2 j + q_3 k$ fo which $q_0 = 0$ with a vector $q_1 e_1 + q_2 e_2 + q_3 e_3$, the same rule applies for addition and scala multiplication as in the case of vectors. In this sense, a quaternion is an extension of a 3 vector described in Chapter 2. In view of this, we call q_0 in Eq. (4.1) the *scalar part* of and $q_1 i + q_2 j + q_3 k$ its *vector part*. We also say that a quaternion is a *scalar* if its vect part is 0 and a *vector* if its scalar part is 0.

If we put the scalar part q_0 of Eq. (4.1) to be $\alpha = q_0$ and its vector part to be $\boldsymbol{a}$ = $q_1 i + q_2 j + q_3 k$, the quaternion q is expressed as a formal sum of the scalar α and the vect $\boldsymbol{a}$ in the form $q = \alpha + \boldsymbol{a}$. Then, Eq. (4.6) is rephrased as follows:

Proposition 4.1 (Quaternion product) *The product of quaternions*

$$q = \alpha + \boldsymbol{a}, \qquad q' = \beta + \boldsymbol{b} \tag{4.8}$$

In particular, the product of their vector parts $\boldsymbol{a}$ *and* $\boldsymbol{b}$ *is*

$$\boldsymbol{ab} = -\langle \boldsymbol{a}, \boldsymbol{b} \rangle + \boldsymbol{a} \times \boldsymbol{b}. \tag{4.10}$$

Here, $\langle \boldsymbol{a}, \boldsymbol{b} \rangle$ and $\boldsymbol{a} \times \boldsymbol{b}$ in Eqs. (4.9) and (4.10) are the inner and the vector products, respectively, computed by identifying $\{i, j, k\}$ with the orthonormal basis $\{e_1, e_2, e_3\}$ in Chapter 2. Equation (4.10) allows us to interpret the quaternion product $\boldsymbol{ab}$ to be simultaneous computation of their inner product $\langle \boldsymbol{a}, \boldsymbol{b} \rangle$ and vector product $\boldsymbol{a} \times \boldsymbol{b}$.

4.3 CONJUGATE, NORM, AND INVERSE

From Eq. (4.4), we can interpret the quaternion q to be an extended complex number with three imaginary units i, j, and k. In view of this, we define the *conjugate* $q^\dagger$ of the quaternion q of Eq. (4.1) by

$$q^\dagger = q_0 - q_1 i - q_2 j - q_3 k. \tag{4.11}$$

Evidently, the conjugate of the conjugate is the original quaternion. For the quaternions q and q' in Eq. (4.8), we can see from Eq. (4.9) that

$$(qq')^\dagger = (\alpha\beta - \langle \boldsymbol{a}, \boldsymbol{b} \rangle) - \alpha\boldsymbol{b} - \beta\boldsymbol{a} - \boldsymbol{a} \times \boldsymbol{b}. \tag{4.12}$$

On the other hand, we see from $q^\dagger = \alpha - \boldsymbol{a}$ and $q'^\dagger = \beta - \boldsymbol{b}$ that

$$q'^\dagger q^\dagger = (\beta\alpha - \langle \boldsymbol{b}, \boldsymbol{a} \rangle) - \beta\boldsymbol{a} - \alpha\boldsymbol{b} + \boldsymbol{b} \times \boldsymbol{a}. \tag{4.13}$$

Hence, we observe that

Proposition 4.2 (Conjugate of quaternion) *The following identities hold:*

$$q^{\dagger\dagger} = q, \qquad (qq')^\dagger = q'^\dagger q^\dagger. \tag{4.14}$$

From the definition of the conjugate, we conclude that

Proposition 4.3 (Classification of quaternion) *A quaternion* q *is a scalar or a vector if and only if*

$$q^\dagger = q, \qquad q^\dagger = -q, \tag{4.15}$$

respectively.

From the rule of Eq. (4.6), the product $qq^\dagger$ can be written as

$$\begin{aligned} qq^\dagger &= (q_0^2 + q_1^2 + q_2^2 + q_3^2) + (-q_0q_1 + q_1q_0 - q_2q_3 + q_3q_2)i \\ &\quad +(-q_0q_2 + q_2q_0 - q_3q_1 + q_1q_3)j + (-q_0q_3 + q_3q_0 - q_1q_2 + q_2q_1)k \\ &= q_0^2 + q_1^2 + q_2^2 + q_3^2 \ (= q^\dagger q). \end{aligned} \tag{4.16}$$

We define the *norm* of quaternion q by

$$\|q\| = \sqrt{qq^\dagger} = \sqrt{q^\dagger q} = \sqrt{q_0^2 + q_1^2 + q_2^2 + q_3^2}. \tag{4.17}$$

For a vector $\boldsymbol{a}$ regarded as a quaternion, its quaternion norm $\|\boldsymbol{a}\|$ coincides with its vector norm $\|\boldsymbol{a}\|$.

$$q\left(\frac{q^\dagger}{\|q\|^2}\right) = \left(\frac{q^\dagger}{\|q\|^2}\right)q = 1, \qquad (4.1$$

which means that $q^\dagger/\|q\|^2$ is the *inverse* q^{-1} of q:

$$q^{-1} = \frac{q^\dagger}{\|q\|^2}, \qquad qq^{-1} = q^{-1}q = 1. \qquad (4.1$$

In other words, every nonzero quaternion q has its inverse q^{-1}, which allows division b quaternions. An algebra is called a *field* if every nonzero element of it has its inverse and it is closed under division by nonzero elements. The set of all real numbers is a field; so the set of all quaternions.

The existence of an inverse that admits division means that $qq' = qq''$ for $q \neq 0$ impli $q' = q''$. This property, however, does not hold for the inner product and the vector produ of vectors. In fact, $\langle \boldsymbol{a}, \boldsymbol{b}\rangle = \langle \boldsymbol{a}, \boldsymbol{c}\rangle$ for $\boldsymbol{a} \neq 0$ does not imply $\boldsymbol{b} = \boldsymbol{c}$, because we can add to any vector that is orthogonal to $\boldsymbol{a}$. Similarly, $\boldsymbol{a} \times \boldsymbol{b} = \boldsymbol{a} \times \boldsymbol{c}$ for $\boldsymbol{a} \neq 0$ does not imply $\boldsymbol{b}$ $\boldsymbol{c}$, because we can add to $\boldsymbol{b}$ any vector that is parallel to $\boldsymbol{a}$. If we recall that the quaternic product of vectors can be regarded as simultaneous computation of the inner and the vect products, as shown in Eq. (4.10), the existence of an inverse is evident: if $\langle \boldsymbol{a}, \boldsymbol{b}\rangle = \langle \boldsymbol{a},$ and $\boldsymbol{a} \times \boldsymbol{b} = \boldsymbol{a} \times \boldsymbol{c}$ at the same time for $\boldsymbol{a} \neq 0$, we must have $\boldsymbol{b} = \boldsymbol{c}$.

4.4 REPRESENTATION OF ROTATION BY QUATERNION

We call a quaternion q of unit norm a *unit quaternion*. Consider the product $q\boldsymbol{a}q^\dagger$ for unit quaternion q and a vector $\boldsymbol{a}$ regarded as a quaternion. Its conjugate is

$$(q\boldsymbol{a}q^\dagger)^\dagger = q^{\dagger\dagger}\boldsymbol{a}^\dagger q^\dagger = -q\boldsymbol{a}q^\dagger. \qquad (4.20$$

Hence, $q\boldsymbol{a}q^\dagger$ is a vector from Proposition 4.3. If we let

$$\boldsymbol{a}' = q\boldsymbol{a}q^\dagger, \qquad (4.2$$

its square norm is

$$\|\boldsymbol{a}'\|^2 = \boldsymbol{a}'\boldsymbol{a}'^\dagger = q\boldsymbol{a}q^\dagger q\boldsymbol{a}^\dagger q^\dagger = q\boldsymbol{a}\|q\|^2\boldsymbol{a}^\dagger q^\dagger = q\boldsymbol{a}\boldsymbol{a}^\dagger q^\dagger = q\|\boldsymbol{a}\|^2 q^\dagger = \|\boldsymbol{a}\|^2 qq^\dagger = \|\boldsymbol{a}\|^2. \qquad (4.2$$

Equation (4.21) defines a linear mapping of $\boldsymbol{a}$ that preserves the norm. Hence, it is eithe a pure rotation or a composition of a rotation and a reflection. We now show that this is pure rotation. Since $q_0^2 + q_1^2 + q_2^2 + q_3^2 = 1$, there exists an angle Ω such that

$$q_0 = \cos\frac{\Omega}{2}, \qquad \sqrt{q_1^2 + q_2^2 + q_3^2} = \sin\frac{\Omega}{2}. \qquad (4.2$$

Hence, we can write a unit quaternion q as the sum of its scalar and vector parts in th form

$$q = \cos\frac{\Omega}{2} + \boldsymbol{l}\sin\frac{\Omega}{2}, \qquad (4.24$$

where $\boldsymbol{l}$ is a unit vector ($\|\boldsymbol{l}\| = 1$). Then,

$$\begin{aligned}
\boldsymbol{a}' = q\boldsymbol{a}q^\dagger &= (\cos\frac{\Omega}{2} + \boldsymbol{l}\sin\frac{\Omega}{2})\boldsymbol{a}(\cos\frac{\Omega}{2} - \boldsymbol{l}\sin\frac{\Omega}{2}) \\
&= \boldsymbol{a}\cos^2\frac{\Omega}{2} - \boldsymbol{a}\boldsymbol{l}\cos\frac{\Omega}{2}\sin\frac{\Omega}{2} + \boldsymbol{l}\boldsymbol{a}\sin\frac{\Omega}{2}\cos\frac{\Omega}{2} + \boldsymbol{l}\boldsymbol{a}\boldsymbol{l}\sin^2\frac{\Omega}{2} \\
&= \boldsymbol{a}\cos^2\frac{\Omega}{2} + (\boldsymbol{l}\boldsymbol{a} - \boldsymbol{a}\boldsymbol{l})\cos\frac{\Omega}{2}\sin\frac{\Omega}{2} - \boldsymbol{l}\boldsymbol{a}\boldsymbol{l}\sin^2\frac{\Omega}{2}.
\end{aligned} \qquad (4.25$$

$$= -\langle \boldsymbol{a}, \boldsymbol{l} \rangle \boldsymbol{l} + \|\boldsymbol{l}\|^2 \boldsymbol{a} - \langle \boldsymbol{l}, \boldsymbol{a} \rangle \boldsymbol{l} = \boldsymbol{a} - 2\langle \boldsymbol{a}, \boldsymbol{l} \rangle \boldsymbol{l}, \tag{4.26}$$

where we have noted that $\langle \boldsymbol{l}, \boldsymbol{a} \times \boldsymbol{l} \rangle = |\boldsymbol{l}, \boldsymbol{a}, \boldsymbol{l}| = 0$ and used Eq. (2.20) of Chapter 2 for the vector triple product. Hence, Eq. (4.25) can be written as

$$\begin{aligned}
\boldsymbol{a}' &= \boldsymbol{a} \cos^2 \frac{\Omega}{2} + 2\boldsymbol{l} \times \boldsymbol{a} \cos \frac{\Omega}{2} \sin \frac{\Omega}{2} - (\boldsymbol{a} - 2\langle \boldsymbol{a}, \boldsymbol{l} \rangle \boldsymbol{l}) \sin^2 \frac{\Omega}{2} \\
&= \boldsymbol{a}(\cos^2 \frac{\Omega}{2} - \sin^2 \frac{\Omega}{2}) + 2\boldsymbol{l} \times \boldsymbol{a} \cos \frac{\Omega}{2} \sin \frac{\Omega}{2} + 2 \sin^2 \frac{\Omega}{2} \langle \boldsymbol{a}, \boldsymbol{l} \rangle \boldsymbol{l} \\
&= \boldsymbol{a} \cos \Omega + \boldsymbol{l} \times \boldsymbol{a} \sin \Omega + \langle \boldsymbol{a}, \boldsymbol{l} \rangle \boldsymbol{l}(1 - \cos \Omega).
\end{aligned} \tag{4.27}$$

This is nothing but the Rodrigues formula of Eq. (2.46) in Chapter 2. Hence, the quaternion $\boldsymbol{q}$ of Eq. (4.24) represents a rotation around axis $\boldsymbol{l}$ by angle Ω. The important thing is that a quaternion acts on vector $\boldsymbol{a}$ as the "sandwich" in the form of Eq. (4.21). In this sense, we call Eq. (4.24) a *rotor* around axis $\boldsymbol{l}$ by angle Ω. We see from Eq. (4.21) that q and $-q$ define the same rotor. In fact,

$$-q = -\cos \frac{\Omega}{2} - \boldsymbol{l} \sin \frac{\Omega}{2} = \cos \frac{2\pi - \Omega}{2} - \boldsymbol{l} \sin \frac{2\pi - \Omega}{2} \tag{4.28}$$

represents a rotation around axis $-\boldsymbol{l}$ by angle $2\pi - \Omega$, which is the same as the rotation around $\boldsymbol{l}$ by Ω.

Let $\mathcal{R}$ symbolically denote the operation of rotating a vector around some axis by some angle. Suppose we perform a rotation $\mathcal{R}$ followed by another rotation $\mathcal{R}'$. We write the resulting rotation, i.e., the composition of the two rotations, as $\mathcal{R}' \circ \mathcal{R}$. If we apply a rotor q to vector $\boldsymbol{a}$ and then apply another rotor q', we obtain

$$\boldsymbol{a}' = q'(q\boldsymbol{a}q^\dagger)q'^\dagger = (q'q)\boldsymbol{a}(q'q)^\dagger. \tag{4.29}$$

Hence, if a rotor q defines rotation $\mathcal{R}$ and another rotor q' rotation $\mathcal{R}'$, their composition $\mathcal{R}' \circ \mathcal{R}$ is defined by the rotor $q'q$. We express this fact by saying that the product of rotors and the composition of rotations are *homomorphic* to each other, meaning that the rule of computation has the same form.

From Eq. (4.24), the composition of a rotation around axis $\boldsymbol{l}$ by angle Ω and a rotation around axis $\boldsymbol{l}'$ by angle Ω' is given by

$$\begin{aligned}
&(\cos \frac{\Omega'}{2} + \boldsymbol{l}' \sin \frac{\Omega'}{2})(\cos \frac{\Omega}{2} + \boldsymbol{l} \sin \frac{\Omega}{2}) \\
&= \cos \frac{\Omega'}{2} \cos \frac{\Omega}{2} + \boldsymbol{l} \cos \frac{\Omega'}{2} \sin \frac{\Omega}{2} + \boldsymbol{l}' \sin \frac{\Omega'}{2} \cos \frac{\Omega}{2} + \boldsymbol{l}'\boldsymbol{l} \sin \frac{\Omega'}{2} \sin \frac{\Omega}{2} \\
&= \cos \frac{\Omega'}{2} \cos \frac{\Omega}{2} + \boldsymbol{l} \cos \frac{\Omega'}{2} \sin \frac{\Omega}{2} + \boldsymbol{l}' \sin \frac{\Omega'}{2} \cos \frac{\Omega}{2} + (-\langle \boldsymbol{l}', \boldsymbol{l} \rangle + \boldsymbol{l}' \times \boldsymbol{l}) \sin \frac{\Omega'}{2} \sin \frac{\Omega}{2} \\
&= \left(\cos \frac{\Omega'}{2} \cos \frac{\Omega}{2} - \langle \boldsymbol{l}', \boldsymbol{l} \rangle \sin \frac{\Omega'}{2} \sin \frac{\Omega}{2}\right) + \boldsymbol{l} \cos \frac{\Omega'}{2} \sin \frac{\Omega}{2} + \boldsymbol{l}' \sin \frac{\Omega'}{2} \cos \frac{\Omega}{2} \\
&\quad + \boldsymbol{l}' \times \boldsymbol{l} \sin \frac{\Omega'}{2} \sin \frac{\Omega}{2}.
\end{aligned} \tag{4.30}$$

From this, we see that the axis $\boldsymbol{l}''$ and the angle Ω of the composite rotation are given by

$$\begin{aligned}
\cos \frac{\Omega''}{2} &= \cos \frac{\Omega'}{2} \cos \frac{\Omega}{2} - \langle \boldsymbol{l}', \boldsymbol{l} \rangle \sin \frac{\Omega'}{2} \sin \frac{\Omega}{2}, \\
\boldsymbol{l}'' \cos \frac{\Omega''}{2} &= \boldsymbol{l} \sin \frac{\Omega'}{2} \sin \frac{\Omega}{2} + \boldsymbol{l}' \sin \frac{\Omega'}{2} \cos \frac{\Omega}{2} + \boldsymbol{l}' \times \boldsymbol{l} \sin \frac{\Omega'}{2} \sin \frac{\Omega}{2}.
\end{aligned} \tag{4.31}$$

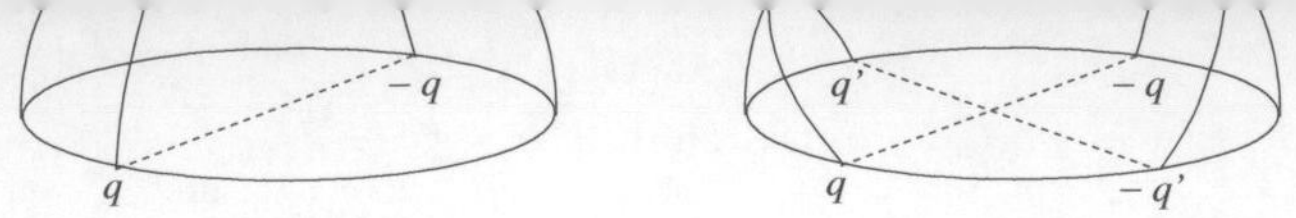

(a) (b)

FIGURE 4.1 The set of all quaternions that represent rotations corresponds to a hemisphere radius 1 in 4D such that all antipodal points q and $-q$ on the boundary are pasted together. (If a closed path that represents continuous variations of rotation reaches the boundary, it appea on the opposite side. This loop cannot be continuously shrunk to a point. (b) If a closed loo passes through the boundary twice, it can be continuously shrunk to a point: we first rotate th diametrical segment connecting q' and $-q'$ so that they coincide with q and $-q$ and then shri the loop to q and $-q$, which represent the same point.

Traditional World 4.1 (Rotation matrix in quaternion) The equation $\boldsymbol{a}' = q\boldsymbol{a}$ defines a linear mapping from $\boldsymbol{a}$ to $\boldsymbol{a}'$. Hence, if $\boldsymbol{a}$ and $\boldsymbol{a}'$ are regarded as vertical arrays their components, it can be expressed as multiplication of $\boldsymbol{a}$ by a matrix. For $q = q_0 + q_1 i$ $q_2 j + q_3 k$, the expression $q\boldsymbol{a}q^\dagger$ is quadratic in q_0, q_1, q_2, and q_3, so the matrix elements a their quadratic forms. It is easy to confirm that $\boldsymbol{a}' = q\boldsymbol{a}q^\dagger$ is equivalent to the following (Exercise 4.1):

$$\boldsymbol{a}' = \boldsymbol{R}\boldsymbol{a},$$
$$\boldsymbol{R} = \begin{pmatrix} q_0^2 + q_1^2 - q_2^2 - q_3^2 & 2(q_1q_2 - q_0q_3) & 2(q_1q_3 + q_0q_2) \\ 2(q_2q_1 + q_0q_3) & q_0^2 - q_1^2 + q_2^2 - q_3^2 & 2(q_2q_3 - q_0q_1) \\ 2(q_3q_1 - q_0q_2) & 2(q_3q_2 + q_0q_1) & q_0^2 - q_1^2 - q_2^2 + q_3^2 \end{pmatrix}. \quad (4.3$$

This matrix $\boldsymbol{R}$ should be the same as the rotation matrix $\boldsymbol{R}$ in Eq. (2.49) in Chapter This means that if q_0, q_1, q_2, and q_3 are expressed in terms of the angle Ω and the ax $\boldsymbol{l}$ using Eq. (4.24), we obtain Eq. (2.49) in Chapter 2. Equation (4.32) is frequently use for parameterizing the matrix $\boldsymbol{R}$ for computing a rotation that satisfies given condition For example, we can search the unit sphere $q_0^2 + q_1^2 + q_2^2 + q_3^2 = 1$ in 4D for a matrix that has some desirable property, e.g., minimizing something. This parameterization is mo convenient than searching the space of $\boldsymbol{l}$ and Ω in the form of Eq. (2.49) in Chapter 2. Or of the main reasons is that Eq. (2.49) in Chapter 2 has a singularity at $\Omega = 0$. In fact, if $= 0$, we have $\boldsymbol{R} = \boldsymbol{I}$ (the identity matrix) whatever the axis $\boldsymbol{l}$ may be, and hence variatio of $\boldsymbol{l}$ are not reflected in the variations of $\boldsymbol{R}$. This type of singularity causes trouble in man numerical search algorithms. In contrast, the four variables q_0, q_1, q_2, and q_3 play almo the same role in Eq. (4.32), causing no such problems.

Traditional World 4.2 (Topology) Equation (4.21) defines a linear mapping such tha $\|\boldsymbol{a}'\| = \|\boldsymbol{a}\|$. Hence, it is either a pure rotation or a combination of a rotation and a reflectio but it can be shown directly that no reflection is involved by a simple consideration withou converting Eq. (4.21) to the Rodrigues formula. Evidently, $q = \pm 1$ acts as the identity, an an arbitrary unit quaternion $q = q_0 + q_1 i + q_2 j + q_3 k$ can be smoothly reduced to $q = \pm$ by varying $q_0 \to \pm 1$, $q_1 \to 0$, $q_2 \to 0$, and $q_3 \to 0$ without violating $q_0^2 + q_1^2 + q_2^2 + q_3^2 = 1$. reflections are involved, q cannot be continuously reduced to the identity. The mathematic study of such reasoning based on "continuous variations" is called *topology*.

Since a set of four numbers q_0, q_1, q_2, q_3 such that $q_0^2 + q_1^2 + q_2^2 + q_3^2 = 1$ represent

between S^3 and the set of rotations is 2 to 1. This can be made 1 to 1 by considering a hemisphere, e.g., the part of S^3 for $q_0 \geq 0$. But then there arises a problem of continuity, because each point on the boundary and its antipode, i.e., the other end of the diametrical segment, represent the same rotation. Hence, if the trajectory of continuously varying rotations reaches the boundary, it appears from the antipode. In order to eliminate such a discontinuity, we need to paste each point on the boundary to its antipode (Fig. 4.1). The resulting space is denoted by $\mathbb{P}^3$ and called in topological terms the 3D *projective space*. Thus, the set of all rotations continuously corresponds 1 to 1 to $\mathbb{P}^3$. We refer to this fact by saying that the set of all rotations is *homeomorphic* to $\mathbb{P}^3$.

It is easy to see that a closed loop on this hemisphere that reaches the boundary and reappears from the antipode cannot be continuously shrunk to a point (Fig. 4.1(a)). A space is said to be *connected* if any two points in it can be connected by a smooth path, and *simply connected* if any closed loop in it can be continuously shrunk to a point. Thus, $\mathbb{P}^3$ is connected but not simply connected. However, it is easy to mentally visualize that a closed loop that passes through the boundary twice (or an even number of times) can be continuously shrunk to a point (Fig. 4.1(b)).

Traditional World 4.3 (Infinitesimal rotations) We have seen that by expressing a unit quaternion q in the form of Eq. (4.24), we can derive the Rodrigues formula. Hence, Eq. (4.24) represents the rotation around axis $\boldsymbol{l}$ by angle Ω. However, we can directly show that a rotation around axis $\boldsymbol{l}$ by angle Ω must have the form of Eq. (4.24). To show this, let us consider infinitesimal rotations close to the identity. Since the identity corresponds to the rotor $q = 1$, the rotor for an infinitesimal rotation has the form

$$q = 1 + \delta q + O(\delta q^2), \qquad q^\dagger = 1 + \delta q^\dagger + O(\delta q^2). \tag{4.33}$$

Since q is a unit quaternion, its square norm

$$\|q\|^2 = qq^\dagger = (1 + \delta q + O(\delta q^2))(1 + \delta q^\dagger + O(\delta q^2)) = 1 + \delta q + \delta q^\dagger + O(\delta q^2) \tag{4.34}$$

must be identically 1. Hence, $\delta q^\dagger = -\delta q$, so δq is a vector from Eq. (4.15). From Eq. (4.10), an infinitesimal rotation of vector $\boldsymbol{a}$ is given by

$$\begin{aligned} \boldsymbol{a}' &= q\boldsymbol{a}q^\dagger = (1 + \delta q)\boldsymbol{a}(1 - \delta q) + O(\delta q^2) = \boldsymbol{a} + \delta q\boldsymbol{a} - \boldsymbol{a}\delta q + O(\delta q^2) \\ &= \boldsymbol{a} + 2\delta q \times \boldsymbol{a} + O(\delta q^2). \end{aligned} \tag{4.35}$$

Comparing this with Eq. (2.47) in Chapter 2, we can express δq in terms of an axis $\boldsymbol{l}$ and an infinitesimal angle $\Delta\Omega$ in the form

$$\delta q = 1 + \frac{\Delta\Omega}{2}\boldsymbol{l}. \tag{4.36}$$

Let $q_l(\Omega)$ be the rotor around axis $\boldsymbol{l}$ by angle Ω. Differentiating it with respect to Ω, we see that

$$\begin{aligned} \frac{dq_l(\Omega)}{d\Omega} &= \lim_{\Delta\Omega\to 0} \frac{q_l(\Omega + \Delta\Omega) - q_l(\Omega)}{\Delta\Omega} = \lim_{\Delta\Omega\to 0} \frac{q_l(\Delta\Omega)q_l(\Omega) - q_l(\Omega)}{\Delta\Omega} \\ &= \lim_{\Delta\Omega\to 0} \frac{q_l(\Delta\Omega) - 1}{\Delta\Omega} q_l(\Omega) = \frac{1}{2}\boldsymbol{l}q_l(\Omega), \end{aligned} \tag{4.37}$$

etc. Since $q_l(0) = 1$, we obtain the Taylor expression around $\Omega = 0$ in the following form:

$$q_l(\Omega) = 1 + \frac{\Omega}{2}\boldsymbol{l} + \frac{1}{2!}\frac{\Omega^2}{4}\boldsymbol{l}^2 + \frac{1}{3!}\frac{\Omega^3}{8}\boldsymbol{l}^3 + \cdots = \sum_{k=1}^{\infty}\frac{1}{k!}\Big(\frac{\Omega}{2}\boldsymbol{l}\Big)^k = \exp\frac{\Omega}{2}\boldsymbol{l}. \tag{4.38}$$

The last term is a symbolic expression of $\sum_{k=1}^{\infty}(\Omega\boldsymbol{l}/2)^k/k!$. Since $\boldsymbol{l}$ is a unit vector, we have $\boldsymbol{l}^2 = -1$ from Eq. (4.10). Hence Eq. (4.38) can be rewritten in the form of Eq. (4.24):

$$\begin{aligned} q_l(\Omega) &= \Big(1 - \frac{1}{2!}\Big(\frac{\Omega}{2}\Big)^2 + \frac{1}{4!}\Big(\frac{\Omega}{2}\Big)^4 + \cdots\Big) + \boldsymbol{l}\Big(\frac{\Omega}{2} - \frac{1}{3!}\Big(\frac{\Omega}{2}\Big)^3 + \frac{1}{5!}\Big(\frac{\Omega}{2}\Big)^5 + \cdots\Big) \\ &= \cos\frac{\Omega}{2} + \boldsymbol{l}\sin\frac{\Omega}{2}. \end{aligned} \tag{4.39}$$

Traditional World 4.4 (Representation of rotation group) A set of elements for which multiplication is defined is called a *group* if 1) the multiplication is associative, 2) there exists a unique *identity* whose multiplication does not change any element, and 3) each element has its *inverse* whose multiplication results in the identity. The set of all rotations is a group, called the *group of rotations* and denoted by $SO(3)$, where multiplication is defined by composition; the identity is the rotation of angle 0 and the inverse is the opposite rotation (sign reversal of the angle).

We have already seen that the action of rotors is homomorphic to the composition of rotations. If we regard vectors as an array of numbers and express a rotation by multiplication by a matrix, the multiplication of matrices is homomorphic to the composition of rotations. Namely, if matrices $\boldsymbol{R}$ and $\boldsymbol{R}'$ represent rotations $\mathcal{R}$ and $\mathcal{R}'$, respectively, the product $\boldsymbol{R}'\boldsymbol{R}$ represents their composition $\mathcal{R}' \circ \mathcal{R}$. In mathematical terms, if each element of a group corresponds to a matrix and if multiplication of group elements is homomorphic to multiplication of the corresponding matrices, we say that this correspondence is a *representation* of the group. The matrix in Eq. (4.32) and the matrix in Eq. (2.49) in Chapter 2 are both representations of $SO(3)$, but there exist many other representations. Among them is the following expression of the quaternion $q = q_0 + q_1 i + q_2 j + q_3 k$ in a 2×2 matrix form with complex components:

$$\boldsymbol{U} = \begin{pmatrix} q_0 - iq_3 & -q_2 - iq_1 \\ q_2 - iq_1 & q_0 + iq_3 \end{pmatrix}. \tag{4.40}$$

To say that this is a representation is equivalent to

$$\begin{pmatrix} q_0'' - iq_3'' & -q_2'' - iq_1'' \\ q_2'' - iq_1'' & q_0'' + iq_3'' \end{pmatrix} = \begin{pmatrix} q_0' - iq_3' & -q_2' - iq_1' \\ q_2' - iq_1' & q_0' + iq_3' \end{pmatrix} \begin{pmatrix} q_0 - iq_3 & -q_2 - iq_1 \\ q_2 - iq_1 & q_0 + iq_3 \end{pmatrix}, \tag{4.41}$$

where the quaternion $q'' = q_0'' + q_1'' i + q_2'' j + q_3'' k$ is the product of the quaternions $q' = q_0' + q_1' i + q_2' j + q_3' k$ and $q = q_0 + q_1 i + q_2 j + q_3 k$. This can be easily confirmed by direct calculation, but the following argument is more informative. We can rewrite Eq. (4.40) in the form

$$\boldsymbol{U} = q_0\boldsymbol{I} + q_1\boldsymbol{S}_1 + q_2\boldsymbol{S}_2 + q_3\boldsymbol{S}_3, \tag{4.42}$$

where we define

$$\boldsymbol{I} = \begin{pmatrix} 1 & 0 \\ 0 & 1 \end{pmatrix}, \quad \boldsymbol{S}_1 = \begin{pmatrix} 0 & -i \\ -i & 0 \end{pmatrix}, \quad \boldsymbol{S}_2 = \begin{pmatrix} 0 & -1 \\ 1 & 0 \end{pmatrix}, \quad \boldsymbol{S}_3 = \begin{pmatrix} -i & 0 \\ 0 & i \end{pmatrix}. \tag{4.43}$$

$$S_2S_3 = S_1, \qquad S_3S_1 = S_2, \qquad S_1S_2 = S_3,$$

$$S_3S_2 = -S_1, \qquad S_1S_3 = -S_2, \qquad S_2S_1 = -S_3. \tag{4.45}$$

In other words, pairwise multiplication of $\boldsymbol{I}$, $\boldsymbol{S}_1$, $\boldsymbol{S}_2$, and $\boldsymbol{S}_3$ has the same form as pairwise multiplication of 1, i, j, and k in Eqs. (4.4) and (4.5). Hence, we can identify the quaternion multiplication of 1, i, j, and k in Eqs. (4.4) and (4.5) with matrix multiplication of $\boldsymbol{I}$, $\boldsymbol{S}_1$, $\boldsymbol{S}_2$, and $\boldsymbol{S}_3$.

Traditional World 4.5 (Stereographic projection) Let us write Eq. (4.40) as

$$\boldsymbol{U} = \begin{pmatrix} \alpha & \beta \\ \gamma & \delta \end{pmatrix}. \tag{4.46}$$

This matrix has the form of Eq. (4.40) with $q_0^2 + q_1^2 + q_2^2 + q_3^2 = 1$ if and only if

$$\gamma = -\beta^\dagger, \qquad \delta = -\alpha^\dagger, \qquad \alpha\delta - \beta\gamma = 1, \tag{4.47}$$

where † denotes the complex conjugate. Thus, a rotation can be specified by four complex numbers α, β, γ, and δ that satisfy Eq. (4.47). They are called the *Cayley–Klein parameters*. The group of matrices in the form of Eq. (4.46), for which Eq. (4.47) holds, is called the *special unitary group* and denoted by $SU(2)$. The term "unitary" refers to the fact that $\boldsymbol{U}$ is a *unitary matrix*, i.e., $\boldsymbol{U}^\dagger\boldsymbol{U} = \boldsymbol{I}$, where † denotes the *Hermitian conjugate*, i.e., the transpose of the complex conjugate. The term "special" refers to the fact that the determinant is 1: $|\boldsymbol{U}| = \alpha\delta - \beta\gamma = 1$.

Equation (4.46) is not only a representation of the group of rotations but also a representation of the group of transformations of the complex plane in the form

$$z' = \frac{\gamma + \delta z}{\alpha + \beta z}. \tag{4.48}$$

A transformation of this form is called a *linear fractional transformation* or a *Möbius transformation*. If this transformation is followed by another transformation specified by α', β', γ', and δ', the resulting transformation is shown to be a linear fractional transformation again (↪ Exercise 4.6). Namely, the set of linear fractional transformations forms a group with respect to composition. Let α'', β'', γ'', and δ'' be the parameters of the above composite transformation, and let $\boldsymbol{U}$ be the 2×2 matrix having them as its elements. Then, $\boldsymbol{U}'' = \boldsymbol{U}'\boldsymbol{U}$ holds, where $\boldsymbol{U}'$ is the 2×2 matrix consisting of α', β', γ', and δ'. In other words, Eq. (4.46) is a representation of the group of linear fractional transformations. This suggests that Eq. (4.48) corresponds to a rotation in some sense. In fact, this correspondence is given by the *stereographic projection* of a unit sphere at the origin: the the point (x, y) obtained by projecting a point on the sphere from its "south pole" onto the xy plane is identified with the complex number $z = x + iy$ (Fig. 4.2). It can be shown that if the sphere rotates by a rotation specified by the Cayley–Klein parameters α, β, γ, and δ, the corresponding transformation of the complex number z is given by the linear fractional transformation in the form of Eq. (4.48).

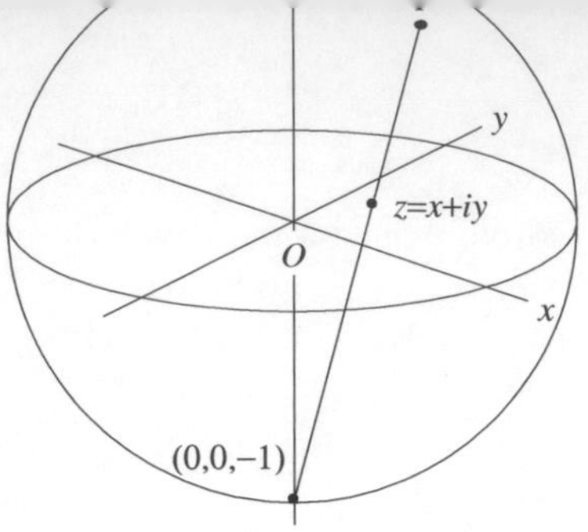

FIGURE 4.2 Stereographic projection of a unit sphere onto a plane. The projection of a point on the sphere from the "south pole" $(0, 0, -1)$ onto the xy plane can be regarded as a compl number $z = x + iy$. If the sphere rotates, the projected point on the xy plane undergoes a line fractional transformation.

4.5 SUPPLEMENTAL NOTE

The quaternion was introduced by Sir William Rowan Hamilton (1805–1865), an Irish mat ematician. He found that complex numbers can be extended by introducing additional ima inary units j and k in such a way that divisions are admitted. The quaternion algebra this chapter is described from the viewpoint of today's vector calculus.

The quaternion algebra is a typical example of describing geometry by means of oper tions on symbols. As stated in Proposition 4.1, a quaternion is a formal sum of a scalar an a vector, and the quaternion product of vectors corresponds to simultaneous computati of the inner and the vector products. It was the American physicist Gibbs, as mention in the supplemental note to Chapter 2, who separated the quaternion product into t inner and vector products. We can see from Eq. (4.6) that the inner product is comput if Eq. (4.4) is replaced by $i^2 = j^2 = k^2 = 1$ and Eq. (4.5) is replaced by $jk = ki = ij$ $kj = ik = ji = 0$, while the vector product is computed if Eq. (4.4) is replaced by i^2 $j^2 = k^2 = 0$. In this way, Gibbs established a system of inner and vector products that convenient for describing physics. At the cost of this, the mathematical structure behi the quaternion algebra was lost. The opposite direction is also possible: a more gener mathematical structure can be defined by introducing additional operations to quaternion resulting in the Clifford algebra to be described in Chapter 6. Thus, Hamilton's quaterni algebra evolved in two opposite directions.

For the group structure of rotations mentioned in Sec. 4.4, see the classical textbook Pontryagin [18]. Kanatani [13] gave an elementary and intuitive explanation intended f computer vision researchers.

4.6 EXERCISES

4.1. Letting $\boldsymbol{q} = q_0 + q_1 i + q_2 j + q_3 k$, $\boldsymbol{a} = a_1 e_1 + a_2 e_2 + a_3 e_3$, and $\boldsymbol{a}' = a_1' e_1 + a_2' e_2 + a_3' e$ in Eq. (4.21), express a_1', a_2', and a_3' in terms of a_1, a_2, and a_3.

4.2. Given a unit quaternion $\boldsymbol{q} = q_0 + q_1 i + q_2 j + q_3 k$, the angle Ω of the rotation represents can be obtained from Eq. (4.23) in two ways:

Which expression is more convenient in actual computation?

4.3. A rotation around an axis by an angle larger than π and smaller than 2π can be regarded as a rotation by an angle less than π in the opposite sense. Given a unit quaternion $\boldsymbol{q} = q_0 + q_1 i + q_2 j + q_3 k$, how can we obtain the axis $\boldsymbol{l}$ and the angle Ω of the rotation it represents in such a way that the rotation is in the right-handed screw sense around $\boldsymbol{l}$ by $0 \leq \Omega \leq \pi$?

4.4. The symbols i, j, and k are themselves unit quaternions. What rotations do they represent? Also, a unit vector $\boldsymbol{l}$ can be identified with a unit quaternion. What rotation does it represent?

4.5. A scalar α is also a quaternion itself. Its conjugate is $\alpha^\dagger = \alpha$. For a vector $\boldsymbol{a}$ regarded as a quaternion, what transformation does $\boldsymbol{a}' = \alpha \boldsymbol{a} \alpha^\dagger$ define? Then, for an arbitrary quaternion $\boldsymbol{q}$ which is not necessarily of unit norm, what transformation does $\boldsymbol{a}' = \boldsymbol{q}\boldsymbol{a}\boldsymbol{q}^\dagger$ represent?

4.6. Show that the composition of linear fractional transformations of the form of Eq. (4.48) is again a linear fractional transformation. Also show that if the parameters are arranged in matrix form as in Eq. (4.46), the parameters after the composition are written as the matrix product $\boldsymbol{U}'' = \boldsymbol{U}'\boldsymbol{U}$.

Grassmann's Outer Product Algebra

This chapter shows how subspaces can be described by introducing an operation called "outer product" on vectors. It becomes the basis of the Clifford algebra to be described in the next chapter. We first specify subspaces of different dimensions (the origin, lines passing through the origin, planes passing through the origin, and the entire 3D space) in terms of the outer product, which makes clear the properties of the outer product operation. Then, we introduce an operation called "contraction" and define the norm and the duality of subspaces. Finally, we show that two methods exist for specifying subspaces: the direct representation and the dual representation.

5.1 SUBSPACES

A *subspace* is a subset of a vector space which is itself a vector space. To be specific, it is the set of all linear combinations of some vectors starting from the origin, or the space *spanned* by these vectors. In 3D, a subspace is of dimension 0, 1, 2, or 3: a 0D subspace is the origin itself; a 1D subspace is a line passing through the origin; a 2D subspace is a plane passing through the origin; a 3D subspace is the entire space itself. We discuss in this chapter how a subspace is specified and how computations involving subspaces are performed. Since we only consider vectors starting from the origin in this chapter, we omit the proviso "starting from the origin." Also, we only consider lines and planes passing through the origin, so we omit the phrase "passing through the origin," and call 1D subspaces, 2D subspaces, and 3D subspaces simply "lines," "planes," and "spaces," respectively. In this chapter, we regard subspaces as having an orientation and a magnitude, where the orientation and the magnitude are combined concepts in the sense that an orientation and a magnitude are the same as the orientation in the opposite direction and the magnitude of opposite sign. Magnitude 0 is interpreted to be "nonexistence," so whatever exists has a nonzero magnitude.

5.1.1 Lines

We define an orientation and a magnitude to a line. Although a line has an infinite length, we think of it as, say, an electric wire through which a current is flowing, and regard the intensity of the flow as the magnitude of that line. We refer to the line spanned by vector $\boldsymbol{a}$ simply as "line $\boldsymbol{a}$"; its orientation is the direction of the vector $\boldsymbol{a}$ and its magnitude is the magnitude of the former. A line $\boldsymbol{a}$ and a line $-\boldsymbol{a}$ have mutually opposite orientations.

If two lines have the same geometric figure, we can define their sum to be the li having the sum of their magnitudes after aligning their orientations by adjusting the sign. For example, if $\boldsymbol{a}$ and $\boldsymbol{a}'$ are collinear, there exists a scalar α such that $\boldsymbol{a}' = \alpha\boldsymbol{a}$, so the su of the two lines is $(1+\alpha)\boldsymbol{a}$. For two lines $\boldsymbol{a}$ and $\boldsymbol{b}$ that are not collinear, we define the sum to be the line generated by $\boldsymbol{a}+\boldsymbol{b}$. From this definition, we can freely compute sum and scalar multiples of lines. Thus, the set of lines forms a vector space.

5.1.2 Planes

We also define an orientation and a magnitude to a plane. Although a plane has an infini area, we think of it as, say, a magnetized iron plate and regard the (signed) strength the magnetism as its magnitude. Defining the orientation of a plane is equivalent to disti guishing its front side from its back. This is done as follows. Two vectors $\boldsymbol{a}$ and $\boldsymbol{b}$ that a not collinear span a plane, which we write $\boldsymbol{a}\wedge\boldsymbol{b}$. The operation $\wedge$ is called the *outer* (*exterior*) *product*, and two vectors combined by $\wedge$ are called a *bivector* or *2-vector*. The sic on which the rotation of $\boldsymbol{a}$ toward $\boldsymbol{b}$ is anticlockwise, i.e., a positive rotation, is defined be the front side, and the magnitude is defined to be the area of the parallelogram define by $\boldsymbol{a}$ and $\boldsymbol{b}$. Equivalently, we may regard the side on which the rotation of $\boldsymbol{a}$ toward $\boldsymbol{b}$ clockwise, i.e., a negative rotation, as the front side and its magnitude as the negative the area of the parallelogram defined by $\boldsymbol{a}$ and $\boldsymbol{b}$.

By definition, the plane spanned by $\alpha\boldsymbol{a}$ and $\beta\boldsymbol{b}$ is, as a geometric figure, the same the plane spanned by $\boldsymbol{a}$ and $\boldsymbol{b}$, but the former has $\alpha\beta$ times the magnitude of the latte Similarly, the plane spanned by $\boldsymbol{a}$ and $\boldsymbol{b}$ is, as a geometric figure, the same as the plar spanned by $\boldsymbol{b}$ and $\boldsymbol{a}$, but the front and back sides are reversed. Evidently, one vector $\boldsymbol{a}$ alo cannot define a plane. In equations, these relations are expressed by

$$(\alpha\boldsymbol{a})\wedge(\beta\boldsymbol{b}) = (\alpha\beta)\boldsymbol{a}\wedge\boldsymbol{b}, \qquad \boldsymbol{a}\wedge\boldsymbol{b} = -\boldsymbol{b}\wedge\boldsymbol{a}, \qquad \boldsymbol{a}\wedge\boldsymbol{a} = 0, \qquad (5.$$

where the number 0 on the right side of the last equation means "nonexistence."

The sum of two planes that are the same as a geometric figure is defined by adding the magnitude after aligning the orientation by adjusting the sign. For example, if bivecto $\boldsymbol{a}\wedge\boldsymbol{b}$ and $\boldsymbol{a}'\wedge\boldsymbol{b}'$ define coplanar planes, there exists a scalar such that $\boldsymbol{a}'\wedge\boldsymbol{b}' = \alpha\boldsymbol{a}\wedge\boldsymbol{b}$, s we define their sum to be $(1+\alpha)\boldsymbol{a}\wedge\boldsymbol{b}$. If vectors $\boldsymbol{a}$, $\boldsymbol{b}$, and $\boldsymbol{c}$ are coplanar, the area of th parallelogram defined by $\boldsymbol{a}$ and $\alpha\boldsymbol{b}+\beta\boldsymbol{c}$ for any α and β is the sum of α times the area the parallelogram defined by $\boldsymbol{a}$ and $\boldsymbol{b}$ and β times the area of the parallelogram defined b $\boldsymbol{a}$ and $\boldsymbol{c}$ (Fig. 5.1(a)). Hence, the following equality holds:

$$\boldsymbol{a}\wedge(\alpha\boldsymbol{b}+\beta\boldsymbol{c}) = \alpha\boldsymbol{a}\wedge\boldsymbol{b}+\beta\boldsymbol{a}\wedge\boldsymbol{c}. \qquad (5.2$$

This states that distributivity of the outer product $\wedge$ holds for coplanar vectors.

Since a plane is defined by its orientation and magnitude, it can be represented b different pairs of vectors. For example, a pair of vectors $\boldsymbol{a}$ and $\boldsymbol{b}$ can span the same plane b another pair $\boldsymbol{a}'$ and $\boldsymbol{b}'$ on it, i.e., $\boldsymbol{a}\wedge\boldsymbol{b} = \boldsymbol{a}'\wedge\boldsymbol{b}'$, if they have the same relative configuratio i.e., the orientation of rotating one toward the other, and the area of the parallelogram the same. For instance, vectors $\boldsymbol{a}$ and $\boldsymbol{b}$ have, for any α, the same relative configuration a $\boldsymbol{a}$ and $\boldsymbol{b}+\alpha\boldsymbol{a}$, and the areas of the parallelograms they define are the same (Fig. 5.1(b) Hence, the following equality holds:

$$\boldsymbol{a}\wedge(\boldsymbol{b}+\alpha\boldsymbol{a}) = \boldsymbol{a}\wedge\boldsymbol{b}. \qquad (5.3$$

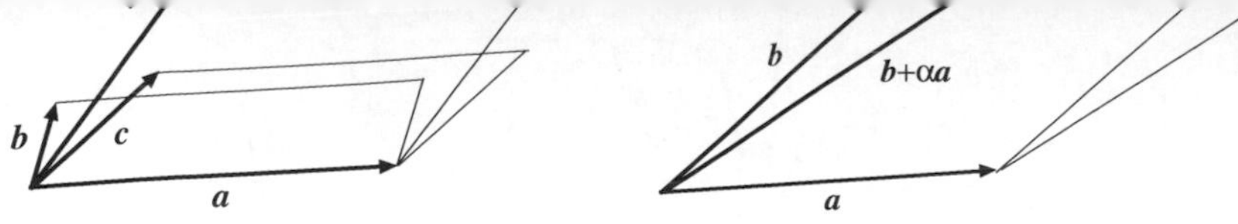

(a) (b)

FIGURE 5.1 (a) If vectors $\boldsymbol{a}$, $\boldsymbol{b}$, and $\boldsymbol{c}$ are coplanar, the area of the parallelogram defined by $\boldsymbol{a}$ and $\alpha\boldsymbol{b} + \beta\boldsymbol{c}$ is the sum of α times the area of the parallelogram defined by $\boldsymbol{a}$ and $\boldsymbol{b}$ and β times the area of the parallelogram defined by $\boldsymbol{a}$ and $\boldsymbol{c}$. (b) The area of the parallelogram defined by vectors $\boldsymbol{a}$ and $\boldsymbol{b}$ equals the area of the parallelogram defined by $\boldsymbol{a}$ and $\boldsymbol{b} + \alpha\boldsymbol{a}$.

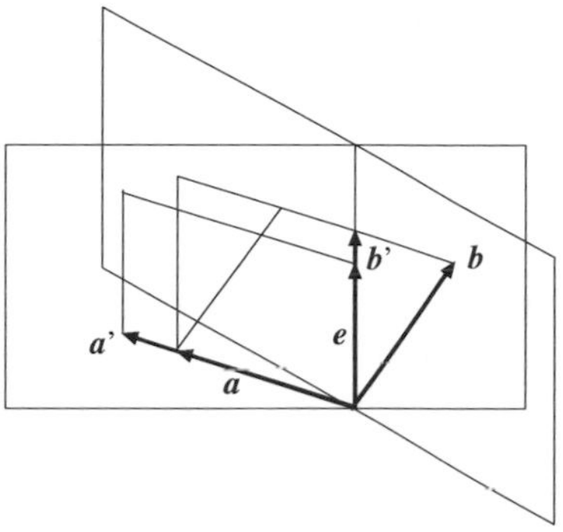

FIGURE 5.2 For the vectors $\boldsymbol{a}$ and $\boldsymbol{b}$, there exists a vector $\boldsymbol{b}'$, having the same orientation as $\boldsymbol{e}$, such that $\boldsymbol{a} \wedge \boldsymbol{b} = \boldsymbol{a} \wedge \boldsymbol{b}'$. Also, there exists a vector $\boldsymbol{a}'$, having the same orientation as $\boldsymbol{a}$, such that $\boldsymbol{a} \wedge \boldsymbol{b}' = \boldsymbol{a}' \wedge \boldsymbol{e}$.

This can also be shown by combining Eqs. (5.2) and (5.1): $\boldsymbol{a} \wedge (\boldsymbol{b} + \alpha\boldsymbol{a}) = \boldsymbol{a} \wedge \boldsymbol{b} + \alpha\boldsymbol{a} \wedge \boldsymbol{a} = \boldsymbol{a} \wedge \boldsymbol{b}$.

For bivectors $\boldsymbol{a} \wedge \boldsymbol{b}$ and $\boldsymbol{c} \wedge \boldsymbol{d}$ that define different planes, let $\boldsymbol{e}$ be the vector along their intersection. Then, there exists a vector $\boldsymbol{a}'$ such that $\boldsymbol{a} \wedge \boldsymbol{b} = \boldsymbol{a}' \wedge \boldsymbol{e}$ (Fig. 5.2). Similarly, there exists a vector $\boldsymbol{c}'$ such that $\boldsymbol{c} \wedge \boldsymbol{d} = \boldsymbol{c}' \wedge \boldsymbol{e}$. We define the sum of the two planes by

$$\boldsymbol{a} \wedge \boldsymbol{b} + \boldsymbol{c} \wedge \boldsymbol{d} = \boldsymbol{a}' \wedge \boldsymbol{e} + \boldsymbol{c}' \wedge \boldsymbol{e} = (\boldsymbol{a}' + \boldsymbol{c}') \wedge \boldsymbol{e}. \tag{5.4}$$

This means that we allow distributivity to hold for non-coplanar vectors as well. Thus, we can freely compute sums and scalar multiples of planes. In other words, the set of all planes forms a vector space.

5.1.3 Spaces

We also regard the entire 3D space as having an orientation and a magnitude. Its orientation is defined by a sign. Although the space has an infinite volume, we imagine that it has some measurable value, which we regard as its magnitude. Non-coplanar vectors $\boldsymbol{a}$, $\boldsymbol{b}$, and $\boldsymbol{c}$ span a 3D space, which we denote by $\boldsymbol{a} \wedge \boldsymbol{b} \wedge \boldsymbol{c}$. Three vectors connected by $\wedge$ are called a *trivector* or *3-vector*. As a geometric figure, it represents the entire space, but its sign is defined to be positive if $\boldsymbol{a}$, $\boldsymbol{b}$, and $\boldsymbol{c}$ are a right-handed system and negative if they are left-handed (↪ Sec. 2.5 of Chapter 2). The magnitude of that space is given by the volume of the

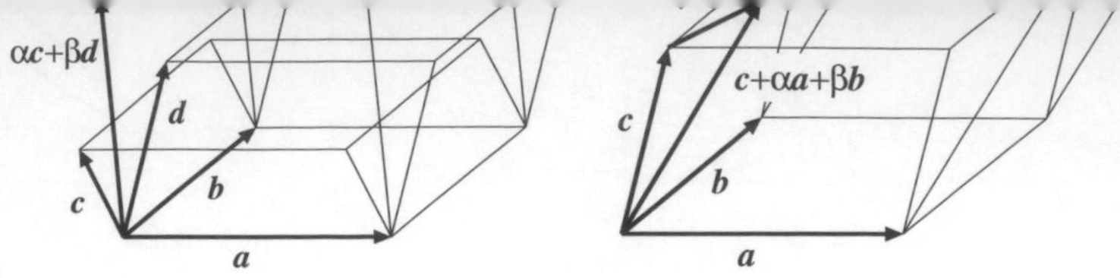

(a) (b)

FIGURE 5.3 (a) For vectors $\boldsymbol{a}$, $\boldsymbol{b}$, $\boldsymbol{c}$, and $\boldsymbol{d}$, the volume of the parallelepiped defined by $\boldsymbol{a}$, $\boldsymbol{b}$, ar $\alpha\boldsymbol{c} + \beta\boldsymbol{d}$ is the sum of α times the volume of the parallelepiped defined by $\boldsymbol{a}$, $\boldsymbol{b}$, and $\boldsymbol{c}$ and times the volume of the parallelepiped defined by $\boldsymbol{a}$, $\boldsymbol{b}$, and $\boldsymbol{d}$. (b) The volume of the parallelepipe defined by $\boldsymbol{a}$, $\boldsymbol{b}$, and $\boldsymbol{c}$ equals the volume of the parallelepiped defined by $\boldsymbol{a}$, $\boldsymbol{b}$, and $\boldsymbol{c} + \alpha\boldsymbol{a} + \beta\boldsymbol{b}$.

parallelepiped defined by $\boldsymbol{a}$, $\boldsymbol{b}$, and $\boldsymbol{c}$ (positive if $\boldsymbol{a}$, $\boldsymbol{b}$, and $\boldsymbol{c}$ are right-handed and negati if left-handed). Equivalently, we may change the signs of the orientation and the magnituc simultaneously.

As in the case of planes, the space spanned by $\alpha\boldsymbol{a}$, $\beta\boldsymbol{b}$, and $\gamma\boldsymbol{c}$ is the same as a geometr figure as the space spanned by $\boldsymbol{a}$, $\boldsymbol{b}$, and $\boldsymbol{c}$, but the former has $\alpha\beta\gamma$ times the magnitude the latter. Similarly, the space spanned by $\boldsymbol{a}$, $\boldsymbol{b}$, and $\boldsymbol{c}$ is the same as the space spanned b $\boldsymbol{b}$, $\boldsymbol{a}$, and $\boldsymbol{c}$ as a geometric figure but they have opposite signs. Evidently, two vectors a unable to define a space. In equations, these relations are expressed by

$$(\alpha\boldsymbol{a}) \wedge (\beta\boldsymbol{b}) \wedge (\beta\boldsymbol{c}) = (\alpha\beta\gamma)\boldsymbol{a} \wedge \boldsymbol{b} \wedge \boldsymbol{c}, \qquad \boldsymbol{a} \wedge \boldsymbol{b} \wedge \boldsymbol{c} = \boldsymbol{b} \wedge \boldsymbol{c} \wedge \boldsymbol{a} = \boldsymbol{c} \wedge \boldsymbol{a} \wedge \boldsymbol{b}, \tag{5.5}$$

$$\boldsymbol{a} \wedge \boldsymbol{b} \wedge \boldsymbol{c} = -\boldsymbol{b} \wedge \boldsymbol{a} \wedge \boldsymbol{c} = -\boldsymbol{c} \wedge \boldsymbol{b} \wedge \boldsymbol{a} = -\boldsymbol{a} \wedge \boldsymbol{c} \wedge \boldsymbol{b}, \tag{5.6}$$

$$\boldsymbol{a} \wedge \boldsymbol{c} \wedge \boldsymbol{c} = \boldsymbol{a} \wedge \boldsymbol{b} \wedge \boldsymbol{a} = \boldsymbol{a} \wedge \boldsymbol{a} \wedge \boldsymbol{c} = 0, \tag{5.7}$$

where the number 0 on the right side of the last equation means "nonexistence."

Any space is the same as a geometric figure, so we define addition of spaces by th sum of their magnitudes, including sign. For two trivectors $\boldsymbol{a} \wedge \boldsymbol{b} \wedge \boldsymbol{c}$ and $\boldsymbol{a}' \wedge \boldsymbol{b}' \wedge \boldsymbol{c}'$, fo example, there exists a scalar α such that $\boldsymbol{a}' \wedge \boldsymbol{b}' \wedge \boldsymbol{c}' = \alpha\boldsymbol{a} \wedge \boldsymbol{b} \wedge \boldsymbol{c}$, so their sum is given b $(1+\alpha)\boldsymbol{a} \wedge \boldsymbol{b} \wedge \boldsymbol{c}$. Thus, we can freely compute sums and scalar multiples of spaces. In othe words, the set of all spaces forms a vector space.

The volume of the parallelepiped defined by $\boldsymbol{a}$, $\boldsymbol{b}$, and $\alpha\boldsymbol{c} + \beta\boldsymbol{d}$ is, for any α and β the sum of α times the volume of the parallelepiped defined by $\boldsymbol{a}$, $\boldsymbol{b}$, and $\boldsymbol{c}$ and β time the volume of the parallelepiped defined by $\boldsymbol{a}$, $\boldsymbol{b}$, and $\boldsymbol{d}$ (Fig. 5.3(a)). Hence, we obtain th equality

$$\boldsymbol{a} \wedge \boldsymbol{b} \wedge (\alpha\boldsymbol{c} + \beta\boldsymbol{d}) = \alpha\boldsymbol{a} \wedge \boldsymbol{b} \wedge \boldsymbol{c} + \beta\boldsymbol{a} \wedge \boldsymbol{b} \wedge \boldsymbol{d}. \tag{5.8}$$

This states that distributivity holds for the outer product $\wedge$.

By definition, vectors $\boldsymbol{a}$, $\boldsymbol{b}$, and $\boldsymbol{c}$ and vectors $\boldsymbol{a}'$, $\boldsymbol{b}'$, and $\boldsymbol{c}'$ span the same space, i.e $\boldsymbol{a} \wedge \boldsymbol{b} \wedge \boldsymbol{c} = \boldsymbol{a}' \wedge \boldsymbol{b}' \wedge \boldsymbol{c}'$, if the parallelepiped defined by $\boldsymbol{a}$, $\boldsymbol{b}$, and $\boldsymbol{c}$ and the parallelepiped define by $\boldsymbol{a}'$, $\boldsymbol{b}'$, and $\boldsymbol{c}'$ have the same volume, including sign. For example, the parallelepipe defined by $\boldsymbol{a}$, $\boldsymbol{b}$, and $\boldsymbol{c}$ and the parallelepiped defined by $\boldsymbol{a}$, $\boldsymbol{b}$, and $\boldsymbol{c} + \alpha\boldsymbol{a} + \beta\boldsymbol{b}$ have, fo any α and β, the same volume, including sign (Fig. 5.3(b)). Hence, we obtain the equalit

$$\boldsymbol{a} \wedge \boldsymbol{b} \wedge (\boldsymbol{c} + \alpha\boldsymbol{a} + \beta\boldsymbol{b}) = \boldsymbol{a} \wedge \boldsymbol{b} \wedge \boldsymbol{c}. \tag{5.9}$$

This can also be obtained from the distributivity of the outer product $\wedge$ and the rule o (5.1): $\boldsymbol{a} \wedge \boldsymbol{b} \wedge (\boldsymbol{c} + \alpha\boldsymbol{a} + \beta\boldsymbol{b}) = \boldsymbol{a} \wedge \boldsymbol{b} \wedge \boldsymbol{c} + \alpha\boldsymbol{a} \wedge \boldsymbol{a} \wedge \boldsymbol{b} + \beta\boldsymbol{a} \wedge \boldsymbol{b} \wedge \boldsymbol{b} = \boldsymbol{a} \wedge \boldsymbol{b} \wedge \boldsymbol{c}$.

magnitude. The orientation is defined by its sign, and the magnitude is given by a signed nonzero number (magnitude 0 is interpreted to be "nonexistence"). Although the origin as a geometric figure does not have a size, we think of it as having, say, electric charge, and regard its (signed) quantity as its magnitude. Hence, we can freely compute sums and scalar multiples of origins of different orientations and magnitudes, defining a vector space.

This suggests that the origin, regarded as a subspace, can be *identified with a scalar*. In this sense, we call a scalar a *0-vector*. Similarly, a vector is also called a *1-vector*. Thus, subspaces of the 3D space are defined by *k-vectors*, $k = 0, 1, 2, 3$, generating scalars (0D subspaces), lines (1D subspaces), planes (2D subspaces), and spaces (3D subspaces), respectively.

5.2 OUTER PRODUCT ALGEBRA

We now summarize the properties of the outer product in the form of axioms and derive expressions in terms of vector components with respect to the basis.

5.2.1 Axioms of outer product

What we described so far is summarized as follows. A subspace has its orientation and magnitude. Addition and scalar multiplication can be defined for subspaces of the same dimension. A subspace generated by different subspaces is defined by the outer product $\wedge$. Two lines $\boldsymbol{a}$ and $\boldsymbol{b}$ generate a 2D subspace (bivector) $\boldsymbol{a}\wedge\boldsymbol{b}$, whose sign depends on the order of composition. A line $\boldsymbol{a}$ and a plane $\boldsymbol{b}\wedge\boldsymbol{c}$ generate a 3D subspace (trivector) $\boldsymbol{a}\wedge(\boldsymbol{b}\wedge\boldsymbol{c})$, which is the same 3D subspace (trivector) $\boldsymbol{a}\wedge\boldsymbol{b}\wedge\boldsymbol{c}$ generated by three lines $\boldsymbol{a}$, $\boldsymbol{b}$, and $\boldsymbol{c}$. This can also be viewed as the 3D subspace (trivector) $(\boldsymbol{a}\wedge\boldsymbol{b})\wedge\boldsymbol{c}$ generated by plane $\boldsymbol{a}\wedge\boldsymbol{b}$ and line $\boldsymbol{c}$. Thus, *associativity holds for the outer product* $\wedge$:

$$\boldsymbol{a}\wedge(\boldsymbol{b}\wedge\boldsymbol{c}) = (\boldsymbol{a}\wedge\boldsymbol{b})\wedge\boldsymbol{c} = \boldsymbol{a}\wedge\boldsymbol{b}\wedge\boldsymbol{c}. \tag{5.10}$$

The sign changes if the order of composition is altered.

Since every subspace contains the origin, the outer product with the origin does not change the subspace as a geometric figure. However, the magnitude is multiplied by the magnitude of that origin. In other words, *the outer product with the origin can be regarded as scalar multiplication*. In view of this, we define the outer product of a scalar α and a vector $\boldsymbol{a}$ by

$$\alpha\wedge\boldsymbol{a} = \boldsymbol{a}\wedge\alpha = \alpha\boldsymbol{a}. \tag{5.11}$$

The outer product of two scalars is identified with the ordinary product: $\alpha\wedge\beta = \alpha\beta$.

In 3D, all subspaces have dimensions up to 3, and no k-vectors for $k > 3$ exist. Hence,

$$\boldsymbol{a}\wedge\boldsymbol{b}\wedge\boldsymbol{c}\wedge\boldsymbol{d} = 0, \qquad \boldsymbol{a}\wedge\boldsymbol{b}\wedge\boldsymbol{c}\wedge\boldsymbol{d}\wedge\boldsymbol{e} = 0, \qquad \ldots, \tag{5.12}$$

where the number 0 on the right sides means "nonexistence." Note that for four or more vectors either they contain duplication or one can be expressed by a linear combination of the rest. Hence, the above result is a consequence of $\boldsymbol{a}\wedge\boldsymbol{a} = 0$.

Thus, all properties of $\wedge$ can be derived from the following rules for arbitrary scalars α and β and arbitrary vectors $\boldsymbol{a}$, $\boldsymbol{b}$, and $\boldsymbol{c}$, which can be viewed as the axiom of the outer product:

associativity: $\boldsymbol{a} \wedge (\boldsymbol{b} \wedge \boldsymbol{c}) = (\boldsymbol{a} \wedge \boldsymbol{b}) \wedge \boldsymbol{c}$, which we write $\boldsymbol{a} \wedge \boldsymbol{b} \wedge \boldsymbol{c}$.

scalar operation: $\alpha \wedge \beta = \alpha\beta, \ \ \alpha \wedge \boldsymbol{a} = \boldsymbol{a} \wedge \alpha = \alpha\boldsymbol{a}$.

5.2.2 Basis expressions

If we express vectors $\boldsymbol{a}$ and $\boldsymbol{b}$ in terms of the basis $\{e_1, e_2, e_3\}$ in the form $\boldsymbol{a} = a_1e_1$ $a_2e_2 + a_3e_3$ and $\boldsymbol{b} = b_1e_1 + b_2e_2 + b_3e_3$ and substitute them into the bivector $\boldsymbol{a} \wedge \boldsymbol{b}$, we ca reduce it, using the distributivity and the antisymmetry of the outer product, ultimately a linear combination of $e_2 \wedge e_3$, $e_3 \wedge e_1$, and $e_1 \wedge e_2$:

$$\begin{aligned}\boldsymbol{a} \wedge \boldsymbol{b} &= (a_1e_1 + a_2e_2 + a_3e_3) \wedge (b_1e_1 + b_2e_2 + b_3e_3) \\ &= a_1b_1e_1 \wedge e_1 + a_1b_2e_1 \wedge e_2 + a_1b_3e_1 \wedge e_3 + a_2b_1e_2 \wedge e_1 + a_2b_2e_2 \wedge e_2 \\ &\quad + a_2b_3e_2 \wedge e_3 + a_3b_1e_3 \wedge e_1 + a_3b_2e_3 \wedge e_2 + a_3b_3e_3 \wedge e_3 \\ &= (a_2b_3 - a_3b_2)e_2 \wedge e_3 + (a_3b_1 - a_1b_3)e_3 \wedge e_1 + (a_1b_2 - a_2b_1)e_1 \wedge e_2. \qquad (5.1\end{aligned}$$

Similarly, substituting $\boldsymbol{c} = c_1e_1 + c_2e_2 + c_3e_3$ and expanding the trivector $\boldsymbol{a} \wedge \boldsymbol{b} \wedge \boldsymbol{c}$, w obtain in the end, using the distributivity and the antisymmetry of the outer product, scalar multiple of $e_1 \wedge e_2 \wedge e_3$:

$$\begin{aligned}\boldsymbol{a} \wedge \boldsymbol{b} \wedge \boldsymbol{c} &= (a_1e_1 + a_2e_2 + a_3e_3) \wedge (b_1e_1 + b_2e_2 + b_3e_3) \wedge (c_1e_1 + c_2c_2 + c_3e_3) \\ &= a_1b_2c_3e_1 \wedge e_2 \wedge e_3 + a_2b_3c_1e_2 \wedge e_3 \wedge e_1 + a_3b_1c_2e_3 \wedge e_1 \wedge e_2 \\ &\quad + a_1b_3c_2e_1 \wedge e_3 \wedge e_2 + a_2b_1c_3e_2 \wedge e_1 \wedge e_3 + a_3b_2c_1e_3 \wedge e_2 \wedge e_1 \\ &= (a_1b_2c_3 + a_2b_3c_1 + a_3b_1c_2 - a_1b_3c_2 - a_2b_1c_3 - a_3b_2c_1)e_1 \wedge e_2 \wedge e_3. \qquad (5.14\end{aligned}$$

In summary,

Proposition 5.1 (Outer product of vectors) *The outer product of vectors* $\boldsymbol{a} = \sum_{i=1}^{3} a_ie_i$ *and* $\boldsymbol{b} = \sum_{i=1}^{3} b_ie_i$ *is given by*

$$\boldsymbol{a} \wedge \boldsymbol{b} = (a_2b_3 - a_3b_2)e_2 \wedge e_3 + (a_3b_1 - a_1b_3)e_3 \wedge e_1 + (a_1b_2 - a_2b_1)e_1 \wedge e_2, \qquad (5.1$$

and the outer product of vectors $\boldsymbol{a} = \sum_{i=1}^{3} a_ie_i$, $\boldsymbol{b} = \sum_{i=1}^{3} b_ie_i$, *and* $\boldsymbol{c} = \sum_{i=1}^{3} c_ie_i$ *is give by*

$$\boldsymbol{a} \wedge \boldsymbol{b} \wedge \boldsymbol{c} = (a_1b_2c_3 + a_2b_3c_1 + a_3b_1c_2 - a_1b_3c_2 - a_2b_1c_3 - a_3b_2c_1)e_1 \wedge e_2 \wedge e_3. \qquad (5.16$$

Traditional World 5.1 (Antisymmetrization of indices) Grassmann algebra can als be defined in traditional tensor calculus, where vectors are regarded as arrays of number For numerical vectors, the outer product is defined not by the symbol $\wedge$ but by "antisym metrization" of indices. We write the bivector constructed from contravariant vectors a and b^i ($\hookrightarrow$ Traditional World 3.2 in Chapter 3) as $a^{[i}b^{j]}$, and the trivector constructed fro contravariant vectors a^i, b^i, and c^i as $a^{[i}b^jc^{k]}$, where the square brackets $[\cdots]$ surroundin indices denote an operation called *antisymmetrization*: we add all terms with permuted in dices multiplied by their signatures and then divide the sum by the number of permutations

$$a^{[i}b^{j]} = \frac{1}{2}(a^i b^j - a^j b^i),$$

$$a^{[i}b^j c^{k]} = \frac{1}{6}(a^i b^j c^k + a^j b^k c^i + a^k b^i c^j - a^k b^j c^i - a^j b^i c^k - a^i b^k c^j). \tag{5.17}$$

This operation is called "antisymmetrization" because by definition exchanging two indices in $[\cdots]$ an odd number of times changes the sign. Although multiplied by 1/2 and 1/6, the above expressions have the same meaning as Eqs. (5.15) and (5.16). Consider $a^{[i}b^{j]}$, for example, and write this in the form $a^{[i}b^{j]}e_i \wedge e_j$, where Einstein's summation convention ($\hookrightarrow$ Sec. 3.3 in Chapter 3) is used. Since $a^{[i}b^{j]}$ and $e_i \wedge e_j$ are both antisymmetric with respect to the indices, the term for $i = 1$ and $j = 2$ and the term for $i = 2$ and $i = 1$ in the summation are the same, so

$$a^{[1}b^{2]}e_1 \wedge e_2 + a^{[2}b^{1]}e_2 \wedge e_1 = \frac{1}{2}(a^1b^2 - a^2b^1)e_1 \wedge e_2 - \frac{1}{2}(a^2b^1 - a^1b^2)e_1 \wedge e_2$$
$$= (a^1b^2 - a^2b^1)e_1 \wedge e_2. \tag{5.18}$$

The same applies to the terms of $e_2 \wedge e_3$ and $e_3 \wedge e_1$, so we obtain Eq. (5.15). Similarly, consider the expression $a^{[i}b^j c^{k]}e_i \wedge e_j \wedge e_k$. It sums six terms, but since $a^{[i}b^j c^{k]}$ and $e_i \wedge e_j \wedge e_k$ are both antisymmetric with respect to the indices, having the same signature for the same permutation, all terms are equal. Hence, the sum is six times $a^{[1}b^2c^{3]}e_1 \wedge e_2 \wedge e_3$, resulting in Eq. (5.16). A geometric object with antisymmetric indices is called an *antisymmetric tensor*. The bivector $a^{[i}b^{j]}$ and the trivector $a^{[i}b^j c^{k]}$ are antisymmetric tensors.

5.3 CONTRACTION

Contraction is an operation that reduces a kD subspace to a subspace of a lower dimension. To be specific, we multiply a k-vector representing a kD subspace by some j-vector, $j \leq k$, to generate a $(k-j)$-vector representing a $(k-j)$D subspace.

5.3.1 Contraction of a line

Lowering the dimensionality of a line results in the origin, i.e., a scalar. We denote by $\boldsymbol{x} \cdot \boldsymbol{a}$ the scalar obtained by contracting the line $\boldsymbol{a}$ by vector $\boldsymbol{x}$. It represents the origin as a geometric figure, but it has a (signed) magnitude, which we define by the inner product

$$\boldsymbol{x} \cdot \boldsymbol{a} = \langle \boldsymbol{x}, \boldsymbol{a} \rangle. \tag{5.19}$$

Geometrically, we can interpret this as computing the intersection of the line $\boldsymbol{a}$ with the plane whose surface normal is $\boldsymbol{x}$ (Fig. 5.4). We can imagine that the intersection of a line and a plane is "strong" if they meet nearly perpendicularly and "weak" if they are nearly parallel. If they are exactly parallel, i.e., if the line $\boldsymbol{a}$ lies on that plane, no intersection exists, so it is 0 (= nonexistence).

5.3.2 Contraction of a plane

Lowering the dimensionality of a plane results in either a line or the origin. We denote by $\boldsymbol{x} \cdot (\boldsymbol{a} \wedge \boldsymbol{b})$ the line obtained by contracting the plane $\boldsymbol{a} \wedge \boldsymbol{b}$ by vector $\boldsymbol{x}$. Hereafter, we regard the outer product $\wedge$ as a stronger operation than contraction and omit the parentheses to

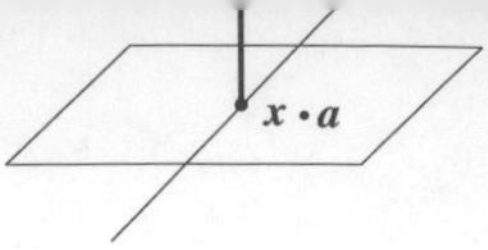

FIGURE 5.4 The point $\boldsymbol{x} \cdot \boldsymbol{a}$ obtained by contracting the line $\boldsymbol{a}$ by vector $\boldsymbol{x}$ is the intersection the line $\boldsymbol{a}$ with the plane whose surface normal is $\boldsymbol{x}$.

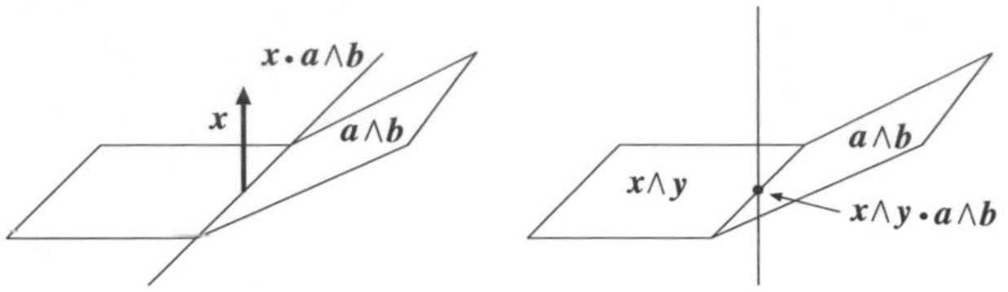

(a) (b)

FIGURE 5.5 (a) The line $\boldsymbol{x} \cdot \boldsymbol{a} \wedge \boldsymbol{b}$ obtained by contracting the plane $\boldsymbol{a} \wedge \boldsymbol{b}$ by vector $\boldsymbol{x}$ is th intersection of the plane $\boldsymbol{a} \wedge \boldsymbol{b}$ with the plane whose surface normal is $\boldsymbol{x}$. (b) The point $\boldsymbol{x} \wedge \boldsymbol{y} \cdot \boldsymbol{a} \wedge$ obtained by contracting the plane $\boldsymbol{a} \wedge \boldsymbol{b}$ by bivector $\boldsymbol{x} \wedge \boldsymbol{y}$ is the intersection of the plane $\boldsymbol{a} \wedge \boldsymbol{b}$ wit the surface normal to the plane $\boldsymbol{x} \wedge \boldsymbol{y}$.

write simply $\boldsymbol{x} \cdot \boldsymbol{a} \wedge \boldsymbol{b}$ for $\boldsymbol{x} \cdot (\boldsymbol{a} \wedge \boldsymbol{b})$. We now introduce the rule that *the inner product successively computed by alternately changing the sign*:

$$\boldsymbol{x} \cdot \boldsymbol{a} \wedge \boldsymbol{b} = \langle \boldsymbol{x}, \boldsymbol{a} \rangle \wedge \boldsymbol{b} - \boldsymbol{a} \wedge \langle \boldsymbol{x}, \boldsymbol{b} \rangle = \langle \boldsymbol{x}, \boldsymbol{a} \rangle \boldsymbol{b} - \langle \boldsymbol{x}, \boldsymbol{b} \rangle \boldsymbol{a}. \quad (5.2\text{(}$$

Since this is a linear combination of $\boldsymbol{a}$ and $\boldsymbol{b}$, the resulting line is contained in the plar $\boldsymbol{a} \wedge \boldsymbol{b}$ unless $\langle \boldsymbol{x}, \boldsymbol{a} \rangle = \langle \boldsymbol{x}, \boldsymbol{b} \rangle = 0$. We can see that this line is orthogonal to $\boldsymbol{x}$:

$$\langle \boldsymbol{x}, \langle \boldsymbol{x}, \boldsymbol{a} \rangle \boldsymbol{b} - \langle \boldsymbol{x}, \boldsymbol{b} \rangle \boldsymbol{a} \rangle = \langle \boldsymbol{x}, \boldsymbol{a} \rangle \langle \boldsymbol{x}, \boldsymbol{b} \rangle - \langle \boldsymbol{x}, \boldsymbol{b} \rangle \langle \boldsymbol{x}, \boldsymbol{a} \rangle = 0. \quad (5.21)$$

Hence, contraction of the line $\boldsymbol{a} \wedge \boldsymbol{b}$ by vector $\boldsymbol{x}$ can be geometrically interpreted as *computin the intersection of the plane $\boldsymbol{a} \wedge \boldsymbol{b}$ with the plane whose surface normal is $\boldsymbol{x}$* (Fig. 5.5(a)); th intersection is "strong" if they meet nearly perpendicularly and "weak" if they are nearl parallel.

We denote by $\boldsymbol{x} \wedge \boldsymbol{y} \cdot \boldsymbol{a} \wedge \boldsymbol{b}$ (the parentheses in $(\boldsymbol{x} \wedge \boldsymbol{y}) \cdot (\boldsymbol{a} \wedge \boldsymbol{b})$ are omitted) the scala obtained by contracting the plane $\boldsymbol{a} \wedge \boldsymbol{b}$ by bivector $\boldsymbol{x} \wedge \boldsymbol{y}$. Again, we introduce the rule tha *the inner product is successively computed from inside by alternately changing the sign*:

$$\begin{aligned} \boldsymbol{x} \wedge \boldsymbol{y} \cdot \boldsymbol{a} \wedge \boldsymbol{b} &= \boldsymbol{x} \cdot (\boldsymbol{y} \cdot \boldsymbol{a} \wedge \boldsymbol{b}) = \boldsymbol{x} \cdot (\langle \boldsymbol{y}, \boldsymbol{a} \rangle \boldsymbol{b} - \langle \boldsymbol{y}, \boldsymbol{b} \rangle \boldsymbol{a}) \\ &= \langle \boldsymbol{y}, \boldsymbol{a} \rangle \langle \boldsymbol{x}, \boldsymbol{b} \rangle - \langle \boldsymbol{y}, \boldsymbol{b} \rangle \langle \boldsymbol{x}, \boldsymbol{a} \rangle = \langle \boldsymbol{x}, \boldsymbol{b} \rangle \langle \boldsymbol{y}, \boldsymbol{a} \rangle - \langle \boldsymbol{x}, \boldsymbol{a} \rangle \langle \boldsymbol{y}, \boldsymbol{b} \rangle. \end{aligned} \quad (5.22)$$

This is a scalar, representing the origin. Geometrically, we can interpret this as *computin the intersection of the plane $\boldsymbol{a} \wedge \boldsymbol{b}$ with the surface normal to the plane $\boldsymbol{x} \wedge \boldsymbol{y}$* (Fig. 5.5(b)); th intersection is "strong" if they meet nearly perpendicularly and "weak" if they are nearl parallel.

5.3.3 Contraction of a space

Lowering the dimensionality of a space results in a plane, a line, or the origin. We denote b $\boldsymbol{x} \cdot \boldsymbol{a} \wedge \boldsymbol{b} \wedge \boldsymbol{c}$ (parentheses omitted) the plane obtained by contracting the space $\boldsymbol{a} \wedge \boldsymbol{b} \wedge \boldsymbol{c}$ b

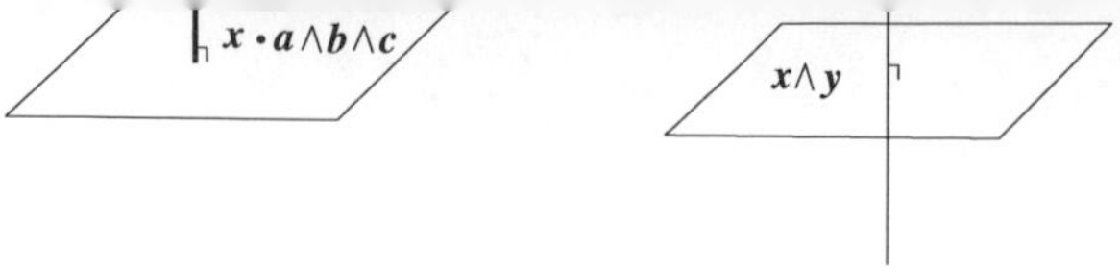

(a) (b)

FIGURE 5.6 (a) The plane $\boldsymbol{x} \cdot \boldsymbol{a} \wedge \boldsymbol{b} \wedge \boldsymbol{c}$ obtained by contracting the space $\boldsymbol{a} \wedge \boldsymbol{b} \wedge \boldsymbol{c}$ by vector $\boldsymbol{x}$ is a plane whose surface normal is $\boldsymbol{x}$. (b) The line $\boldsymbol{x} \wedge \boldsymbol{y} \cdot \boldsymbol{a} \wedge \boldsymbol{b}$ obtained by contracting the space $\boldsymbol{a} \wedge \boldsymbol{b} \wedge \boldsymbol{c}$ by bivector $\boldsymbol{x} \wedge \boldsymbol{y}$ is the surface normal to the surface $\boldsymbol{x} \wedge \boldsymbol{y}$.

vector $\boldsymbol{x}$. Here, too, we introduce the rule that *the inner product is successively computed by alternately changing the sign*:

$$\begin{aligned} \boldsymbol{x} \cdot \boldsymbol{a} \wedge \boldsymbol{b} \wedge \boldsymbol{c} &= \langle \boldsymbol{x}, \boldsymbol{a} \rangle \wedge \boldsymbol{b} \wedge \boldsymbol{c} - \boldsymbol{a} \wedge \langle \boldsymbol{x}, \boldsymbol{b} \rangle \wedge \boldsymbol{c} + \boldsymbol{a} \wedge \boldsymbol{b} \wedge \langle \boldsymbol{x}, \boldsymbol{c} \rangle \\ &= \langle \boldsymbol{x}, \boldsymbol{a} \rangle \boldsymbol{b} \wedge \boldsymbol{c} + \langle \boldsymbol{x}, \boldsymbol{b} \rangle \boldsymbol{c} \wedge \boldsymbol{a} + \langle \boldsymbol{x}, \boldsymbol{c} \rangle \boldsymbol{a} \wedge \boldsymbol{b}. \end{aligned} \tag{5.23}$$

The resulting plane is orthogonal to $\boldsymbol{x}$. In fact, if we contract this plane by vector $\boldsymbol{x}$, we obtain from Eq. (5.20)

$$\langle \boldsymbol{x}, \boldsymbol{a} \rangle (\langle \boldsymbol{x}, \boldsymbol{b} \rangle \boldsymbol{c} - \langle \boldsymbol{x}, \boldsymbol{c} \rangle \boldsymbol{b}) + \langle \boldsymbol{x}, \boldsymbol{b} \rangle (\langle \boldsymbol{x}, \boldsymbol{c} \rangle \boldsymbol{a} - \langle \boldsymbol{x}, \boldsymbol{a} \rangle \boldsymbol{c}) + \langle \boldsymbol{x}, \boldsymbol{c} \rangle (\langle \boldsymbol{x}, \boldsymbol{a} \rangle \boldsymbol{b} - \langle \boldsymbol{x}, \boldsymbol{b} \rangle \boldsymbol{a}) = 0. \tag{5.24}$$

We see from Eq. (5.20) that a plane whose contraction by $\boldsymbol{x}$ is 0 must be orthogonal to $\boldsymbol{x}$, since otherwise a line orthogonal to $\boldsymbol{x}$ would result. Thus, the contraction of Eq. (5.23) can be geometrically interpreted as *computing the intersection of the space $\boldsymbol{a} \wedge \boldsymbol{b} \wedge \boldsymbol{c}$ with the plane whose surface normal is $\boldsymbol{x}$* (Fig. 5.6(a)).

We denote by $\boldsymbol{x} \wedge \boldsymbol{y} \cdot \boldsymbol{a} \wedge \boldsymbol{b} \wedge \boldsymbol{c}$ (parentheses omitted) the line obtained by contracting the space $\boldsymbol{a} \wedge \boldsymbol{b} \wedge \boldsymbol{c}$ by bivector $\boldsymbol{x} \wedge \boldsymbol{y}$. As before, *the inner product is successively computed from inside by alternately changing the sign*:

$$\begin{aligned} &\boldsymbol{x} \wedge \boldsymbol{y} \cdot \boldsymbol{a} \wedge \boldsymbol{b} \wedge \boldsymbol{c} = \boldsymbol{x} \cdot (\boldsymbol{y} \cdot \boldsymbol{a} \wedge \boldsymbol{b} \wedge \boldsymbol{c}) \boldsymbol{x} \cdot (\langle \boldsymbol{y}, \boldsymbol{a} \rangle \boldsymbol{b} \wedge \boldsymbol{c} + \langle \boldsymbol{y}, \boldsymbol{b} \rangle \boldsymbol{c} \wedge \boldsymbol{a} + \langle \boldsymbol{y}, \boldsymbol{c} \rangle \boldsymbol{a} \wedge \boldsymbol{b}) \\ &= \langle \boldsymbol{y}, \boldsymbol{a} \rangle (\langle \boldsymbol{x}, \boldsymbol{b} \rangle \boldsymbol{c} - \langle \boldsymbol{x}, \boldsymbol{c} \rangle \boldsymbol{b}) + \langle \boldsymbol{y}, \boldsymbol{b} \rangle (\langle \boldsymbol{x}, \boldsymbol{c} \rangle \boldsymbol{a} - \langle \boldsymbol{x}, \boldsymbol{a} \rangle \boldsymbol{c}) + \langle \boldsymbol{y}, \boldsymbol{c} \rangle (\langle \boldsymbol{x}, \boldsymbol{a} \rangle \boldsymbol{b} - \langle \boldsymbol{x}, \boldsymbol{b} \rangle \boldsymbol{a}) \\ &= (\langle \boldsymbol{x}, \boldsymbol{c} \rangle \langle \boldsymbol{y}, \boldsymbol{b} \rangle - \langle \boldsymbol{x}, \boldsymbol{b} \rangle \langle \boldsymbol{y}, \boldsymbol{c} \rangle) \boldsymbol{a} + (\langle \boldsymbol{x}, \boldsymbol{a} \rangle \langle \boldsymbol{y}, \boldsymbol{c} \rangle \\ &\quad - \langle \boldsymbol{x}, \boldsymbol{c} \rangle \langle \boldsymbol{y}, \boldsymbol{a} \rangle) \boldsymbol{b} + (\langle \boldsymbol{x}, \boldsymbol{b} \rangle \langle \boldsymbol{y}, \boldsymbol{a} \rangle - \langle \boldsymbol{x}, \boldsymbol{a} \rangle \langle \boldsymbol{y}, \boldsymbol{b} \rangle) \boldsymbol{c}. \end{aligned} \tag{5.25}$$

This line is orthogonal to both $\boldsymbol{x}$ and $\boldsymbol{y}$. In fact, the inner product of Eq. (5.25) with $\boldsymbol{x}$ is

$$\begin{aligned} &(\langle \boldsymbol{x}, \boldsymbol{c} \rangle \langle \boldsymbol{y}, \boldsymbol{b} \rangle - \langle \boldsymbol{x}, \boldsymbol{b} \rangle \langle \boldsymbol{y}, \boldsymbol{c} \rangle) \langle \boldsymbol{x}, \boldsymbol{a} \rangle + (\langle \boldsymbol{x}, \boldsymbol{a} \rangle \langle \boldsymbol{y}, \boldsymbol{c} \rangle - \langle \boldsymbol{x}, \boldsymbol{c} \rangle \langle \boldsymbol{y}, \boldsymbol{a} \rangle) \langle \boldsymbol{x}, \boldsymbol{b} \rangle \\ &+ (\langle \boldsymbol{x}, \boldsymbol{b} \rangle \langle \boldsymbol{y}, \boldsymbol{a} \rangle - \langle \boldsymbol{x}, \boldsymbol{a} \rangle \langle \boldsymbol{y}, \boldsymbol{b} \rangle) \langle \boldsymbol{x}, \boldsymbol{c} \rangle = 0. \end{aligned} \tag{5.26}$$

Similarly, the inner product with $\boldsymbol{y}$ is

$$\begin{aligned} &(\langle \boldsymbol{x}, \boldsymbol{c} \rangle \langle \boldsymbol{y}, \boldsymbol{b} \rangle - \langle \boldsymbol{x}, \boldsymbol{b} \rangle \langle \boldsymbol{y}, \boldsymbol{c} \rangle) \langle \boldsymbol{y}, \boldsymbol{a} \rangle + (\langle \boldsymbol{x}, \boldsymbol{a} \rangle \langle \boldsymbol{y}, \boldsymbol{c} \rangle - \langle \boldsymbol{x}, \boldsymbol{c} \rangle \langle \boldsymbol{y}, \boldsymbol{a} \rangle) \langle \boldsymbol{y}, \boldsymbol{b} \rangle \\ &+ (\langle \boldsymbol{x}, \boldsymbol{b} \rangle \langle \boldsymbol{y}, \boldsymbol{a} \rangle - \langle \boldsymbol{x}, \boldsymbol{a} \rangle \langle \boldsymbol{y}, \boldsymbol{b} \rangle) \langle \boldsymbol{y}, \boldsymbol{c} \rangle = 0. \end{aligned} \tag{5.27}$$

Geometrically, we can interpret the contraction of Eq. (5.25) to be *computing the intersection of the space $\boldsymbol{a} \wedge \boldsymbol{b} \wedge \boldsymbol{c}$ with the surface normal to the plane $\boldsymbol{x} \wedge \boldsymbol{y}$* (Fig. 5.6(b)).

We denote by $\boldsymbol{x} \wedge \boldsymbol{y} \wedge \boldsymbol{z} \cdot \boldsymbol{a} \wedge \boldsymbol{b} \wedge \boldsymbol{c}$ (parentheses omitted) the scalar obtained by contracting

$$\begin{aligned}
\boldsymbol{x}\wedge\boldsymbol{y}\wedge\boldsymbol{z}\cdot\boldsymbol{a}\wedge\boldsymbol{b}\wedge\boldsymbol{c} &= \boldsymbol{x}\cdot(\boldsymbol{y}\cdot(\boldsymbol{z}\cdot\boldsymbol{a}\wedge\boldsymbol{b}\wedge\boldsymbol{c}))\\
&= \boldsymbol{x}\cdot\Big((\langle\boldsymbol{y},\boldsymbol{c}\rangle\langle\boldsymbol{z},\boldsymbol{b}\rangle-\langle\boldsymbol{y},\boldsymbol{b}\rangle\langle\boldsymbol{z},\boldsymbol{c}\rangle)\boldsymbol{a}+(\langle\boldsymbol{y},\boldsymbol{a}\rangle\langle\boldsymbol{z},\boldsymbol{c}\rangle-\langle\boldsymbol{y},\boldsymbol{c}\rangle\langle\boldsymbol{z},\boldsymbol{a}\rangle)\boldsymbol{b}\\
&\quad+(\langle\boldsymbol{y},\boldsymbol{b}\rangle\langle\boldsymbol{z},\boldsymbol{a}\rangle-\langle\boldsymbol{y},\boldsymbol{a}\rangle\langle\boldsymbol{z},\boldsymbol{b}\rangle)\boldsymbol{c}\Big)\\
&= (\langle\boldsymbol{y},\boldsymbol{c}\rangle\langle\boldsymbol{z},\boldsymbol{b}\rangle-\langle\boldsymbol{y},\boldsymbol{b}\rangle\langle\boldsymbol{z},\boldsymbol{c}\rangle)\langle\boldsymbol{x},\boldsymbol{a}\rangle+(\langle\boldsymbol{y},\boldsymbol{a}\rangle\langle\boldsymbol{z},\boldsymbol{c}\rangle-\langle\boldsymbol{y},\boldsymbol{c}\rangle\langle\boldsymbol{z},\boldsymbol{a}\rangle)\langle\boldsymbol{x},\boldsymbol{b}\rangle\\
&\quad+(\langle\boldsymbol{y},\boldsymbol{b}\rangle\langle\boldsymbol{z},\boldsymbol{a}\rangle-\langle\boldsymbol{y},\boldsymbol{a}\rangle\langle\boldsymbol{z},\boldsymbol{b}\rangle)\langle\boldsymbol{x},\boldsymbol{c}\rangle\\
&= \langle\boldsymbol{x},\boldsymbol{a}\rangle\langle\boldsymbol{y},\boldsymbol{c}\rangle\langle\boldsymbol{z},\boldsymbol{b}\rangle+\langle\boldsymbol{x},\boldsymbol{b}\rangle\langle\boldsymbol{y},\boldsymbol{a}\rangle\langle\boldsymbol{z},\boldsymbol{c}\rangle+\langle\boldsymbol{x},\boldsymbol{c}\rangle\langle\boldsymbol{y},\boldsymbol{b}\rangle\langle\boldsymbol{z},\boldsymbol{a}\rangle\\
&\quad-\langle\boldsymbol{x},\boldsymbol{a}\rangle\langle\boldsymbol{y},\boldsymbol{b}\rangle\langle\boldsymbol{z},\boldsymbol{c}\rangle-\langle\boldsymbol{x},\boldsymbol{b}\rangle\langle\boldsymbol{y},\boldsymbol{c}\rangle\langle\boldsymbol{z},\boldsymbol{a}\rangle-\langle\boldsymbol{x},\boldsymbol{c}\rangle\langle\boldsymbol{y},\boldsymbol{a}\rangle\langle\boldsymbol{z},\boldsymbol{b}\rangle. \qquad (5.28)
\end{aligned}$$

This scalar represents the origin, so geometrically the above contraction can be interprete as *computing the intersection of the space $\boldsymbol{a}\wedge\boldsymbol{b}\wedge\boldsymbol{c}$ with the origin, which is orthogonal the space $\boldsymbol{x}\wedge\boldsymbol{y}\wedge\boldsymbol{z}$.*

5.3.4 Summary of contraction

Contracting a subspace by a k-vector, $k = 1, 2, 3$, lowers the dimensionality by k. Subspac of dimension less than 0 do not exist, so, for example, $\boldsymbol{x}\wedge\boldsymbol{y}\cdot\boldsymbol{a} = 0$. The origin is a 0 subspace. Hence, its dimension is not lowered by a scalar α (= 0-vector), but its magnitu is multiplied by α. We define

$$\alpha\cdot\beta=\alpha\beta,\qquad \alpha\cdot\boldsymbol{a}=\alpha\boldsymbol{a},\qquad \alpha\cdot\boldsymbol{a}\wedge\boldsymbol{b}=\alpha\boldsymbol{a}\wedge\boldsymbol{b},\qquad \alpha\cdot\boldsymbol{a}\wedge\boldsymbol{b}\wedge\boldsymbol{c}=\alpha\boldsymbol{a}\wedge\boldsymbol{b}\wedge\boldsymbol{c}. \qquad (5.29)$$

Since no subspace has dimension less than 0, contraction of a scalar by anything other tha a scalar is 0 (= nonexistence):

$$\boldsymbol{x}\cdot\alpha=0,\qquad \boldsymbol{x}\wedge\boldsymbol{y}\cdot\alpha=0,\qquad \boldsymbol{x}\wedge\boldsymbol{y}\wedge\boldsymbol{z}\cdot\alpha=0. \qquad (5.30)$$

All the results we have obtained so far are summarized as follows:

Proposition 5.2 (Contraction computation) *Subspaces are contracted by a k-vecto $k = 0, 1, 2, 3$, in the following form:*

Contraction by scalar α:

$$\alpha\cdot\beta=\alpha\beta,\qquad \alpha\cdot\boldsymbol{a}=\alpha\boldsymbol{a},\qquad \alpha\cdot\boldsymbol{a}\wedge\boldsymbol{b}=\alpha\boldsymbol{a}\wedge\boldsymbol{b},\qquad \alpha\cdot\boldsymbol{a}\wedge\boldsymbol{b}\wedge\boldsymbol{c}=\alpha\boldsymbol{a}\wedge\boldsymbol{b}\wedge\boldsymbol{c}. \qquad (5.31)$$

Contraction by vector $\boldsymbol{x}$:

$$\boldsymbol{x}\cdot\alpha=0,\qquad \boldsymbol{x}\cdot\boldsymbol{a}=\langle\boldsymbol{x},\boldsymbol{a}\rangle,\qquad \boldsymbol{x}\cdot\boldsymbol{a}\wedge\boldsymbol{b}=\langle\boldsymbol{x},\boldsymbol{a}\rangle\boldsymbol{b}-\langle\boldsymbol{x},\boldsymbol{b}\rangle\boldsymbol{a},$$

$$\boldsymbol{x}\cdot\boldsymbol{a}\wedge\boldsymbol{b}\wedge\boldsymbol{c}=\langle\boldsymbol{x},\boldsymbol{a}\rangle\boldsymbol{b}\wedge\boldsymbol{c}+\langle\boldsymbol{x},\boldsymbol{b}\rangle\boldsymbol{c}\wedge\boldsymbol{a}+\langle\boldsymbol{x},\boldsymbol{c}\rangle\boldsymbol{a}\wedge\boldsymbol{b}. \qquad (5.32)$$

Contraction by bivector $\boldsymbol{x}\wedge\boldsymbol{y}$:

$$\boldsymbol{x}\wedge\boldsymbol{y}\cdot\alpha=0,\qquad \boldsymbol{x}\wedge\boldsymbol{y}\cdot\boldsymbol{a}=0,\qquad \boldsymbol{x}\wedge\boldsymbol{y}\cdot\boldsymbol{a}\wedge\boldsymbol{b}=\langle\boldsymbol{x},\boldsymbol{b}\rangle\langle\boldsymbol{y},\boldsymbol{a}\rangle-\langle\boldsymbol{x},\boldsymbol{a}\rangle\langle\boldsymbol{y},\boldsymbol{b}\rangle,$$

$$\begin{aligned}
\boldsymbol{x}\wedge\boldsymbol{y}\cdot\boldsymbol{a}\wedge\boldsymbol{b}\wedge\boldsymbol{c} &= (\langle\boldsymbol{x},\boldsymbol{c}\rangle\langle\boldsymbol{y},\boldsymbol{b}\rangle-\langle\boldsymbol{x},\boldsymbol{b}\rangle\langle\boldsymbol{y},\boldsymbol{c}\rangle)\boldsymbol{a}+(\langle\boldsymbol{x},\boldsymbol{a}\rangle\langle\boldsymbol{y},\boldsymbol{c}\rangle-\langle\boldsymbol{x},\boldsymbol{c}\rangle\langle\boldsymbol{y},\boldsymbol{a}\rangle)\boldsymbol{b}\\
&\quad+(\langle\boldsymbol{x},\boldsymbol{b}\rangle\langle\boldsymbol{y},\boldsymbol{a}\rangle-\langle\boldsymbol{x},\boldsymbol{a}\rangle\langle\boldsymbol{y},\boldsymbol{b}\rangle)\boldsymbol{c}. \qquad (5.33)
\end{aligned}$$

$$\boldsymbol{x}\wedge\boldsymbol{y}\wedge\boldsymbol{z}\cdot\boldsymbol{a}\wedge\boldsymbol{b}\wedge\boldsymbol{c} = \langle\boldsymbol{x},\boldsymbol{a}\rangle\langle\boldsymbol{y},\boldsymbol{c}\rangle\langle\boldsymbol{z},\boldsymbol{b}\rangle + \langle\boldsymbol{x},\boldsymbol{b}\rangle\langle\boldsymbol{y},\boldsymbol{a}\rangle\langle\boldsymbol{z},\boldsymbol{c}\rangle + \langle\boldsymbol{x},\boldsymbol{c}\rangle\langle\boldsymbol{y},\boldsymbol{b}\rangle\langle\boldsymbol{z},\boldsymbol{a}\rangle$$
$$-\langle\boldsymbol{x},\boldsymbol{a}\rangle\langle\boldsymbol{y},\boldsymbol{b}\rangle\langle\boldsymbol{z},\boldsymbol{c}\rangle - \langle\boldsymbol{x},\boldsymbol{b}\rangle\langle\boldsymbol{y},\boldsymbol{c}\rangle\langle\boldsymbol{z},\boldsymbol{a}\rangle - \langle\boldsymbol{x},\boldsymbol{c}\rangle\langle\boldsymbol{y},\boldsymbol{a}\rangle\langle\boldsymbol{z},\boldsymbol{b}\rangle. \quad (5.34)$$

We have also observed the following:

Proposition 5.3 (Geometric meaning of contraction) *The contraction of a subspace by a k-vector is included in that subspace, having a dimension smaller by k, and is orthogonal to the subspace specified by that k-vector.*

For a kD subspace, $k = 0, 1, 2, 3$, the $(3-k)$D subspace orthogonal to it is called its *orthogonal complement*. We have seen that contraction by a k-vector can be interpreted as computing *the intersection of the subspace with the orthogonal complement of the subspace specified by that k-vector*; the magnitude of the intersection depends not only on the magnitudes of the two subspaces but also on their angle of intersection. In particular, if the k-vector is orthogonal to the subspace, the contraction by that k-vector is 0. Hence, we observe

Proposition 5.4 (Contraction and orthogonality) *Contraction of a subspace by a subspace is 0 if and only if the two subspaces are orthogonal:*

$$(\cdots)\cdot(\cdots) = 0 \quad \Leftrightarrow \quad (\cdots)\perp(\cdots). \quad (5.35)$$

The inner product can be viewed as a special case of contraction, i.e., contraction of a line by a line.

Traditional World 5.2 (Tensor calculus and contraction) In traditional tensor calculus, where an array of numbers is regarded as a vector, contraction means juxtaposing a k-vector constructed from contravariant vectors and a j-vector constructed from covariant vectors, which are summed over corresponding upper and lower indices. For example, contraction of a bivector $a^{[i}b^{j]}$ constructed from contravariant vectors a^i and b^i by covariant vector x_i is

$$x_i a^{[i}b^{j]} = x_i\Big(\frac{1}{2}(a^i b^j - a^j b^i)\Big) = \frac{1}{2}((x_i a^i)b^j - (x_i b^i)a^j). \quad (5.36)$$

Since $x_i a^i$ and $x_i b^i$ (Einstein's summation convention is used) mean the inner products $\langle\boldsymbol{x},\boldsymbol{a}\rangle$ and $\langle\boldsymbol{x},\boldsymbol{b}\rangle$, respectively, the above expression describes the same thing as Eq. (5.20) apart from the multiplier $1/2$. The operation "alternately changing the sign" corresponds to antisymmetrization of indices. Contraction of $a^{[i}b^{j]}$ by the bivector $x_{[i}y_{j]}$ obtained by antisymmetrization of covariant vectors x_i and y_i is

$$x_{[j}y_{i]}a^{[i}b^{j]} = x_j y_i a^{[i}b^{j]} = x_j y_i\Big(\frac{1}{2}(a^i b^j - a^j b^i)\Big) = \frac{1}{2}(x_i a^i b^j - x_i b^i a^j)$$
$$= \frac{1}{2}\Big((x_j b^j)(y_i a^i) - (x_j a^j)(y_i b^i)\Big), \quad (5.37)$$

which describes the same thing as Eq. (5.22) apart from the multiplier $1/2$. The operation "successively from inside" corresponds to summing over the "nearest" indices of the same letter first. Note that $x_{[j}y_{i]}a^{[i}b^{j]}$ can be replaced by $x_j y_i a^{[i}b^{j]}$, because for summing two

In fact,

$$x_j y_i a^{[i}b^{j]} = \frac{1}{2}(x_j y_i a^{[i}b^{j]} + x_i y_j a^{[j}b^{i]}) = \frac{1}{2}(x_j y_i a^{[i}b^{j]} - x_i y_j a^{[i}b^{j]})$$
$$= \frac{1}{2}(x_j y_i - x_i y_j) a^{[i}b^{j]} = x_{[j}y_{i]} a^{[i}b^{j]}, \qquad (5.3$$

so if one set of indices is antisymmetric, the other indices to be summed are automatical antisymmetrized. Contraction of a trivector $a^{[i}b^j c^{k]}$ by x_i is

$$x_i a^{[i}b^j c^{k]} = x_i\Big(\frac{1}{6}(a^i b^j c^k + a^j b^k c^i + a^k b^i c^j - a^k b^j c^i - a^j b^i c^k - a^i b^k c^j)\Big)$$
$$= \frac{1}{6}\Big(x_i a^i (b^j c^k - b^k c^j) + x_i b^i (c^j a^k - c^k a^j) + x_i c^i (a^j b^k - a^k b^j)\Big)$$
$$= \frac{1}{3}\Big((x_i a^i) b^{[j}c^{k]} + (x_i b^i) c^{[j}a^{k]} + (x_i c^i) a^{[j}b^{k]}\Big), \qquad (5.3$$

which describes the same thing as Eq. (5.23) apart from the multiplier 1/3. Contraction trivector $a^{[i}b^j c^{k]}$ by $x_{[i}y_{j]}$ is

$$x_{[j}y_{i]} a^{[i}b^j c^{k]} = x_j y_i\Big(\frac{1}{6}(a^i b^j c^k + a^j b^k c^i + a^k b^i c^j - a^k b^j c^i - a^j b^i c^k - a^i b^k c^j)\Big)$$
$$= \frac{1}{6}\Big(((x_j c^j)(y_i b^i) - (x_j b^j)(y_i c^i))a^k + ((x_j a^j)(y_i c^i) - (x_j c^j)(y_i a^i))b^k$$
$$+((x_j b^j)(y_i a^i) - (x_j a^j)(y_i b^i))c^k\Big), \qquad (5.4$$

which describes the same thing as Eq. (5.25) apart from the multiplier 1/6. We can repla $x_{[j}y_{i]}$ by $x_j y_i$, because, as explained earlier, the indices of the term to be multiplied a antisymmetric. Contraction of trivector $a^{[i}b^j c^{k]}$ by $x_{[i}y_j z_{k]}$ is

$$x_{[k}y_j z_{i]} a^{[i}b^j c^{k]} = x_k y_j z_i\Big(\frac{1}{6}(a^i b^j c^k + a^j b^k c^i + a^k b^i c^j - a^k b^j c^i - a^j b^i c^k - a^i b^k c^j)\Big)$$
$$= \frac{1}{6}\Big((x_k c^k)(y_j b^j)(z_i a^i) + (x_k b^k)(y_j a^j)(z_i c^i) + (x_k a^k)(y_j c^j)(z_i b^i)$$
$$-(x_k a^k)(y_j b^j)(z_i c^i) - (x_k c^k)(y_j a^j)(z_i b^i) - (x_k b^k)(y_j c^j)(z_i a^i)\Big), \qquad (5.41$$

which describes the same thing as Eq. (5.28) apart from the multiplier 1/6. We can repla $x_{[k}y_j z_{i]}$ by $x_k y_j z_i$, because the indices of the corresponding term are antisymmetric.

5.4 NORM

We define the *norm* of a k-vector, $k = 0, 1, 2, 3$, in such a way that its square equals i contraction by its *reversal* (the reversal of $\boldsymbol{a}\wedge\boldsymbol{b}$ is $\boldsymbol{b}\wedge\boldsymbol{a}$, the reversal of $\boldsymbol{a}\wedge\boldsymbol{b}\wedge\boldsymbol{c}$ is $\boldsymbol{c}\wedge\boldsymbol{b}\wedge\boldsymbol{a}$ etc.).

If we contract a scalar α by itself, we have from the first equation in Eq. (5.31)

$$\|\alpha\|^2 = \alpha\cdot\alpha = \alpha^2. \qquad (5.42$$

Hence, the norm $\|\alpha\|$ equals $|\alpha|$.

which coincides with definition of the usual vector norm.

If we contract a bivector $\boldsymbol{a} \wedge \boldsymbol{b}$ by its reversal $\boldsymbol{b} \wedge \boldsymbol{a}$, we have from the third equation in Eq. (5.33)

$$\|\boldsymbol{a} \wedge \boldsymbol{b}\|^2 = \boldsymbol{b} \wedge \boldsymbol{a} \cdot \boldsymbol{a} \wedge \boldsymbol{b} = \langle \boldsymbol{b}, \boldsymbol{b}\rangle\langle \boldsymbol{a}, \boldsymbol{a}\rangle - \langle \boldsymbol{b}, \boldsymbol{a}\rangle^2 = \|\boldsymbol{a}\|^2\|\boldsymbol{b}\|^2 - \langle \boldsymbol{a}, \boldsymbol{b}\rangle^2. \tag{5.44}$$

This equals the square of the area of the parallelogram defined by $\boldsymbol{a}$ and $\boldsymbol{b}$. In fact, if we let θ be the angle made by $\boldsymbol{a}$ and $\boldsymbol{b}$, its area S is $\|\boldsymbol{a}\|\|\boldsymbol{b}\|\sin\theta$, and hence

$$\begin{aligned} S^2 &= \|\boldsymbol{a}\|^2\|\boldsymbol{b}\|^2\sin^2\theta = \|\boldsymbol{a}\|^2\|\boldsymbol{b}\|^2(1-\cos^2\theta) = \|\boldsymbol{a}\|^2\|\boldsymbol{b}\|^2\Big(1 - \Big(\frac{\langle \boldsymbol{a}, \boldsymbol{b}\rangle}{\|\boldsymbol{a}\|\|\boldsymbol{b}\|}\Big)^2\Big) \\ &= \|\boldsymbol{a}\|^2\|\boldsymbol{b}\|^2 - \langle \boldsymbol{a}, \boldsymbol{b}\rangle^2. \end{aligned} \tag{5.45}$$

If we contract a trivector $\boldsymbol{a} \wedge \boldsymbol{b} \wedge \boldsymbol{c}$ by its reversal $\boldsymbol{c} \wedge \boldsymbol{b} \wedge \boldsymbol{a}$, we have from the fourth equation in Eq. (5.34)

$$\begin{aligned} \|\boldsymbol{a} \wedge \boldsymbol{b} \wedge \boldsymbol{c}\|^2 &= \boldsymbol{c} \wedge \boldsymbol{b} \wedge \boldsymbol{a} \cdot \boldsymbol{a} \wedge \boldsymbol{b} \wedge \boldsymbol{c} \\ &= \langle \boldsymbol{c}, \boldsymbol{a}\rangle\langle \boldsymbol{b}, \boldsymbol{c}\rangle\langle \boldsymbol{a}, \boldsymbol{b}\rangle + \langle \boldsymbol{c}, \boldsymbol{b}\rangle\langle \boldsymbol{b}, \boldsymbol{a}\rangle\langle \boldsymbol{a}, \boldsymbol{c}\rangle + \langle \boldsymbol{c}, \boldsymbol{c}\rangle\langle \boldsymbol{b}, \boldsymbol{b}\rangle\langle \boldsymbol{a}, \boldsymbol{a}\rangle \\ &\quad \langle \boldsymbol{c}, \boldsymbol{a}\rangle\langle \boldsymbol{b}, \boldsymbol{b}\rangle\langle \boldsymbol{a}, \boldsymbol{c}\rangle - \langle \boldsymbol{c}, \boldsymbol{b}\rangle\langle \boldsymbol{b}, \boldsymbol{c}\rangle\langle \boldsymbol{a}, \boldsymbol{a}\rangle - \langle \boldsymbol{c}, \boldsymbol{c}\rangle\langle \boldsymbol{b}, \boldsymbol{a}\rangle\langle \boldsymbol{a}, \boldsymbol{b}\rangle \\ &= \|\boldsymbol{a}\|^2\|\boldsymbol{b}\|^2\|\boldsymbol{c}\|^2 + 2\langle \boldsymbol{b}, \boldsymbol{c}\rangle\langle \boldsymbol{c}, \boldsymbol{a}\rangle\langle \boldsymbol{a}, \boldsymbol{b}\rangle - \|\boldsymbol{a}\|^2\langle \boldsymbol{b}, \boldsymbol{c}\rangle^2 - \|\boldsymbol{b}\|^2\langle \boldsymbol{c}, \boldsymbol{a}\rangle^2 - \|\boldsymbol{c}\|^2\langle \boldsymbol{a}, \boldsymbol{b}\rangle^2. \end{aligned} \tag{5.46}$$

It can be shown that this equals the square of the volume of the parallelogram defined by $\boldsymbol{a}$, $\boldsymbol{b}$, and $\boldsymbol{c}$. The above results are summarized as follows:

Proposition 5.5 (Norm of a k-vector) *The norm of scalar α is $\|\alpha\| = |\alpha|$, i.e., its absolute value. The norm $\|\boldsymbol{a}\|$ of vector $\boldsymbol{a}$ is its length. The norm of bivector $\boldsymbol{a} \wedge \boldsymbol{b}$ is*

$$\|\boldsymbol{a} \wedge \boldsymbol{b}\| = \sqrt{\|\boldsymbol{a}\|^2\|\boldsymbol{b}\|^2 - \langle \boldsymbol{a}, \boldsymbol{b}\rangle^2}, \tag{5.47}$$

which equals the area of the parallelogram defined by $\boldsymbol{a}$ and $\boldsymbol{b}$. The norm of trivector $\boldsymbol{a} \wedge \boldsymbol{b} \wedge \boldsymbol{c}$ is

$$\|\boldsymbol{a} \wedge \boldsymbol{b} \wedge \boldsymbol{c}\| = \sqrt{\|\boldsymbol{a}\|^2\|\boldsymbol{b}\|^2\|\boldsymbol{c}\|^2 + 2\langle \boldsymbol{b}, \boldsymbol{c}\rangle\langle \boldsymbol{c}, \boldsymbol{a}\rangle\langle \boldsymbol{a}, \boldsymbol{b}\rangle - \|\boldsymbol{a}\|^2\langle \boldsymbol{b}, \boldsymbol{c}\rangle^2 - \|\boldsymbol{b}\|^2\langle \boldsymbol{c}, \boldsymbol{a}\rangle^2 - \|\boldsymbol{c}\|^2\langle \boldsymbol{a}, \boldsymbol{b}\rangle^2}, \tag{5.48}$$

which equals the volume of the parallelepiped defined by $\boldsymbol{a}$, $\boldsymbol{b}$, and $\boldsymbol{c}$.

If we write $\boldsymbol{a} = \sum_{k=1}^{3} a_i e_i$ using the basis, Eq. (5.43) reads

$$\|\boldsymbol{a}\|^2 = a_1^2 + a_2^2 + a_3^2. \tag{5.49}$$

Let $\boldsymbol{b} = \sum_{k=1}^{3} b_i e_i$. Since Eq. (5.44) equals the square of the area of the parallelogram defined by $\boldsymbol{a}$ and $\boldsymbol{b}$, we can also write

$$\|\boldsymbol{a} \wedge \boldsymbol{b}\|^2 = (a_2 b_3 - a_3 b_2)^2 + (a_3 b_1 - a_1 b_3)^2 + (a_1 b_2 - a_2 b_1)^2 \tag{5.50}$$

($\hookrightarrow$ Eq. (2.18) and Exercise 2.6 in Chapter 2). Let $\boldsymbol{c} = \sum_{k=1}^{3} c_i e_i$. Since Eq. (5.46) is the square of the volume of the parallelepiped defined by $\boldsymbol{a}$, $\boldsymbol{b}$, and $\boldsymbol{c}$, we can also write

$$\|\boldsymbol{a} \wedge \boldsymbol{b} \wedge \boldsymbol{c}\|^2 = (a_1 b_2 c_3 + a_2 b_3 c_1 + a_3 b_1 c_2 - a_1 b_3 c_2 - a_2 b_1 c_3 - a_3 b_2 c_1)^2 \tag{5.51}$$

($\hookrightarrow$ Eq. (2.28) and Exercise 2.9 in Chapter 2).

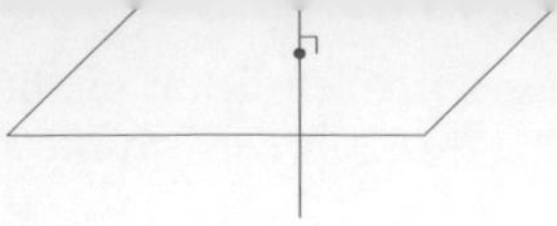

FIGURE 5.7 The orthogonal complement of a line is a plane orthogonal to it, and the orthogon complement of a plane is its surface normal. The orthogonal complement of the entire space is th origin, whose orthogonal complement is the entire space.

Traditional World 5.3 (Determinant and volume) The fact that (5.48) gives the vo ume of the parallelepiped is easily seen if we invoke traditional linear algebra. Regard $\boldsymbol{a}$, $\boldsymbol{b}$ and $\boldsymbol{c}$ as columns obtained by vertically aligning their components, and let $\boldsymbol{A} = (\boldsymbol{a}, \boldsymbol{b}, \boldsymbol{c})$ be the matrix consisting of these columns. As is well known in linear algebra, the (signed volume of the parallelepiped defined by $\boldsymbol{a}$, $\boldsymbol{b}$, and $\boldsymbol{c}$ (positive if they are right-handed, an negative if left-handed) is given by the determinant $|\boldsymbol{A}|$. Since a matrix $\boldsymbol{A}$ and its transpos $\boldsymbol{A}^\top$ has the same determinant, and the determinant of the product of two matrices equa the product of their determinants, we see that

$$|\boldsymbol{A}|^2 = |\boldsymbol{A}^\top||\boldsymbol{A}| = |\boldsymbol{A}^\top\boldsymbol{A}|. \qquad (5.52)$$

The transpose $\boldsymbol{A}^\top$ is given by vertically aligning, from top to bottom, the rows $\boldsymbol{a}^\top$, $\boldsymbol{b}^\top$ and $\boldsymbol{c}^\top$ obtained by transposing the columns $\boldsymbol{a}$, $\boldsymbol{b}$, and $\boldsymbol{c}$, so we can write $\boldsymbol{A}^\top = \begin{pmatrix} \boldsymbol{a}^\top \\ \boldsymbol{b}^\top \\ \boldsymbol{c}^\top \end{pmatrix}$

According to the matrix multiplication rule, we see that

$$\boldsymbol{A}^\top\boldsymbol{A} = \begin{pmatrix} \boldsymbol{a}^\top \\ \boldsymbol{b}^\top \\ \boldsymbol{c}^\top \end{pmatrix} (\boldsymbol{a}, \boldsymbol{b}, \boldsymbol{c}) = \begin{pmatrix} \boldsymbol{a}^\top\boldsymbol{a} & \boldsymbol{b}^\top\boldsymbol{a} & \boldsymbol{c}^\top\boldsymbol{a} \\ \boldsymbol{b}^\top\boldsymbol{a} & \boldsymbol{b}^\top\boldsymbol{b} & \boldsymbol{c}^\top\boldsymbol{b} \\ \boldsymbol{c}^\top\boldsymbol{a} & \boldsymbol{b}^\top\boldsymbol{c} & \boldsymbol{c}^\top\boldsymbol{c} \end{pmatrix} = \begin{pmatrix} \|\boldsymbol{a}\|^2 & \langle \boldsymbol{a}, \boldsymbol{b} \rangle & \langle \boldsymbol{c}, \boldsymbol{a} \rangle \\ \langle \boldsymbol{a}, \boldsymbol{b} \rangle & \|\boldsymbol{b}\|^2 & \langle \boldsymbol{b}, \boldsymbol{c} \rangle \\ \langle \boldsymbol{c}, \boldsymbol{a} \rangle & \langle \boldsymbol{b}, \boldsymbol{c} \rangle & \|\boldsymbol{c}\|^2 \end{pmatrix}. \qquad (5.53)$$

Evaluating this determinant, we obtain Eq. (5.46). Hence, Eq. (5.48) is the volume of th parallelepiped.

5.5 DUALITY

A k-vector, $k = 0$, 1, 2, 3, specifies a subspace of dimension k. However, it can also b specified by its orthogonal complement, i.e., the $(n-k)$D subspace orthogonal to it. W call the $(n-k)$-vector that specifies the orthogonal complement the *dual* of the origin k-vector and use the asterisk $*$ to denote it. We first describe orthogonal complements fro a geometric consideration and then express them in terms of the basis.

5.5.1 Orthogonal complements

The orthogonal complement of a line is a plane orthogonal to it, and the orthogonal com plement of a plane is its surface normal. The orthogonal complement of the entire space i the origin, whose orthogonal complement is the entire space (Fig. 5.7). These orthogona complements are specified by the following k-vectors, $k = 0$, 1, 2, 3:

(i) The orthogonal complement of line $\boldsymbol{a}$ is a plane orthogonal it. We define the dual $\boldsymbol{a}$ to be a bivector $\boldsymbol{b} \wedge \boldsymbol{c}$ that specifies that plane and has the same magnitude as $\boldsymbol{a}$.

magnitude as $\boldsymbol{a} \wedge \boldsymbol{b}$.

(iii) The orthogonal complement of the space specified by trivector $\boldsymbol{a}\wedge\boldsymbol{b}\wedge\boldsymbol{c}$ is the origin (a scalar). We define the dual $(\boldsymbol{a}\wedge\boldsymbol{b}\wedge\boldsymbol{c})^*$ to be a scalar α that has the same magnitude as $\boldsymbol{a}\wedge\boldsymbol{b}\wedge\boldsymbol{c}$.

(iv) The orthogonal complement of the origin, i.e., a scalar α, is the entire space. We define the dual α^* to be a trivector $\boldsymbol{a}\wedge\boldsymbol{b}\wedge\boldsymbol{c}$ that has magnitude α.

However, we need to determine the sign of the k-vector that specifies the orthogonal complement, since a k-vector and its sign reversal define the same subspace (with magnitudes of opposite signs). To determine the sign, we let I be a trivector with magnitude 1 and call it the *volume element*. For example, if $\boldsymbol{a}$, $\boldsymbol{b}$, and $\boldsymbol{c}$ are right-handed and the parallelepiped they define has volume 1, we can let $I = \boldsymbol{a}\wedge\boldsymbol{b}\wedge\boldsymbol{c}$. Using this, we define the dual $(\cdots)^*$ of a k-vector $(\cdots)$ by

$$(\cdots)^* = -(\cdots)\cdot I. \tag{5.54}$$

As stated at the end of Sec. 5.3.4, the right side computes the intersection of the orthogonal complement of $(\cdots)$ with the entire space specified by I. Hence, it specifies the orthogonal complement, including its sign.

Consider a vector $\boldsymbol{a}$, for example. Let $\boldsymbol{b}$ and $\boldsymbol{c}$ be vectors orthogonal to $\boldsymbol{a}$ such that $\boldsymbol{a}$, $\boldsymbol{b}$, and $\boldsymbol{c}$ are right-handed and $\|\boldsymbol{b}\wedge\boldsymbol{c}\|$ is equal to $\|\boldsymbol{a}\|$. Then, $\boldsymbol{a}\wedge\boldsymbol{b}\wedge\boldsymbol{c}/\|\boldsymbol{a}\|^2$ is the volume element I, so the dual $\boldsymbol{a}^*$ is given, from the fourth equation in Eq. (5.32), by

$$\boldsymbol{a}^* = -\boldsymbol{a}\cdot I = -\boldsymbol{a}\cdot\frac{\boldsymbol{a}\wedge\boldsymbol{b}\wedge\boldsymbol{c}}{\|\boldsymbol{a}\|^2} = -\frac{\langle\boldsymbol{a},\boldsymbol{a}\rangle\boldsymbol{b}\wedge\boldsymbol{c}+\langle\boldsymbol{a},\boldsymbol{b}\rangle\boldsymbol{c}\wedge\boldsymbol{a}+\langle\boldsymbol{a},\boldsymbol{c}\rangle\boldsymbol{a}\wedge\boldsymbol{b}}{\|\boldsymbol{a}\|^2} = -\boldsymbol{b}\wedge\boldsymbol{c}. \tag{5.55}$$

Consider a bivector $\boldsymbol{a}\wedge\boldsymbol{b}$. Let $\boldsymbol{c}$ be a vector orthogonal to $\boldsymbol{a}$ and $\boldsymbol{b}$ such that $\boldsymbol{a}$, $\boldsymbol{b}$, and $\boldsymbol{c}$ are right-handed and $\|\boldsymbol{c}\|$ equals $\|\boldsymbol{a}\wedge\boldsymbol{b}\|$. Then, $\boldsymbol{a}\wedge\boldsymbol{b}\wedge\boldsymbol{c}/\|\boldsymbol{a}\wedge\boldsymbol{b}\|^2$ is the volume element I, so the dual $(\boldsymbol{a}\wedge\boldsymbol{b})^*$ is given, from Eq. (5.44) and the fourth equation in Eq. (5.33), by

$$\begin{aligned}(\boldsymbol{a}\wedge\boldsymbol{b})^* &= -\boldsymbol{a}\wedge\boldsymbol{b}\cdot I = -\frac{\boldsymbol{a}\wedge\boldsymbol{b}\cdot\boldsymbol{a}\wedge\boldsymbol{b}\wedge\boldsymbol{c}}{\|\boldsymbol{a}\wedge\boldsymbol{b}\|^2}\\ &= -\frac{1}{\|\boldsymbol{a}\wedge\boldsymbol{b}\|^2}\Big((\langle\boldsymbol{a},\boldsymbol{c}\rangle\langle\boldsymbol{b},\boldsymbol{b}\rangle-\langle\boldsymbol{a},\boldsymbol{b}\rangle\langle\boldsymbol{b},\boldsymbol{c}\rangle)\boldsymbol{a}+(\langle\boldsymbol{a},\boldsymbol{a}\rangle\langle\boldsymbol{b},\boldsymbol{c}\rangle-\langle\boldsymbol{a},\boldsymbol{c}\rangle\langle\boldsymbol{b},\boldsymbol{a}\rangle)\boldsymbol{b}\\ &\quad+(\langle\boldsymbol{a},\boldsymbol{b}\rangle\langle\boldsymbol{b},\boldsymbol{a}\rangle-\langle\boldsymbol{a},\boldsymbol{a}\rangle\langle\boldsymbol{b},\boldsymbol{b}\rangle)\boldsymbol{c}\Big) = -\frac{\langle\boldsymbol{a},\boldsymbol{b}\rangle^2-\|\boldsymbol{a}\|^2\|\boldsymbol{b}\|^2}{\|\boldsymbol{a}\wedge\boldsymbol{b}\|^2}\boldsymbol{c} = \boldsymbol{c}.\end{aligned} \tag{5.56}$$

Consider a trivector $\boldsymbol{a}\wedge\boldsymbol{b}\wedge\boldsymbol{c}$. If $\boldsymbol{a}$, $\boldsymbol{b}$, and $\boldsymbol{c}$ are right-handed, then $\boldsymbol{a}\wedge\boldsymbol{b}\wedge\boldsymbol{c}/\|\boldsymbol{a}\wedge\boldsymbol{b}\wedge\boldsymbol{c}\|$ is the volume element I. From the definition of the norm $\|\boldsymbol{a}\wedge\boldsymbol{b}\wedge\boldsymbol{c}\|$, the dual $(\boldsymbol{a}\wedge\boldsymbol{b}\wedge\boldsymbol{c})^*$ is given by

$$\begin{aligned}(\boldsymbol{a}\wedge\boldsymbol{b}\wedge\boldsymbol{c})^* &= -\boldsymbol{a}\wedge\boldsymbol{b}\wedge\boldsymbol{c}\cdot I = -\frac{\boldsymbol{a}\wedge\boldsymbol{b}\wedge\boldsymbol{c}\cdot\boldsymbol{a}\wedge\boldsymbol{b}\wedge\boldsymbol{c}}{\|\boldsymbol{a}\wedge\boldsymbol{b}\wedge\boldsymbol{c}\|}\\ &= \frac{\boldsymbol{a}\wedge\boldsymbol{b}\wedge\boldsymbol{c}\cdot\boldsymbol{c}\wedge\boldsymbol{b}\wedge\boldsymbol{a}}{\|\boldsymbol{a}\wedge\boldsymbol{b}\wedge\boldsymbol{c}\|} = \frac{\|\boldsymbol{a}\wedge\boldsymbol{b}\wedge\boldsymbol{c}\|^2}{\|\boldsymbol{a}\wedge\boldsymbol{b}\wedge\boldsymbol{c}\|} = \|\boldsymbol{a}\wedge\boldsymbol{b}\wedge\boldsymbol{c}\|.\end{aligned} \tag{5.57}$$

If $\boldsymbol{a}$, $\boldsymbol{b}$, and $\boldsymbol{c}$ are left-handed, $-\boldsymbol{a}\wedge\boldsymbol{b}\wedge\boldsymbol{c}$ is the volume element I, so the sign is reversed, and we have $(\boldsymbol{a}\wedge\boldsymbol{b}\wedge\boldsymbol{c})^* = -\|\boldsymbol{a}\wedge\boldsymbol{b}\wedge\boldsymbol{c}\|$.

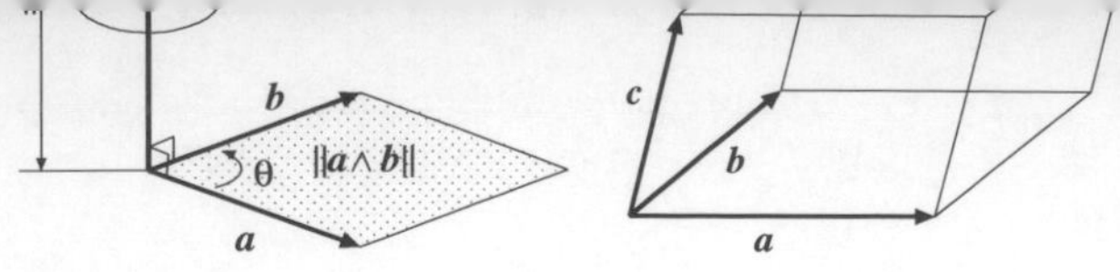

(a) (b)

FIGURE 5.8 (a) The dual $(\boldsymbol{a}\wedge\boldsymbol{b})^*$ of bivector $\boldsymbol{a}\wedge\boldsymbol{b}$ is a vector orthogonal to $\boldsymbol{a}$ and $\boldsymbol{b}$ in the directio of the right-handed screw movement of rotating $\boldsymbol{a}$ toward $\boldsymbol{b}$. The length equals the area $\|\boldsymbol{a}\wedge\boldsymbol{b}\|$ the parallelogram defined by $\boldsymbol{a}$ and $\boldsymbol{b}$. (b) If $\boldsymbol{a}$, $\boldsymbol{b}$, and $\boldsymbol{c}$ are right-handed, the dual $(\boldsymbol{a}\wedge\boldsymbol{b}\wedge\boldsymbol{c})^*$ trivector $\boldsymbol{a}\wedge\boldsymbol{b}\wedge\boldsymbol{c}$ equals the volume α of the parallelepiped defined by $\boldsymbol{a}$, $\boldsymbol{b}$, and $\boldsymbol{c}$.

For a scalar α, we see that

$$\alpha^* = -\alpha \cdot I = -\alpha I, \tag{5.58}$$

which is the entire space with magnitude $-\alpha$.

The important thing to note is that *the dual of the dual is sign reversal*. Consider bivector $\boldsymbol{a}\wedge\boldsymbol{b}$, for example. Its dual $\boldsymbol{c} = (\boldsymbol{a}\wedge\boldsymbol{b})^*$ is a vector orthogonal to $\boldsymbol{a}$ and $\boldsymbol{b}$ in th direction of the right-handed screw movement of rotating $\boldsymbol{a}$ toward $\boldsymbol{b}$ with the length equa to the area $\|\boldsymbol{a}\wedge\boldsymbol{b}\|$ of the parallelogram defined by $\boldsymbol{a}$ and $\boldsymbol{b}$ (Fig. 5.8(a)). However, the du of that $\boldsymbol{c}$ is $\boldsymbol{c}^* = -\boldsymbol{a}\wedge\boldsymbol{b}$. Similarly, if $\boldsymbol{a}$, $\boldsymbol{b}$, and $\boldsymbol{c}$ are right-handed, and if the parallelepipe they define has volume α, we have $(\boldsymbol{a}\wedge\boldsymbol{b}\wedge\boldsymbol{c})^* = \alpha$ (Fig. 5.8(b)), but $\alpha^* = -\boldsymbol{a}\wedge\boldsymbol{b}\wedge\boldsymbol{c}$.

5.5.2 Basis expression

Duality is a linear operation for which distributivity holds, so if a k-vector, $k = 0, 1, 2, 3,$ expressed using the basis, its dual is obtained by computing the dual of the basis. For th orthonormal basis $\{e_1, e_2, e_3\}$, the volume element is given by

$$I = c_1 \wedge e_2 \wedge e_3. \tag{5.59}$$

A scalar is a multiple of 1, and the dual of 1 is

$$1^* = -1 \cdot I = -e_1 \wedge e_2 \wedge e_3. \tag{5.60}$$

A vector is given by a linear combination of e_1, e_2, and e_3. We see that

$$e_1^* = -e_1 \cdot I = -e_1 \cdot e_1 \wedge e_2 \wedge e_3 = -\langle e_1, e_1\rangle e_2 \wedge e_3 = -e_2 \wedge e_3, \tag{5.61}$$

and e_2^* and e_3^* are similarly given. A bivector is a linear combination of $e_2 \wedge e_3$, $e_3 \wedge e$ $e_1 \wedge e_2$. We see that

$$\begin{aligned}(e_2 \wedge e_3)^* &= -e_2 \wedge e_3 \cdot I = -e_2 \wedge e_3 \cdot e_1 \wedge e_2 \wedge e_3 \\ &= -e_2 \cdot (e_3 \cdot e_1 \wedge e_2 \wedge e_3) = -e_2 \cdot (e_1 \wedge e_2) = e_1,\end{aligned} \tag{5.62}$$

and $(e_3 \wedge e_1)^*$ and $(e_1 \wedge e_2)^*$ are similarly given. A trivector is a scalar multiple of $e_1 \wedge e_2 \wedge e_3$ and we see that

$$\begin{aligned}(e_1 \wedge e_2 \wedge e_3)^* &= -e_1 \wedge e_2 \wedge e_3 \cdot e_1 \wedge e_2 \wedge e_3 = -e_1 \cdot (e_2 \cdot (e_3 \cdot e_1 \wedge e_2 \wedge e_3)) \\ &= -e_1 \cdot (e_2 \cdot (e_1 \wedge e_2)) = e_1 \cdot e_1 = 1.\end{aligned} \tag{5.63}$$

In summary,

$$(e_1 \wedge e_2 \wedge e_3)^* = 1, \qquad 1^* = e_1 \wedge e_2 \wedge e_3, \tag{5.64}$$

$$(e_2 \wedge e_3)^* = e_1, \qquad (e_3 \wedge e_1)^* = e_2, \qquad (e_1 \wedge e_2)^* = e_3, \tag{5.65}$$

$$e_1^* = -e_2 \wedge e_3, \qquad e_2^* = -e_3 \wedge e_1, \qquad e_3^* = -e_1 \wedge e_2. \tag{5.66}$$

Hence, for a vector $\boldsymbol{a} = \sum_{i=1}^{3} a_i e_i$, we obtain

$$\boldsymbol{a} = a_1 e_1 + a_2 e_2 + a_3 e_3, \qquad \boldsymbol{a}^* = -a_1 e_2 \wedge e_3 - a_2 e_3 \wedge e_1 - a_3 e_1 \wedge e_2. \tag{5.67}$$

The square norm is given by

$$\|\boldsymbol{a}\|^2 = \|\boldsymbol{a}^*\|^2 = a_1^2 + a_2^2 + a_3^2. \tag{5.68}$$

For vectors $\boldsymbol{a} = \sum_{i=1}^{3} a_i e_i$ and $\boldsymbol{b} = \sum_{i=1}^{3} b_i e_i$, we obtain from Eq. (5.15)

$$\boldsymbol{a} \wedge \boldsymbol{b} = (a_2 b_3 - a_3 b_2) e_2 \wedge e_3 + (a_3 b_1 - a_1 b_3) e_3 \wedge e_1 + (a_1 b_2 - a_2 b_1) e_1 \wedge e_2,$$

$$(\boldsymbol{a} \wedge \boldsymbol{b})^* = (a_2 b_3 - a_3 b_2) e_1 + (a_3 b_1 - a_1 b_3) e_2 + (a_1 b_2 - a_2 b_1) e_3. \tag{5.69}$$

The square norm is given by

$$\|\boldsymbol{a} \wedge \boldsymbol{b}\|^2 = \|(\boldsymbol{a} \wedge \boldsymbol{b})^*\|^2 = (a_2 b_3 - a_3 b_2)^2 + (a_3 b_1 - a_1 b_3)^2 + (a_1 b_2 - a_2 b_1)^2. \tag{5.70}$$

For vectors $\boldsymbol{a} = \sum_{i=1}^{3} a_i e_i$, $\boldsymbol{b} = \sum_{i=1}^{3} b_i e_i$, and $\boldsymbol{c} = \sum_{i=1}^{3} c_i e_i$, we obtain from Eq. (5.16)

$$\boldsymbol{a} \wedge \boldsymbol{b} \wedge \boldsymbol{c} = (a_1 b_2 c_3 + a_2 b_3 c_1 + a_3 b_1 c_2 - a_1 b_3 c_2 - a_2 b_1 c_3 - a_3 b_2 c_1) e_1 \wedge e_2 \wedge e_3,$$

$$(\boldsymbol{a} \wedge \boldsymbol{b} \wedge \boldsymbol{c})^* = a_1 b_2 c_3 + a_2 b_3 c_1 + a_3 b_1 c_2 - a_1 b_3 c_2 - a_2 b_1 c_3 - a_3 b_2 c_1. \tag{5.71}$$

The square norm is given by

$$\|\boldsymbol{a} \wedge \boldsymbol{b} \wedge \boldsymbol{c}\|^2 = \|(\boldsymbol{a} \wedge \boldsymbol{b} \wedge \boldsymbol{c})^*\|^2 = (a_1 b_2 c_3 + a_2 b_3 c_1 + a_3 b_1 c_2 - a_1 b_3 c_2 - a_2 b_1 c_3 - a_3 b_2 c_1)^2. \tag{5.72}$$

From this, we obtain the correspondence to the notations in Chapter 2 as follows:

Proposition 5.7 (Vector product and scalar triple product) *The dual* $(\boldsymbol{a} \wedge \boldsymbol{b})^*$ *of bivector* $\boldsymbol{a} \wedge \boldsymbol{b}$ *equals the vector product* $\boldsymbol{a} \times \boldsymbol{b}$*, and the dual* $(\boldsymbol{a} \wedge \boldsymbol{b} \wedge \boldsymbol{c})^*$ *of trivector* $\boldsymbol{a} \wedge \boldsymbol{b} \wedge \boldsymbol{c}$ *equals the scalar triple product* $|\boldsymbol{a}, \boldsymbol{b}, \boldsymbol{c}|$*:*

$$(\boldsymbol{a} \wedge \boldsymbol{b})^* = \boldsymbol{a} \times \boldsymbol{b}, \qquad (\boldsymbol{a} \wedge \boldsymbol{b} \wedge \boldsymbol{c})^* = |\boldsymbol{a}, \boldsymbol{b}, \boldsymbol{c}|. \tag{5.73}$$

Traditional World 5.4 (Duality in tensor calculus) In traditional tensor calculus, where vectors are identified with arrays of numbers, the dual of a bivector $a^{[i}b^{j]}$ constructed from contravariant vectors a^i and b^i is defined, via the permutation signature ϵ_{ijk}, to be the covariant vector

$$c_i = \epsilon_{ijk} a^j b^k. \tag{5.74}$$

The right side means $\epsilon_{ijk} a^{[j} b^{k]}$, but since the indices j and k of ϵ_{ijk} are antisymmetric, we do not need the antisymmetrization $[\cdots]$ for a^j and b^k. We can see that Eq. (5.74) describes the same thing as Eq. (5.69). Similarly the dual of a trivector $a^{[i}b^j b^{k]}$ is defined to be a scalar

$$\alpha = \epsilon_{ijk} a^i b^j c^k. \tag{5.75}$$

as Eq. (5.71). However, one big difference arises by regarding arrays of numbers as vector
In this chapter, we have defined the duality via the volume element I. Usually, we use right-handed xyz coordinate system, but if we use a left-handed coordinate system, th volume element I changes its sign. Yet the definition of duality defined via the permutatic signature ϵ_{ijk} is unchanged whether the coordinate system is right-handed or left-hande Instead, the *geometric interpretation* changes depending on whether the coordinate syste is right-handed or left-handed. In traditional tensor calculus, vectors and scalars whos interpretation depends on the sense of the coordinate system are called *axial vectors* (o *pseudovectors*) (↪ Traditional World 3.3 in Chapter 3) and *pseudoscalars*, respectively.

5.6 DIRECT AND DUAL REPRESENTATIONS

By the "equation" of an object, we mean an equality that a position vector $\boldsymbol{x}$ satisfies and only if it belongs to that object. If the equation has the form

$$\boldsymbol{x} \wedge (\cdots) = 0, \tag{5.76}$$

we call $(\cdots)$ the *direct representation* of the object. If the equation has the form

$$\boldsymbol{x} \cdot (\cdots) = 0, \tag{5.77}$$

we call $(\cdots)$ the *dual representation* of the object.

From the definition of the outer product $\wedge$, the direct representation of lines, planes, th space, and the origin is given as follows:

(i) A position vector $\boldsymbol{x}$ is on the line in the direction of $\boldsymbol{a}$ if and only if $\boldsymbol{x} \wedge \boldsymbol{a} = 0$. Henc $\boldsymbol{a}$ is the direct representation of that line.

(ii) A position vector $\boldsymbol{x}$ is on the plane spanned by $\boldsymbol{a}$ and $\boldsymbol{b}$ if and only if $\boldsymbol{x} \wedge \boldsymbol{a} \wedge \boldsymbol{b} = 0$ Hence, $\boldsymbol{a} \wedge \boldsymbol{b}$ is the direct representation of that plane.

(iii) Any position vector $\boldsymbol{x}$ is in the space spanned by $\boldsymbol{a}$, $\boldsymbol{b}$, and $\boldsymbol{c}$. This fact is written a $\boldsymbol{x} \wedge \boldsymbol{a} \wedge \boldsymbol{b} \wedge \boldsymbol{c} = 0$. Hence, $\boldsymbol{a} \wedge \boldsymbol{b} \wedge \boldsymbol{c}$ is the direct representation of that space.

(iv) A position vector $\boldsymbol{x}$ is at the origin if and only if $\boldsymbol{x} \wedge \alpha\ (= \alpha\boldsymbol{x}) = 0$ for a nonzero α Hence, a nonzero scalar α is the direct representation of the origin.

From the definition of the contraction operation $\cdot$, the dual representation of lines, planes the space, and the origin is given as follows:

(i) From the second equation of Eq. (5.32), a position vector $\boldsymbol{x}$ satisfies $\boldsymbol{x} \cdot \boldsymbol{a} = 0$ and only if $\langle \boldsymbol{x}, \boldsymbol{a} \rangle = 0$, i.e., when $\boldsymbol{x}$ is in the orthogonal complement of $\boldsymbol{a}$ (the plan perpendicular to $\boldsymbol{a}$). Hence, $\boldsymbol{a}$ is its dual representation.

(ii) From the third equation of Eq. (5.32), a position vector $\boldsymbol{x}$ satisfies $\boldsymbol{x} \cdot \boldsymbol{a} \wedge \boldsymbol{b} = 0$ if an only if $\langle \boldsymbol{x}, \boldsymbol{a} \rangle = \langle \boldsymbol{x}, \boldsymbol{b} \rangle = 0$, i.e., when $\boldsymbol{x}$ is in the orthogonal complement of $\boldsymbol{a} \wedge \boldsymbol{b}$ (th surface normal to the plane $\boldsymbol{a} \wedge \boldsymbol{b}$). Hence, $\boldsymbol{a} \wedge \boldsymbol{b}$ is its dual representation.

(iii) From the fourth equation of Eq. (5.32), a position vector $\boldsymbol{x}$ satisfies $\boldsymbol{x} \cdot \boldsymbol{a} \wedge \boldsymbol{b} \wedge \boldsymbol{c} = 0$ if and only if $\langle \boldsymbol{x}, \boldsymbol{a} \rangle = \langle \boldsymbol{x}, \boldsymbol{b} \rangle = \langle \boldsymbol{x}, \boldsymbol{c} \rangle = 0$, i.e., when $\boldsymbol{x}$ is at the origin, which is th orthogonal complement of $\boldsymbol{a} \wedge \boldsymbol{b} \wedge \boldsymbol{c}$. Hence, $\boldsymbol{a} \wedge \boldsymbol{b} \wedge \boldsymbol{c}$ is its dual representation.

subspace	direct representation	dual representation
origin	scalar α	trivector $\boldsymbol{a} \wedge \boldsymbol{b} \wedge \boldsymbol{c}$
line	vector $\boldsymbol{a}$	bivector $\boldsymbol{b} \wedge \boldsymbol{c}$
plane	bivector $\boldsymbol{a} \wedge \boldsymbol{b}$	vector $\boldsymbol{c}$
space	trivector $\boldsymbol{a} \wedge \boldsymbol{b} \wedge \boldsymbol{c}$	scalar α
equation	$\boldsymbol{x} \wedge (\cdots) = 0$	$\boldsymbol{x} \cdot (\cdots) = 0$

(iv) From the first equation of Eq. (5.32), a position vector $\boldsymbol{x}$ always satisfies $\boldsymbol{x} \cdot \alpha = 0$ for any nonzero scalar α. Hence, a nonzero scalar α is the dual representation of the space $\boldsymbol{a} \wedge \boldsymbol{b} \wedge \boldsymbol{c}$.

The above observations are summarized in Table 5.1.

From the definition of the duality, the outer product $\boldsymbol{x} \wedge (\cdots)$ is 0 if and only if its dual is 0. From the rule of the contraction computation, we can rewrite the dual of $\boldsymbol{x} \wedge (\cdots)$ in the form

$$(\boldsymbol{x} \wedge (\cdots))^* = -\boldsymbol{x} \wedge (\cdots) \cdot I = -\boldsymbol{x} \cdot ((\cdots) \cdot I) = \boldsymbol{x} \cdot (\cdots)^*. \tag{5.78}$$

Thus, we observe

Proposition 5.8 (Direct and dual representations) *The following equivalence relations hold:*

$$\boldsymbol{x} \wedge (\cdots) = 0 \quad \Leftrightarrow \quad \boldsymbol{x} \cdot (\cdots)^* = 0,$$
$$\boldsymbol{x} \cdot (\cdots) = 0 \quad \Leftrightarrow \quad \boldsymbol{x} \wedge (\cdots)^* = 0. \tag{5.79}$$

Hence, the dual of the direct representation is the dual representation, and the dual of the dual representation is the direct representation.

Recall that the dual specifies the orthogonal complement. Hence, the equation $\boldsymbol{x} \wedge (\cdots) = 0$ means that $\boldsymbol{x}$ *belongs to* that subspace, and the equation $\boldsymbol{x} \cdot (\cdots) = 0$ means that $\boldsymbol{x}$ *is orthogonal to* that subspace.

5.7 SUPPLEMENTAL NOTE

The Grassmann algebra was introduced by *Hermann Günther Grassmann* (1809–1877), a German mathematician, who introduced the outer product $\wedge$ and described the algebra of subspaces without using vector components. Today's readers who are familiar with vector calculus would find his original formulation a little confusing, wondering why things that could be easily explained in terms of vector products and scalar triple products are described in rather complicated forms. This is reasonable. As mentioned in the supplemental note to Chapter 2, today's vector calculus was established by Gibbs, who combined and simplified Hamilton's quaternion algebra in Chapter 4 and the Grassmann algebra in this chapter. In short, today's vector calculus is a reformulation of the Grassmann algebra made easy in terms of vector products and scalar triple products.

Today, we specify the area and orientation of the parallelogram defined by vectors $\boldsymbol{a}$ and $\boldsymbol{b}$ by the vector product $\boldsymbol{a} \times \boldsymbol{b}$ and the volume and sign of the parallelepiped defined by vectors $\boldsymbol{a}$, $\boldsymbol{b}$, and $\boldsymbol{c}$ by the scalar triple product $|\boldsymbol{a}, \boldsymbol{b}, \boldsymbol{c}|$. As shown in Eq. (5.73), the dual $(\boldsymbol{a} \wedge \boldsymbol{b})^*$ is nothing but the vector product $\boldsymbol{a} \times \boldsymbol{b}$, and the dual $(\boldsymbol{a} \wedge \boldsymbol{b} \wedge \boldsymbol{c})^*$ is the scalar triple product $|\boldsymbol{a}, \boldsymbol{b}, \boldsymbol{c}|$. Today, as shown in Chapter 2, all quantities involving lines and

the Grassmann algebra is seldom taught to students of physics and engineering today.

However, vector calculus has a crucial restriction: *it can be used only in 3D*. In gener nD, two vectors $\boldsymbol{a}$ and $\boldsymbol{b}$ span a 2D subspace. However, there exist infinitely many directio orthogonal to it, so we cannot define a unique surface normal to it. Hence, the only way specify that plane is to write just $\boldsymbol{a} \wedge \boldsymbol{b}$. However, the same plane can be specified in ma different ways, so various rules that the outer product $\wedge$ should satisfy are obtained.

Although this chapter only deals with 3D, all the descriptions can be extended to n almost as is. By *almost*, we mean that a few specialties of 3D are involved here. The fir is when we convert the sum $\boldsymbol{a} \wedge \boldsymbol{b} + \boldsymbol{c} \wedge \boldsymbol{d}$ of two bivectors into one. This is a process call *factorization* ($\hookrightarrow$ Exercise 5.1), which involves the vector $\boldsymbol{e}$ along the intersection betwe the two planes specified by $\boldsymbol{a} \wedge \boldsymbol{b}$ and $\boldsymbol{c} \wedge \boldsymbol{d}$. In higher dimensions, however, two plan passing through the origin generally have no intersection other than the origin, so the su $\boldsymbol{a} \wedge \boldsymbol{b} + \boldsymbol{c} \wedge \boldsymbol{d}$ cannot be reduced any further; we cannot but retain it as a formal sum. Th type of formal sum of k-vectors is also called "k-vector." However, the distinction is ma in many textbooks of geometric algebra [2, 3, 4, 5, 12, 16], where a single term, i.e., t outer product of k vectors, is called a *simple k-vector* or a *blade* of *grade k*, and their form sum is called a *k-vector*. In this chapter, however, we use the more classical and intuiti term "k-vector" for both a single term and their formal sum, because we concentrate 3D, where no confusion should arise.

Another specialty of 3D is Eq. (5.54); the right side would be multiplied $(-1)^{n(n-1)/2}(\cdots) \cdot I$ in nD. In this chapter, we use the concept of a "right-handed sy tem" for defining the 3D volume element I. In general nD, we cannot define the surfa normal or the movement of a right-handed screw. For n basis elements $e_1, \ldots, e_n$, there exi two types of their outer product: one is obtained by interchanging two vectors in $e_1 \wedge \cdots \wedge$ an even number of times, all defining the same n-vector; the other by interchanging the an odd number of times, all defining the n-vector with the opposite sign. Either one can defined to have a "positive orientation," the other a "negative orientation." The positi outer product is called the "volume element" and denoted by I. Hence, the sign depends the coordinate system; the sign flips if two coordinate axes are interchanged. Such a scal is a *pseudoscalar*, as mentioned in the Traditional World 5.4. For this reason, the volum element I itself is called the *unit pseudoscalar* or simply the *pseudoscalar* in many textboo of geometric algebra [2, 3, 4, 5, 12, 16]. This book, however, uses the more classical an intuitive term "volume element."

In many textbooks [4, 12, 16], contraction is simply called "inner product" (althoug the result is not a scalar) and written as $(\cdots) \cdot (\cdots)$, using the dot '$\cdot$'; no distinction made between the vector inner product. Perwass [16] writes $(\cdots) * (\cdots)$ for the vect inner product and $(\cdots) \cdot (\cdots)$ for the contraction. Dorst et al. [5] write $(\cdots) \cdot (\cdots)$ f the vector inner product and $(\cdots) \rfloor (\cdots)$ for the contraction. Bayro-Corrochano [3] writ $(\cdots) \cdot (\cdots)$ for the vector inner product and writes $(\cdots) \dashv (\cdots)$ for the contraction (b uses the dot as well). In this book, we write $\langle \cdots, \cdots \rangle$ for the vector inner product an $(\cdots) \cdot (\cdots)$ for contraction. In [4, 12], contraction of and by a scalar is defined to be whatever the other element is. However, this book follows the definition in Dorst et al. [5

We use an upper right asterisk $*$ to denote the dual and write $(\cdots)^*$, but anoth definition of duality exists for specifying a subspace by its orthogonal complement, writte as $*(\cdots)$ with an asterisk before the expression. This preceding asterisk $*$ is called t *Hodge star operator*. Its definition is slightly different from ours: one first defines the inn product $\langle \cdots, \cdots \rangle$ of two k-vectors in nD and then defines the dual in such a way that t dual coincides with the original expression.

A notable application of the Grassmann algebra is its use in calculus. An infinitesimal variation of a continuously defined physical quantity in the neighborhood of each point is expressed by a differential $\omega = a_1 dx + a_2 dy + a_3 dz$, where a_1, a_2, and a_3 are the change per unit length for infinitesimal movement in the x, y, and z directions, respectively. Such an expression is called a *differential form of degree 1* or simply a *1-form*, and the set of all 1-forms defines a vector space at each point with the origin at that point. Then, we can construct the Grassmann algebra by introducing the outer product $\wedge$ between 1-forms in this vector space. The important consequence is that, by introducing a differential operator called *exterior derivative* $d\omega$, we can systematically derive many integral theorems such as Green's theorem, Gauss's divergence theorem, and Stoke's theorem, where the Hodge star operator is used for duality. Using this algebra, one can describe topological properties of the space in consideration, e.g., whether or not it is simply connected, and differential geometric properties of surfaces such as curvatures. A well known textbook is Flanders [8].

5.8 EXERCISES

5.1. Show that in 3D the sum of arbitrary bivectors can be *factorized*, i.e., for any vectors $\boldsymbol{a}_1, \ldots, \boldsymbol{a}_n$ and $\boldsymbol{b}_1, \ldots, \boldsymbol{b}_n$, there exist $\boldsymbol{a}$ and $\boldsymbol{b}$ such that

$$\boldsymbol{a}_1 \wedge \boldsymbol{b}_1 + \cdots + \boldsymbol{a}_n \wedge \boldsymbol{b}_n = \boldsymbol{a} \wedge \boldsymbol{b}.$$

5.2. Using the duality relationships $(\boldsymbol{a} \wedge \boldsymbol{b})^* = \boldsymbol{a} \times \boldsymbol{b}$, $(\boldsymbol{a} \wedge \boldsymbol{b} \wedge \boldsymbol{c})^* = |\boldsymbol{a}, \boldsymbol{b}, \boldsymbol{c}|$ (Eq. (5.73)) and $\boldsymbol{x} \cdot (\cdots)^* = (\boldsymbol{x} \wedge (\cdots))^*$ (Eq. (5.78)), rewrite the following identities in terms of vector products without using outer products.

(a) The third equation of Eq. (5.32)

$$\boldsymbol{x} \cdot \boldsymbol{a} \wedge \boldsymbol{b} = \langle \boldsymbol{x}, \boldsymbol{a} \rangle \boldsymbol{b} - \langle \boldsymbol{x}, \boldsymbol{b} \rangle \boldsymbol{a}.$$

(b) The third equation of Eq. (5.33)

$$\boldsymbol{x} \wedge \boldsymbol{y} \cdot \boldsymbol{a} \wedge \boldsymbol{b} = \langle \boldsymbol{x}, \boldsymbol{b} \rangle \langle \boldsymbol{y}, \boldsymbol{a} \rangle - \langle \boldsymbol{x}, \boldsymbol{a} \rangle \langle \boldsymbol{y}, \boldsymbol{b} \rangle.$$

5.3. For coplanar vectors $\boldsymbol{a}$, $\boldsymbol{b}$, $\boldsymbol{c}$, and $\boldsymbol{d}$, the bivectors $\boldsymbol{a} \wedge \boldsymbol{b}$ and $\boldsymbol{c} \wedge \boldsymbol{d}$ specify the same plane except for orientation and magnitude. Hence, there exists a scalar γ such that $\boldsymbol{c} \wedge \boldsymbol{d} = \gamma \boldsymbol{a} \wedge \boldsymbol{b}$. Let us regard this γ as the "quotient" of $\boldsymbol{c} \wedge \boldsymbol{d}$ over $\boldsymbol{a} \wedge \boldsymbol{b}$ and write

$$\gamma = \frac{\boldsymbol{c} \wedge \boldsymbol{d}}{\boldsymbol{a} \wedge \boldsymbol{b}}.$$

For four points A, B, C, and D on a line l, their *cross-ratio* (Fig. 5.9(a)) is defined by

$$[A, B; C, D] = \frac{\overrightarrow{OA} \wedge \overrightarrow{OC}}{\overrightarrow{OB} \wedge \overrightarrow{OC}} \Bigg/ \frac{\overrightarrow{OA} \wedge \overrightarrow{OD}}{\overrightarrow{OB} \wedge \overrightarrow{OD}}.$$

(a) Show that the cross-ratio $[A, B; C, D]$ can be rewritten in the form

$$[A, B; C, D] = \frac{AC}{BC} \Bigg/ \frac{AD}{BD},$$

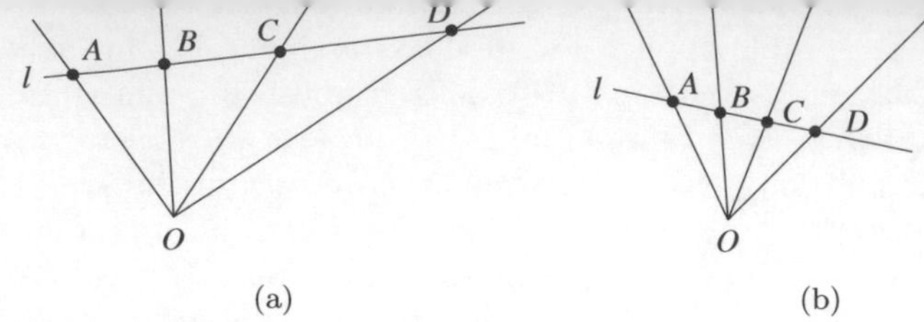

FIGURE 5.9 (a) Cross-ratio of four points. (b) Invariance of cross-ratio.

where we assume that the line is given a direction and AC, etc., are the sign distance between points A and C, etc., measured in that direction (hence CA $-AC$, etc.).

(b) Consider the plane that passes through the origin O and the above line l. Let be an arbitrary line on it, and let A, B', C', and D' be the intersections of line with lines OA, OB, OC, and OD, respectively. Show that the following identi holds (Fig. 5.9(b)):

$$[A, B; C, D] = [A', B'; C', D'].$$

Geometric Product and Clifford Algebra

This chapter describes the "Clifford algebra" that integrates the Hamilton algebra in Chapter 4 and the Grassmann algebra in Chapter 5, using a new operation called "geometric product." We first state the operational rule of the geometric product to show that the inner and outer products of vectors and the quaternion product can be computed using the geometric product. The important fact is that vectors and k-vectors have their inverse with respect to the geometric product. We show how the projection, rejection, reflection, and rotation of vectors are described using the geometric product and point out that orthogonal transformations of the space can be described in the form of "versors."

6.1 GRASSMANN ALGEBRA OF MULTIVECTORS

We have seen in the preceding chapter that bivector $\boldsymbol{a} \wedge \boldsymbol{b}$ specifies the plane spanned by vectors $\boldsymbol{a}$ and $\boldsymbol{b}$ and the orientation and the magnitude of the rotation of $\boldsymbol{a}$ toward $\boldsymbol{b}$. Hence, the bivector $\boldsymbol{a} \wedge \boldsymbol{b}$ can be viewed as specifying a rotation. Then, what vector results if a vector $\boldsymbol{x}$ is rotated by the rotation specified by $\boldsymbol{a} \wedge \boldsymbol{b}$? For such a computation, we need to introduce to the Grassmann algebra a new operation other than the outer product $\wedge$. For this purpose, we consider not individual k-vectors, $k = 0, 1, 2, 3$, but their formal sum

$$\mathcal{A} = \alpha + \boldsymbol{a} + \boldsymbol{b} \wedge \boldsymbol{c} + \boldsymbol{d} \wedge \boldsymbol{e} \wedge \boldsymbol{f}. \tag{6.1}$$

We call this type of formal sum of k-vectors of different k a *multivector*. The set of all multivectors forms an 8D vector space with respect to addition/subtraction and scalar multiplication. For if vectors are expressed in the orthonormal basis $\{e_1, e_2, e_3\}$, an arbitrary multivector is reduced, due to the antisymmetry of the outer product, to a linear combination of eight basis elements 1, e_1, e_2, e_3, $e_2 \wedge e_3$, $e_3 \wedge e_1$, and $e_1 \wedge e_2$, $e_1 \wedge e_2 \wedge e_3$.

Furthermore, the set of multivectors is closed under the outer product $\wedge$, because a product of multivectors can be reduced, after expanding the formal sums and computing the outer product of individual terms, to a linear combination of 1, e_1, e_2, e_3, $e_2 \wedge e_3$, $e_3 \wedge e_1$, $e_1 \wedge e_2$, and $e_1 \wedge e_2 \wedge e_3$. Consider the outer product of more than three symbols, for example, $e_1 \wedge e_2 \wedge e_3 \wedge e_1$. *Interchanging successive terms and changing the sign*, using the antisymmetry of the outer product, we see that

$$e_1 \wedge e_2 \wedge \underbrace{e_3 \wedge e_1} = -e_1 \wedge \underbrace{e_2 \wedge e_1} \wedge e_3 = \underbrace{e_1 \wedge e_1}_{0} \wedge e_2 \wedge e_3 = 0, \tag{6.2}$$

1	1	e_1	e_2	e_3
e_1	e_1	0	$e_1 \wedge e_2$	$-e_3 \wedge e_1$
e_2	e_2	$-e_1 \wedge e_2$	0	$e_2 \wedge e_3$
e_3	e_3	$e_3 \wedge e_1$	$-e_2 \wedge e_3$	0
$e_2 \wedge e_3$	$e_2 \wedge e_3$	$e_1 \wedge e_2 \wedge e_3$	0	0
$e_3 \wedge e_1$	$e_3 \wedge e_1$	0	$e_3 \wedge e_1 \wedge e_2$	0
$e_1 \wedge e_2$	$e_1 \wedge e_2$	0	0	$e_1 \wedge e_2 \wedge e_3$
$e_1 \wedge e_2 \wedge e_3$	$e_1 \wedge e_2 \wedge e_3$	0	0	0

	$e_2 \wedge e_3$	$e_3 \wedge e_1$	$e_1 \wedge e_2$	$e_1 \wedge e_2 \wedge e_3$
1	$e_2 \wedge e_3$	$e_3 \wedge e_1$	$e_1 \wedge e_2$	$e_1 \wedge e_2 \wedge e_3$
e_1	$e_1 \wedge e_2 \wedge e_3$	0	0	0
e_2	0	0	0	0
e_3	0	0	0	0
$e_2 \wedge e_3$	0	0	0	0
$e_3 \wedge e_1$	0	0	0	0
$e_1 \wedge e_2$	0	0	0	0
$e_1 \wedge e_2 \wedge e_3$	0	0	0	0

i.e., *the same symbols ultimately adjoin each other*, resulting in 0 according to the operatio rule. Computing the outer product for all pairs of the basis elements according to th rule, we obtain Table 6.1. The set of multivectors for which addition/subtraction, scal multiplication, and outer product are defined is formally called the *Grassmann algebr* This algebra is formally described as follows:

Proposition 6.1 (Grassmann algebra) *The Grassmann algebra is an algebra generat from 1 and symbols* e_1, e_2, *and* e_3 *by a multiplication operation (outer product) that associative and subject to the following rule:*

$$e_1 \wedge e_1 = e_2 \wedge e_2 = e_3 \wedge e_3 = 0, \tag{6.3}$$

$$e_2 \wedge e_3 = -e_3 \wedge e_2, \qquad e_3 \wedge e_1 = -e_1 \wedge e_3, \qquad e_1 \wedge e_2 = -e_2 \wedge e_1. \tag{6.4}$$

This algebra is an 8D vector space with respect to addition/subtraction and scalar multipl cation.

If we regard the Grassmann algebra as an algebra generated from symbols in this wa the algebra of quaternions in Chapter 4, which we call the *Hamilton algebra*, is formall described as follows:

Proposition 6.2 (Hamilton algebra) *The Hamilton algebra is an algebra generated fro 1 and symbols* i, j, *and* k *by a multiplication operation (quaternion product) that is ass ciative and subject to the following rule:*

$$i^2 = j^2 = k^2 = -1, \tag{6.5}$$

$$jk = i, \qquad ki = j, \qquad ij = k,$$
$$kj = -i, \qquad ik = -j, \qquad ji = -k. \tag{6.6}$$

This algebra is a 4D vector space with respect to addition/subtraction and scalar multipl cation.

	(a)			
	1	i	j	k
1	1	i	j	k
i	i	-1	k	$-j$
j	j	$-k$	-1	i
k	k	j	$-i$	-1

	(b)	
	1	i
1	1	i
i	i	-1

The quaternion product for all pairs of the basis elements is shown in Table 6.2(a). If we consider only those elements for which the coefficients of j and k are 0, they themselves form an algebra, as shown in Table 6.2(b). In other words,

Proposition 6.3 (Complex numbers) *The set of complex numbers is an algebra generated from 1 and symbol i by a multiplication operation (complex product) that is associative and subject to the rule $i^2 = -1$. This algebra is a 2D vector space with respect to addition/subtraction and scalar multiplication.*

Of course, those elements for which the coefficient of i is 0 themselves form an algebra, i.e., the set of real numbers, which is a 1D vector space.

6.2 CLIFFORD ALGEBRA

The *Clifford algebra* is an algebra generated from 1 and symbols e_1, e_2, and e_3, just as the Grassmann algebra, but we define a new product. We do not introduce a new operation symbol but juxtapose elements and call it the *geometric product* or the *Clifford product*. If no confusion will arise, we call it simply the "product." Its operational rule is given as follows:

Proposition 6.4 (Clifford algebra) *The Clifford algebra is an algebra generated from 1 and symbols e_1, e_2, and e_3 by a multiplication operation (geometric product) that is associative and subject to the following rule:*

$$e_1^2 = e_2^2 = e_3^2 = 1, \tag{6.7}$$

$$e_2e_3 = -e_3e_2, \qquad e_3e_1 = -e_1e_3, \qquad e_1e_2 = -e_2e_1. \tag{6.8}$$

This algebra is an 8D vector space with respect to addition/subtraction and scalar multiplication.

The reason that the Clifford algebra is an 8D vector space is that the geometric product of however many elements 1, e_1, e_2, and e_3 reduces to one of the eight terms 1, e_1, e_2, e_3, e_2e_3, e_3e_1, e_1e_2, and $e_1e_2e_3$ or their sign reversals. Consider the geometric product of more than three symbols, for example, $e_1e_2e_3e_1$. *Interchanging successive terms and changing the sign*, using the antisymmetry of the geometric product for different symbols, we see that

$$e_1e_2\underbrace{e_3e_1} = -e_1\underbrace{e_2e_1}e_3 = \underbrace{e_1^2}_{1}e_2e_3 = e_2e_3, \tag{6.9}$$

1	1	e_1	e_2	e_3	e_2e_3	e_3e_1	e_1e_2	$e_1e_2e_3$
e_1	e_1	1	e_1e_2	$-e_3e_1$	$e_1e_2e_3$	$-e_3$	e_2	e_2e_3
e_2	e_2	$-e_1e_2$	1	e_2e_3	e_3	$e_1e_2e_3$	$-e_2$	e_3e_1
e_3	e_3	e_3e_1	$-e_2e_3$	1	$-e_2$	e_1	$e_1e_2e_3$	e_1e_2
e_2e_3	e_2e_3	$e_1e_2e_3$	$-e_3$	e_2	-1	$-e_1e_2$	e_3e_1	$-e_1$
e_3e_1	e_3e_1	e_3	$e_3e_1e_2$	$-e_1$	e_1e_2	-1	$-e_2e_3$	$-e_2$
e_1e_2	e_1e_2	$-e_2$	e_1	$e_1e_2e_3$	$-e_3e_1$	e_2e_3	-1	$-e_3$
$e_1e_2e_3$	$e_1e_2e_3$	e_2e_3	e_3e_1	e_1e_2	$-e_1$	$-e_2$	$-e_3$	-1

i.e., *the same symbols ultimately adjoin each other*, resulting in 1 according to the operatic rule. Computing the geometric product for all pairs of the basis elements according to th rule, we obtain Table 6.3. Hence, an element of the Clifford algebra has the form

$$\mathcal{C} = \alpha + a_1e_1 + a_2e_2 + a_3e_3 + b_1e_2e_3 + b_2e_3e_1 + b_3e_1e_2 + ce_1e_2e_3, \qquad (6.1$$

and this is also called a *multivector*. We call α the *scalar part*, $a_1e_1 + a_2e_2 + a_3e_3$ the *vecto part*, $b_1e_2e_3 + b_2e_3e_1 + b_3e_1e_2$ the *bivector part*, and $ce_1e_2e_3$ the *trivector part*. The numbe of symbols in the product is called its *grade*. Namely, the scalar part, the vector part, th bivector part, and the trivector part have grades 0, 1, 2, and 3, respectively.

6.3 PARITY OF MULTIVECTORS

According to the operational rule of the geometric product, the *parity* of the grade preserved in the Clifford algebra in the following sense. A multivector consisting of term of odd grades

$$\mathcal{A} = a_1e_1 + a_2e_2 + a_3e_3 + ce_1e_2e_3 \qquad (6.1$$

is called an *odd multivector*, and a multivector consisting of terms of even grades

$$\mathcal{B} = \alpha + b_1e_2e_3 + b_2e_3e_1 + b_3e_1e_2 \qquad (6.1$$

an *even multivector*. It is easily seen that the product of two even multivectors and th product of two odd multivectors are even multivectors, and the product of even and od multivectors is an odd multivector. For the number of symbols in the product is the su of the symbols in the two terms if no annihilation occurs and is reduced *by two* each tim annihilation occurs. To be specific, we obtain from the rule of Table 6.2 the following result

Proposition 6.5 (Geometric product of multivectors) *The product of odd multivec tors*

$$\begin{aligned} \mathcal{A} &= a_1e_1 + a_2e_2 + a_3e_3 + ce_1e_2e_3, \\ \mathcal{A}' &= a_1'e_1 + a_2'e_2 + a_3'e_3 + c'e_1e_2e_3, \end{aligned} \qquad (6.13$$

is the following even multivector:

$$\begin{aligned} \mathcal{A}\mathcal{A}' = {}& a_1a_{1'} + a_2a_{2'} + a_3a_{3'} - cc + (a_2a_3' - a_3a_2' + ca_1 + c'a_1)e_2e_3 \\ & +(a_3a_1' - a_1a_3' + ca_2' + c'a_2)e_3e_1 + (a_1a_2' - a_2a_3' + ca_3' + c'a_3)e_1e_2. \end{aligned} \qquad (6.14$$

$$\mathcal{B}' = \alpha' + b_1' e_2 e_3 + b_2' e_3 e_1 + b_3' e_1 e_2, \tag{6.15}$$

is the following even multivector:

$$\begin{aligned}\mathcal{B}\mathcal{B}' &= \alpha\alpha' - b_1 b_1' - b_2 b_2' - b_3 b_3' + (\alpha b_1' + \alpha' b_1 - b_2 b_3' + b_3 b_2') e_2 e_3 \\ &+ (\alpha' b_2 + \alpha' b_2 - b_3 b_1' + b_1 b_3') e_3 e_1 + (\alpha' b_3 + \alpha' b_3 - b_1 b_2' + b_2 b_1') e_1 e_2.\end{aligned} \tag{6.16}$$

The product of the odd multivector $\mathcal{A}$ and the even multivector $\mathcal{B}$ is the following odd multivector:

$$\begin{aligned}\mathcal{A}\mathcal{B} &= (\alpha a_1 + a_3 b_2 - a_2 b_3 - c b_1) e_1 + (\alpha a_2 + a_1 b_3 - a_3 b_1 - c b_2) e_2 \\ &+ (\alpha a_3 + a_2 b_1 - a_1 b_2 - c b_3) e_3 + (\alpha c + a_1 b_1 + a_2 b_2 + a_3 b_3) e_1 e_2 e_3.\end{aligned} \tag{6.17}$$

The product of the even multivector $\mathcal{B}$ and the odd multivector $\mathcal{A}$ is the following odd multivector:

$$\begin{aligned}\mathcal{B}\mathcal{A} &= (\alpha a_1 + b_3 a_2 - b_2 a_3 - b_1 c) e_1 + (\alpha a_2 + b_1 a_3 - b_3 a_1 - b_2 c) e_2 \\ &+ (\alpha a_3 + b_2 a_1 - b_1 a_2 - b_3 c) e_3 + (\alpha c + b_1 a_1 + b_2 a_2 + b_3 a_3) e_1 e_2 e_3.\end{aligned} \tag{6.18}$$

For general multivectors, we write them as sums of odd and even multivectors and compute the product separately in the form $(\mathcal{A}+\mathcal{B})(\mathcal{A}'+\mathcal{B}') = (\mathcal{A}\mathcal{A}'+\mathcal{B}\mathcal{B}') + (\mathcal{A}\mathcal{B}'+\mathcal{B}\mathcal{A}')$. Note that sums and scalar multiples of even multivectors are even multivectors and their products are also even multivectors. This means that the set of even multivectors forms by itself a closed algebra, i.e., a *subalgebra* of the Clifford algebra. In fact, this subalgebra is essentially nothing but the Hamilton algebra. This can be seen if we let

$$i = e_3 e_2 \ (= -e_2 e_3), \qquad j = e_1 e_3 \ (= -e_3 e_1), \qquad k = e_2 e_1 \ (= -e_1 e_2). \tag{6.19}$$

From the multiplication rule of Table 6.2, it is easy to see that Eqs. (6.5) and (6.6) are satisfied (↪ Exercise 6.1). In other words, *the Hamilton algebra is a part of the Clifford algebra*, which also contains the set of complex numbers $\mathbb{C}$ and the set of real numbers $\mathbb{R}$ as subalgebras of the Hamilton algebra.

6.4 GRASSMANN ALGEBRA IN THE CLIFFORD ALGEBRA

We have shown that the Clifford algebra includes the Hamilton algebra via Eq. (6.19). We now show that *the Grassmann algebra is also a part of the Clifford algebra.*

We identify the vector $\boldsymbol{a} = a_1 e_1 + a_2 e_2 + a_3 e_3$ in 3D with an element of the Clifford algebra and define the outer product $\boldsymbol{a} \wedge \boldsymbol{b}$ of vector $\boldsymbol{a}$ and $\boldsymbol{b}$ by the following *antisymmetrization*:

$$\boldsymbol{a} \wedge \boldsymbol{b} = \frac{1}{2}(\boldsymbol{ab} - \boldsymbol{ba}). \tag{6.20}$$

From this definition, we see that

$$\boldsymbol{b} \wedge \boldsymbol{a} = -\boldsymbol{a} \wedge \boldsymbol{b}, \qquad \boldsymbol{a} \wedge \boldsymbol{a} = 0. \tag{6.21}$$

For vectors $\boldsymbol{a}$, $\boldsymbol{b}$, and $\boldsymbol{c}$, we define their outer product by the following antisymmetrization:

$$\boldsymbol{a} \wedge \boldsymbol{b} \wedge \boldsymbol{c} = \frac{1}{6}(\boldsymbol{abc} + \boldsymbol{bca} + \boldsymbol{cab} - \boldsymbol{cba} - \boldsymbol{bac} - \boldsymbol{acb}). \tag{6.22}$$

Finally, the outer product of four or more vectors $\boldsymbol{a} \wedge \boldsymbol{b} \wedge \boldsymbol{c} \wedge \boldsymbol{d} \wedge \cdots$ is defined to be 0. Then all the axioms of the Grassmann outer product $\wedge$ are satisfied. Hence, we can identify th outer product computation as that of the Grassmann algebra.

If we use the basis to write $\boldsymbol{a} = a_1 e_1 + a_2 e_2 + a_3 e_3$ and $\boldsymbol{b} = b_1 e_1 + b_2 e_2 + b_3 e_3$, we se from Eq. (6.14) that the geometric products $\boldsymbol{ab}$ and $\boldsymbol{ba}$ are expressed as

$$\boldsymbol{ab} = a_1 b_1 + a_2 b_2 + a_3 b_3 + (a_2 b_3 - a_3 b_2)e_2 e_3 + (a_3 b_1 - a_1 b_3)e_3 e_1 + (a_1 b_2 - a_2 b_1)e_1 e_2, \quad (6.24)$$

$$\boldsymbol{ba} = b_1 a_1 + b_2 a_2 + b_3 a_3 + (b_2 a_3 - b_3 a_2)e_2 e_3 + (b_3 a_1 - b_1 a_3)e_3 e_1 + (b_1 a_2 - b_2 a_3)e_1 e_2. \quad (6.25)$$

Hence, the outer product $\boldsymbol{a} \wedge \boldsymbol{b}$ has the form

$$\boldsymbol{a} \wedge \boldsymbol{b} = (a_2 b_3 - a_3 b_2)e_2 e_3 + (a_3 b_1 - a_1 b_3)e_3 e_1 + (a_1 b_2 - a_2 b_1)e_1 e_2. \quad (6.26)$$

Vectors $\boldsymbol{a}$, $\boldsymbol{b}$, ... are themselves odd multivectors of grade 1, so the products of thre vectors $\boldsymbol{abc}$, $\boldsymbol{bca}$, ... are all odd multivectors consisting of the grade 1 part and the grad 3 part. It is easy to see that antisymmetrization of Eq. (6.22) cancels out the grade 1 part. Consequently, if we let $\boldsymbol{c} = c_1 e_1 + c_2 e_2 + c_3 e_3$, we can see that in the end

$$\boldsymbol{a} \wedge \boldsymbol{b} \wedge \boldsymbol{c} = (a_1 b_2 c_3 + a_2 b_3 c_1 + a_3 b_1 c_2 - a_1 b_3 c_2 - a_2 b_1 c_3 - a_3 b_2 c_1)e_1 e_2 e_3. \quad (6.27)$$

Thus, identifying Eqs. (6.20) and (6.22) with the Grassmann outer products means identify ing $e_1 e_2$, $e_2 e_3$, $e_3 e_1$, and $e_1 e_2 e_3$ with $e_1 \wedge e_2$, $e_2 \wedge e_3$, $e_3 \wedge e_1$, and $e_1 \wedge e_2 \wedge e_3$, respectively (Eqs. (5.15) and (5.16) in Chapter 5). This is justified from Eqs. (6.7) and (6.8). Recall tha the only difference between the outer and the geometric products is that *the outer produ of the same symbols is 0 while their geometric product is 1*. It follows that for differen symbols the geometric and the outer products follow the same rule, and hence *geometr products of different symbols can be identified with their outer products*. Also note that th square terms $e_1^2 = e_2^2 = e_3^2 = 1$ unique to the geometric product are canceled out by th antisymmetrization operation.

6.5 PROPERTIES OF THE GEOMETRIC PRODUCT

We show how the geometric product is expressed in terms of the contraction (or the inne product) and the outer product. It is then shown that the inverse for the geometric produc exists.

6.5.1 Geometric product and outer product

From Eqs. (6.24) and (6.25), the *symmetrization* of the geometric product $\boldsymbol{ab}$ becomes

$$\frac{1}{2}(\boldsymbol{ab} + \boldsymbol{ba}) = a_1 b_1 + a_2 b_2 + a_3 b_3 = \langle \boldsymbol{a}, \boldsymbol{b} \rangle. \quad (6.28)$$

In particular, if $\boldsymbol{a}$ and $\boldsymbol{b}$ are orthogonal, i.e., $\langle \boldsymbol{a}, \boldsymbol{b} \rangle = 0$, we have $\boldsymbol{ab} = -\boldsymbol{ba}$. Products fo which we can interchange the two terms after sign change are said to be *anticommutative* General geometric products are neither commutative nor anticommutative, but *geometri products of orthogonal vectors are anticommutative*.

where the dot $\cdot$ denotes contraction. The geometric product of vector $\boldsymbol{a}$ and bivector $\boldsymbol{b} \wedge \boldsymbol{c}$ is given by

$$\boldsymbol{a}(\boldsymbol{b} \wedge \boldsymbol{c}) = \boldsymbol{a} \cdot \boldsymbol{b} \wedge \boldsymbol{c} + \boldsymbol{a} \wedge \boldsymbol{b} \wedge \boldsymbol{c}. \tag{6.30}$$

This can be confirmed by expanding both sides. Namely, we replace the left side by $\boldsymbol{a}(\boldsymbol{bc} - \boldsymbol{cb})/2$ and the first term on the right side by $\langle \boldsymbol{a}, \boldsymbol{b} \rangle \boldsymbol{c} - \langle \boldsymbol{a}, \boldsymbol{c} \rangle \boldsymbol{b}$ ($\hookrightarrow$ Eq. (5.32) in Chapter 5). We further replace $\langle \boldsymbol{a}, \boldsymbol{b} \rangle$ and $\langle \boldsymbol{a}, \boldsymbol{c} \rangle$ by $(\boldsymbol{ab} + \boldsymbol{ba})/2$ and $(\boldsymbol{ac} + \boldsymbol{ca})/2$, respectively, and expand the second term on the right side as in Eq. (6.22). Then, both sides of Eq. (6.30) turn out to be the same. Similarly, the geometric product of vector $\boldsymbol{a}$ and trivector $\boldsymbol{b} \wedge \boldsymbol{c} \wedge \boldsymbol{d}$ is given by

$$\boldsymbol{a}(\boldsymbol{b} \wedge \boldsymbol{c} \wedge \boldsymbol{d}) = \boldsymbol{a} \cdot \boldsymbol{b} \wedge \boldsymbol{c} \wedge \boldsymbol{d} + \boldsymbol{a} \wedge \boldsymbol{b} \wedge \boldsymbol{c} \wedge \boldsymbol{d}. \tag{6.31}$$

This can also be confirmed by expansion on both sides. We can summarize Eqs. (6.29), (6.30), and (6.31) as follows:

Proposition 6.6 (Geometric product via contraction and outer product) *The geometric product of vector* $\boldsymbol{a}$ *and a* k*-vector* $(\cdots)$, $k = 0, 1, 2, 3$, *is expressed as the sum of contraction and outer product as follows:*

$$\boldsymbol{a}(\cdots) = \boldsymbol{a} \cdot (\cdots) + \boldsymbol{a} \wedge (\cdots). \tag{6.32}$$

Note this also holds when $(\cdots)$ is a scalar, in which case the first term on the right side is 0 and the second term equals the left side ($\hookrightarrow$ Eqs. (5.11) and (5.30) in Chapter 5). All geometric products are reduced to contractions and outer products by recursively applying this rule.

6.5.2 Inverse

Letting $\boldsymbol{a} = \boldsymbol{b}$ in Eq. (6.28), we see that

$$\|\boldsymbol{a}\|^2 = \boldsymbol{a}^2. \tag{6.33}$$

Hence, if $\|\boldsymbol{a}\| \neq 0$, we have

$$\boldsymbol{a}\frac{\boldsymbol{a}}{\|\boldsymbol{a}\|^2} = \frac{\boldsymbol{a}}{\|\boldsymbol{a}\|^2}\boldsymbol{a} = 1. \tag{6.34}$$

This means that $\boldsymbol{a}/\|\boldsymbol{a}\|^2$ is the *inverse* of $\boldsymbol{a}$:

$$\boldsymbol{a}^{-1} = \frac{\boldsymbol{a}}{\|\boldsymbol{a}\|^2}, \qquad \boldsymbol{a}\boldsymbol{a}^{-1} = \boldsymbol{a}^{-1}\boldsymbol{a} = 1. \tag{6.35}$$

From Eq. (6.29), we can interpret the geometric product to be *computing the inner and outer products simultaneously*. The existence of the inverse that admits division means that $\boldsymbol{ab} = \boldsymbol{ac}$ for $\boldsymbol{a} \neq 0$ implies $\boldsymbol{b} = \boldsymbol{c}$. This does not hold for the inner or outer product. In fact, $\langle \boldsymbol{a}, \boldsymbol{b} \rangle = \langle \boldsymbol{a}, \boldsymbol{c} \rangle$ for $\boldsymbol{a} \neq 0$ does not imply $\boldsymbol{b} = \boldsymbol{c}$, because we can add to $\boldsymbol{b}$ any vector that is orthogonal to $\boldsymbol{a}$. Similarly, $\boldsymbol{a} \wedge \boldsymbol{b} = \boldsymbol{a} \wedge \boldsymbol{c}$ for $\boldsymbol{a} \neq 0$ does not imply $\boldsymbol{b} = \boldsymbol{c}$, because we can add to $\boldsymbol{b}$ any vector that is parallel to $\boldsymbol{a}$. However, if *the inner and outer products are simultaneously considered*, i.e., if $\langle \boldsymbol{a}, \boldsymbol{b} \rangle = \langle \boldsymbol{a}, \boldsymbol{c} \rangle$ *and* $\boldsymbol{a} \times \boldsymbol{b} = \boldsymbol{a} \times \boldsymbol{c}$ for $\boldsymbol{a} \neq 0$, then $\boldsymbol{b} = \boldsymbol{c}$. Thus, the existence of the inverse for the geometric product is a natural consequence.

$$(\boldsymbol{ab})^{-1} = \boldsymbol{b}^{-1}\boldsymbol{a}^{-1}, \quad (\boldsymbol{abc})^{-1} = \boldsymbol{c}^{-1}\boldsymbol{b}^{-1}\boldsymbol{a}^{-1}, \quad (\boldsymbol{abc}\cdots)^{-1} = \cdots \boldsymbol{c}^{-1}\boldsymbol{b}^{-1}\boldsymbol{a}^{-1}. \qquad (6.3$$

This is evident from $\boldsymbol{abc}\cdots\boldsymbol{c}^{-1}\boldsymbol{b}^{-1}\boldsymbol{a}^{-1} = \cdots = \boldsymbol{ab}\underbrace{\boldsymbol{cc}^{-1}}_{1}\boldsymbol{b}^{-1}\boldsymbol{a}^{-1} = \boldsymbol{a}\underbrace{\boldsymbol{bb}^{-1}}_{1}\boldsymbol{a}^{-1} = \underbrace{\boldsymbol{aa}^{-1}}_{1} =$

The inverse of bivector $\boldsymbol{a} \wedge \boldsymbol{b}$ is given by

$$(\boldsymbol{a} \wedge \boldsymbol{b})^{-1} = \frac{\boldsymbol{b} \wedge \boldsymbol{a}}{\|\boldsymbol{a} \wedge \boldsymbol{b}\|^2}, \qquad (6.3$$

which is equivalent to

$$(\boldsymbol{a} \wedge \boldsymbol{b})(\boldsymbol{b} \wedge \boldsymbol{a}) = \|\boldsymbol{a} \wedge \boldsymbol{b}\|^2, \qquad (6.3$$

where the right side is given by Eq. (5.44) in Chapter 5. This is easily confirmed by expandi both sides using Eq. (6.20) (↪ Exercise 6.2), but the following reasoning is much simple We let $\boldsymbol{b}' = \boldsymbol{b} - \alpha\boldsymbol{a}$ and determine α so that $\boldsymbol{a}$ and $\boldsymbol{b}'$ are orthogonal. Since $\boldsymbol{a} \wedge \boldsymbol{b} = \boldsymbol{a} \wedge$ for bivectors (↪ Eq. (5.3) in Chapter 5), we only need to check Eq. (6.38) when $\boldsymbol{a}$ and are orthogonal. If they are orthogonal, then $\boldsymbol{a} \wedge \boldsymbol{b} = \boldsymbol{ab}$ and $\boldsymbol{b} \wedge \boldsymbol{a} = \boldsymbol{ba}$ from Eq. (6.29), we see that

$$(\boldsymbol{a} \wedge \boldsymbol{b})(\boldsymbol{b} \wedge \boldsymbol{a}) = \boldsymbol{abba} = \boldsymbol{ab}^2\boldsymbol{a} = \boldsymbol{a}\|\boldsymbol{b}\|^2\boldsymbol{a} = \boldsymbol{a}^2\|\boldsymbol{b}\|^2 = \|\boldsymbol{a}\|^2\|\boldsymbol{b}\|^2 = \|\boldsymbol{a} \wedge \boldsymbol{b}\|^2. \qquad (6.3$$

Similarly, the inverse of trivector $\boldsymbol{a} \wedge \boldsymbol{b} \wedge \boldsymbol{c}$ is given by

$$(\boldsymbol{a} \wedge \boldsymbol{b} \wedge \boldsymbol{c})^{-1} = \frac{\boldsymbol{c} \wedge \boldsymbol{b} \wedge \boldsymbol{a}}{\|\boldsymbol{a} \wedge \boldsymbol{b} \wedge \boldsymbol{c}\|^2}, \qquad (6.4$$

which is equivalent to

$$(\boldsymbol{a} \wedge \boldsymbol{b} \wedge \boldsymbol{c})(\boldsymbol{c} \wedge \boldsymbol{b} \wedge \boldsymbol{a}) = \|\boldsymbol{a} \wedge \boldsymbol{b} \wedge \boldsymbol{c}\|^2, \qquad (6.4$$

where the right side is given by Eq. (5.46) in Chapter 5. We can show this by expanding bot sides using Eq. (6.22), but this would be tedious. So we let $\boldsymbol{b}' = \boldsymbol{b} - \alpha\boldsymbol{a}$ and $\boldsymbol{c}' = \boldsymbol{c} - \beta\boldsymbol{a} - \gamma$ and determine α, β, and γ so that $\boldsymbol{a}$, $\boldsymbol{b}'$, and $\boldsymbol{c}'$ are orthogonal. Since $\boldsymbol{a} \wedge \boldsymbol{b} \wedge \boldsymbol{c} = \boldsymbol{a} \wedge \boldsymbol{b}' \wedge$ for trivectors (↪ Eq. (5.9) in Chapter 5), we only need to check Eq. (6.41) when $\boldsymbol{a}$, and $\boldsymbol{c}$ are orthogonal. If they are orthogonal, $\boldsymbol{a}$, $\boldsymbol{b}$, and $\boldsymbol{c}$ are mutually anticommutative, $\boldsymbol{a} \wedge \boldsymbol{b} \wedge \boldsymbol{c} = \boldsymbol{abc}$ and $\boldsymbol{c} \wedge \boldsymbol{b} \wedge \boldsymbol{a} = \boldsymbol{cba}$ from Eq. (6.22). Hence, we see that

$$(\boldsymbol{a} \wedge \boldsymbol{b} \wedge \boldsymbol{c})(\boldsymbol{c} \wedge \boldsymbol{b} \wedge \boldsymbol{a}) = \boldsymbol{abccba} = \boldsymbol{abba}\|\boldsymbol{c}\|^2 = \boldsymbol{aa}\|\boldsymbol{b}\|^2\|\boldsymbol{c}\|^2 = \|\boldsymbol{a}\|^2\|\boldsymbol{b}\|^2\|\boldsymbol{c}\|^2 = \|\boldsymbol{a} \wedge \boldsymbol{b} \wedge \boldsymbol{c}\| \qquad (6.4$$

In summary,

Proposition 6.7 (Inverse of k-vectors) *Vector $\boldsymbol{a}$, bivector $\boldsymbol{a} \wedge \boldsymbol{b}$, and trivector $\boldsymbol{a} \wedge \boldsymbol{b} \wedge$ have the following inverses:*

$$\boldsymbol{a}^{-1} = \frac{\boldsymbol{a}}{\|\boldsymbol{a}\|^2}, \qquad (\boldsymbol{a} \wedge \boldsymbol{b})^{-1} = \frac{\boldsymbol{b} \wedge \boldsymbol{a}}{\|\boldsymbol{a} \wedge \boldsymbol{b}\|^2}, \qquad (\boldsymbol{a} \wedge \boldsymbol{b} \wedge \boldsymbol{c})^{-1} = \frac{\boldsymbol{c} \wedge \boldsymbol{b} \wedge \boldsymbol{a}}{\|\boldsymbol{a} \wedge \boldsymbol{b} \wedge \boldsymbol{c}\|^2}. \qquad (6.4$$

The inverse of the geometric product of multiple terms is given by the geometric product their respective inverses in reverse order.

ization well known in linear algebra. Given linearly independent vectors $\boldsymbol{u}_1, \boldsymbol{u}_2, \ldots$, we can construct an orthogonal system $\boldsymbol{v}_1, \boldsymbol{v}_2, \ldots$ as follows. First, let

$$\boldsymbol{v}_1 = \boldsymbol{u}_1. \tag{6.44}$$

Next, let $\boldsymbol{v}_2 = \boldsymbol{u}_2 - c\boldsymbol{v}_1$, and determine the coefficient c so that $\boldsymbol{v}_2$ is orthogonal to $\boldsymbol{v}_1$. From

$$\langle \boldsymbol{v}_1, \boldsymbol{v}_2 \rangle = \langle \boldsymbol{v}_1, \boldsymbol{u}_2 \rangle - c\langle \boldsymbol{v}_1, \boldsymbol{v}_1 \rangle = \langle \boldsymbol{v}_1, \boldsymbol{u}_2 \rangle - c\|\boldsymbol{v}_1\|^2 = 0, \tag{6.45}$$

we obtain $c = \langle \boldsymbol{v}_1, \boldsymbol{u}_2 \rangle / \|\boldsymbol{v}_1\|^2$. Hence,

$$\boldsymbol{v}_2 = \boldsymbol{u}_2 - \frac{\langle \boldsymbol{v}_1, \boldsymbol{u}_2 \rangle}{\|\boldsymbol{v}_1\|^2}\boldsymbol{v}_1. \tag{6.46}$$

Next, let $\boldsymbol{v}_3 = \boldsymbol{u}_3 - c\boldsymbol{v}_1 - c_2\boldsymbol{v}_2$, and determine the coefficients c_1 and c_2 so that $\boldsymbol{v}_3$ is orthogonal to $\boldsymbol{v}_1$ and $\boldsymbol{v}_2$. Since the vectors $\boldsymbol{v}_1$ and $\boldsymbol{v}_2$ are so constructed that they are mutually orthogonal, we have

$$\begin{aligned} \langle \boldsymbol{v}_1, \boldsymbol{v}_3 \rangle &= \langle \boldsymbol{v}_1, \boldsymbol{u}_3 \rangle - c_1\langle \boldsymbol{v}_1, \boldsymbol{v}_1 \rangle - c_2\langle \boldsymbol{v}_1, \boldsymbol{v}_2 \rangle = \langle \boldsymbol{v}_1, \boldsymbol{u}_3 \rangle - c_1\|\boldsymbol{v}_1\|^2 = 0, \\ \langle \boldsymbol{v}_2, \boldsymbol{v}_3 \rangle &= \langle \boldsymbol{v}_2, \boldsymbol{u}_3 \rangle - c_1\langle \boldsymbol{v}_2, \boldsymbol{v}_1 \rangle - c_2\langle \boldsymbol{v}_2, \boldsymbol{v}_2 \rangle = \langle \boldsymbol{v}_2, \boldsymbol{u}_3 \rangle - c_2\|\boldsymbol{v}_2\|^2 = 0, \end{aligned} \tag{6.47}$$

so we obtain $c_1 = \langle \boldsymbol{v}_1, \boldsymbol{u}_3 \rangle / \|\boldsymbol{v}_1\|^2$ and $c_2 = \langle \boldsymbol{v}_2, \boldsymbol{u}_3 \rangle / \|\boldsymbol{v}_2\|^2$. Hence,

$$\boldsymbol{v}_3 = \boldsymbol{u}_3 - \frac{\langle \boldsymbol{v}_1, \boldsymbol{u}_3 \rangle}{\|\boldsymbol{v}_1\|^2}\boldsymbol{v}_1 - \frac{\langle \boldsymbol{v}_2, \boldsymbol{u}_3 \rangle}{\|\boldsymbol{v}_2\|^2}\boldsymbol{v}_2. \tag{6.48}$$

We continue this. After we have obtained an orthogonal system $\boldsymbol{v}_1, \ldots, \boldsymbol{v}_k$, we let $\boldsymbol{v}_{k+1} = \boldsymbol{u}_{k+1} - \sum_{j=1}^{k} c_j\boldsymbol{v}_j$ and determine the coefficients $c_i, \ldots, c_k$ so that $\boldsymbol{v}_{k+1}$ is orthogonal to $\boldsymbol{v}_i$, $i = 1, \ldots, k$. Since $\boldsymbol{v}_i$, $i = 1, \ldots, k$, are so constructed that they are mutually orthogonal, we have

$$\langle \boldsymbol{v}_i, \boldsymbol{u}_{k+1} \rangle = \langle \boldsymbol{v}_i, \boldsymbol{u}_{k+1} \rangle - \sum_{j=1}^{k} c_j\langle \boldsymbol{v}_i, \boldsymbol{v}_j \rangle = \langle \boldsymbol{v}_i, \boldsymbol{u}_{k+1} \rangle - c_i\|\boldsymbol{v}_i\|^2 = 0, \tag{6.49}$$

so we obtain $c_i = \langle \boldsymbol{v}_i, \boldsymbol{u}_{k+1} \rangle / \|\boldsymbol{v}_i\|^2$. Hence,

$$\boldsymbol{v}_{k+1} = \boldsymbol{u}_{k+1} - \sum_{j=1}^{k} \frac{\langle \boldsymbol{v}_j, \boldsymbol{u}_{k+1} \rangle}{\|\boldsymbol{v}_j\|^2}\boldsymbol{v}_j, \tag{6.50}$$

and the process goes on. If we normalize $\boldsymbol{v}_i$, $i = 1, \ldots, k$, into unit vectors $\boldsymbol{e}_i = \boldsymbol{v}_i / \|\boldsymbol{v}_i\|$, Eq. (6.50) can be written in the form

$$\boldsymbol{v}_{k+1} = \boldsymbol{u}_{k+1} - \sum_{j=1}^{k} \langle \boldsymbol{e}_j, \boldsymbol{u}_{k+1} \rangle \boldsymbol{e}_j. \tag{6.51}$$

Since the projected length of the vector $\boldsymbol{u}_{k+1}$ onto the direction along the unit vector $\boldsymbol{e}_j$ is $\langle \boldsymbol{e}_j, \boldsymbol{u}_{k+1} \rangle$, the sum $\sum_{j=1}^{k} \langle \boldsymbol{e}_j, \boldsymbol{u}_{k+1} \rangle \boldsymbol{e}_j$ is the projection of the vector $\boldsymbol{u}_{k+1}$ onto the space spanned by $\boldsymbol{e}_1, \ldots, \boldsymbol{e}_k$. In other words, Eq. (6.51) is nothing but the rejection of the vector $\boldsymbol{u}_{k+1}$ from the space spanned by $\boldsymbol{e}_1, \ldots, \boldsymbol{e}_k$. Thus, the Schmidt orthogonalization can be

dependent, u_{k+1} is expressed as a linear combination of u_1, ..., u_k and hence a line
combination of v_1, ..., v_k. In other words, it is included in the space spanned by e ..., e_k, so Eq. (6.51) is 0, and the orthogonalization process stops. Evidently, we cann orthogonalize more than n vectors in nD.

The Schmidt orthogonalization goes in the same way whether the given vectors are a stractly defined symbols or arrays of numbers. If in particular they are columns of vertical aligned n numbers, we can define an $n\times n$ matrix consisting of linearly independent colum u_1, ..., u_n. If v_1, ..., v_n are the result of their Schmidt orthogonalization, the determina of the $n \times n$ matrix consisting of them is equal to that before the orthogonalization:

$$|\boldsymbol{u}_1, \boldsymbol{u}_2, \ldots, \boldsymbol{u}_n| = |\boldsymbol{v}_1, \boldsymbol{v}_2, \ldots, \boldsymbol{v}_n|. \quad (6.52)$$

This is the consequence of the well-known fact that the determinant is unchanged if w subtract from one column a constant multiple of another column; the Schmidt orthogona ization simply repeats this process. We can translate this consideration into the Grassman algebra terminologies: identifying vectors with elements of the Grassmann algebra, we s that for an arbitrary $k = 1, \ldots, n$

$$\boldsymbol{u}_1 \wedge \boldsymbol{u}_2 \wedge \cdots \wedge \boldsymbol{u}_k = \boldsymbol{v}_1 \wedge \boldsymbol{v}_2 \wedge \cdots \wedge \boldsymbol{v}_k. \quad (6.53)$$

6.6 PROJECTION, REJECTION, AND REFLECTION

A line l along vector $\boldsymbol{a}$ has the unit direction vector $\boldsymbol{a}/\|\boldsymbol{a}\|$. Hence, the projection $\boldsymbol{x}_{\|}$ vector $\boldsymbol{x}$ onto line l is given as follows (↪ Sec. 2.6 in Chapter 2):

$$\boldsymbol{x}_{\|} = \langle \boldsymbol{x}, \frac{\boldsymbol{a}}{\|\boldsymbol{a}\|} \rangle \frac{\boldsymbol{a}}{\|\boldsymbol{a}\|}. \quad (6.54)$$

Noting that $\boldsymbol{a}^{-1} = \boldsymbol{a}/\|\boldsymbol{a}\|^2$, we can write this in the form

$$\boldsymbol{x}_{\|} = \langle \boldsymbol{x}, \boldsymbol{a} \rangle \boldsymbol{a}^{-1}. \quad (6.55)$$

On the other hand, we have $\boldsymbol{x}\boldsymbol{a} = \langle \boldsymbol{x}, \boldsymbol{a} \rangle + \boldsymbol{x} \wedge \boldsymbol{a}$ from Eq. (6.29). Multiplying $\boldsymbol{a}^{-1}$ from th right on both sides, we have

$$\boldsymbol{x} = \langle \boldsymbol{x}, \boldsymbol{a} \rangle \boldsymbol{a}^{-1} + (\boldsymbol{x} \wedge \boldsymbol{a}) \boldsymbol{a}^{-1}. \quad (6.56)$$

Since the first term on the right is the projection $\boldsymbol{x}_{\|} = \langle \boldsymbol{x}, \boldsymbol{a} \rangle \boldsymbol{a}^{-1}$, the second terms shou be the rejection $\boldsymbol{x}_{\perp} = (\boldsymbol{x} \wedge \boldsymbol{a})\boldsymbol{a}^{-1}$ orthogonal to it (Fig. 6.1(a)). The reflection $\boldsymbol{x}_{\top}$ of vect $\boldsymbol{x}$ with respect to line l is obtained by subtracting from $\boldsymbol{x}$ twice the rejection $\boldsymbol{x}_{\perp}$. Henc we obtain

$$\begin{aligned} \boldsymbol{x}_{\top} &= (\boldsymbol{x}_{\|} + \boldsymbol{x}_{\perp}) - 2\boldsymbol{x}_{\perp} = \boldsymbol{x}_{\|} - \boldsymbol{x}_{\perp} = \langle \boldsymbol{x}, \boldsymbol{a} \rangle \boldsymbol{a}^{-1} - (\boldsymbol{x} \wedge \boldsymbol{a})\boldsymbol{a}^{-1} \\ &= (\langle \boldsymbol{x}, \boldsymbol{a} \rangle - \boldsymbol{x} \wedge \boldsymbol{a})\boldsymbol{a}^{-1} = (\langle \boldsymbol{a}, \boldsymbol{x} \rangle + \boldsymbol{a} \wedge \boldsymbol{x})\boldsymbol{a}^{-1} = \boldsymbol{a}\boldsymbol{x}\boldsymbol{a}^{-1}. \end{aligned} \quad (6.57)$$

This is summarized as follows:

Proposition 6.8 (Projection, rejection, and reflection for a line) *The projection $\boldsymbol{x}$ the rejection $\boldsymbol{x}_{\perp}$, and the reflection $\boldsymbol{x}_{\top}$ of vector $\boldsymbol{x}$ for a line in the direction $\boldsymbol{a}$ are given b*

$$\boldsymbol{x}_{\|} = \langle \boldsymbol{x}, \boldsymbol{a} \rangle \boldsymbol{a}^{-1}, \qquad \boldsymbol{x}_{\perp} = (\boldsymbol{x} \wedge \boldsymbol{a})\boldsymbol{a}^{-1}, \qquad \boldsymbol{x}_{\top} = \boldsymbol{a}\boldsymbol{x}\boldsymbol{a}^{-1}. \quad (6.58)$$

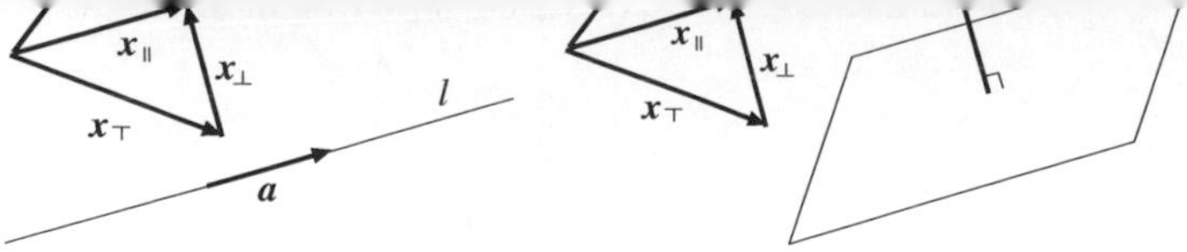

(a) (b)

FIGURE 6.1 (a) The projection $\boldsymbol{x}_{\parallel}$, the rejection $\boldsymbol{x}_{\perp}$, and the reflection $\boldsymbol{x}_{\top}$ of vector $\boldsymbol{x}$ for a line in the direction $\boldsymbol{a}$. (b) The projection $\boldsymbol{x}_{\parallel}$, the rejection $\boldsymbol{x}_{\perp}$, and the reflection $\boldsymbol{x}_{\top}$ of vector $\boldsymbol{x}$ for a plane with unit surface normal $\boldsymbol{n}$.

The projection $\boldsymbol{x}_{\parallel}$ of vector $\boldsymbol{x}$ onto a plane with unit surface normal $\boldsymbol{n}$ equals its rejection from the surface normal (Fig. 6.1(b)), so $\boldsymbol{x}_{\parallel} = (\boldsymbol{x} \wedge \boldsymbol{n})\boldsymbol{n}^{-1}$ from Eq. (6.58). Conversely, the rejection $\boldsymbol{x}_{\perp}$ from this plane equals its projection onto the surface normal, so $\boldsymbol{x}_{\perp} = \langle \boldsymbol{x}, \boldsymbol{n} \rangle \boldsymbol{n}^{-1}$ from Eq. (6.58). The reflection $\boldsymbol{x}_{\top}$ of vector $\boldsymbol{x}$ with respect to this plane is obtained by subtracting from $\boldsymbol{x}$ twice the projection $\boldsymbol{x}_{\perp}$. Hence, we obtain

$$\begin{aligned} \boldsymbol{x}_{\top} &= (\boldsymbol{x}_{\parallel} + \boldsymbol{x}_{\perp}) - 2\boldsymbol{x}_{\perp} = \boldsymbol{x}_{\parallel} - \boldsymbol{x}_{\perp} = (\boldsymbol{x} \wedge \boldsymbol{n})\boldsymbol{n}^{-1} - \langle \boldsymbol{x}, \boldsymbol{n} \rangle \boldsymbol{n}^{-1} \\ &= -(\langle \boldsymbol{n}, \boldsymbol{x} \rangle + \boldsymbol{n} \wedge \boldsymbol{x})\boldsymbol{n}^{-1} = -\boldsymbol{n}\boldsymbol{x}\boldsymbol{n}^{-1}. \end{aligned} \tag{6.59}$$

This is summarized as follows:

Proposition 6.9 (Projection, rejection, and reflection for a plane) *The projection $\boldsymbol{x}_{\parallel}$, the rejection $\boldsymbol{x}_{\perp}$, and the reflection $\boldsymbol{x}_{\top}$ of vector $\boldsymbol{x}$ with respect to a plane with unit surface normal $\boldsymbol{n}$ are given by*

$$\boldsymbol{x}_{\parallel} = (\boldsymbol{x} \wedge \boldsymbol{n})\boldsymbol{n}^{-1}, \qquad \boldsymbol{x}_{\perp} = \langle \boldsymbol{x}, \boldsymbol{n} \rangle \boldsymbol{n}^{-1}, \qquad \boldsymbol{x}_{\top} = -\boldsymbol{n}\boldsymbol{x}\boldsymbol{n}^{-1}. \tag{6.60}$$

6.7 ROTATION AND GEOMETRIC PRODUCT

One of the most significant roles of the geometric product in practical applications is its ability to represent rotations of vectors and planes in a systematic manner. First, we show that a rotation can be represented by a composition of two successive reflections. Then, we show how it is expressed in terms of the surface element of the plane, which leads to an expression in the exponential function of the surface element.

6.7.1 Representation by reflections

Consider a rotation around an axis $\boldsymbol{l}$. Let $\boldsymbol{a}$ and $\boldsymbol{b}$ be two vectors orthogonal to $\boldsymbol{l}$. We now show that rotation of vector $\boldsymbol{x}$ around $\boldsymbol{l}$ is achieved by composition of two successive reflections. To see this, let $\tilde{\boldsymbol{x}}$ be the reflection of $\boldsymbol{x}$ with respect to a plane orthogonal to $\boldsymbol{a}$, and $\boldsymbol{x}'$ the reflection of $\tilde{\boldsymbol{x}}$ with respect to a plane orthogonal to $\boldsymbol{b}$ (Fig. 6.2(a)). Successive reflections do not alter the norm of vectors, and the axis $\boldsymbol{l}$ remains unchanged. Hence, $\boldsymbol{x}'$ is a rotation of $\boldsymbol{x}$ around $\boldsymbol{l}$. Since $\tilde{\boldsymbol{x}} = -\boldsymbol{a}\boldsymbol{x}\boldsymbol{a}^{-1}$ from Eq. (6.60), the vector $\boldsymbol{x}'$ is expressed as follows:

$$\boldsymbol{x}' = -\boldsymbol{b}\tilde{\boldsymbol{x}}\boldsymbol{b}^{-1} = -\boldsymbol{b}(-\boldsymbol{a}\boldsymbol{x}\boldsymbol{a}^{-1})\boldsymbol{b}^{-1} = (\boldsymbol{b}\boldsymbol{a})\boldsymbol{x}(\boldsymbol{b}\boldsymbol{a})^{-1}. \tag{6.61}$$

This can be written as

$$\boldsymbol{x}' = \mathcal{R}\boldsymbol{x}\mathcal{R}^{-1}, \tag{6.62}$$

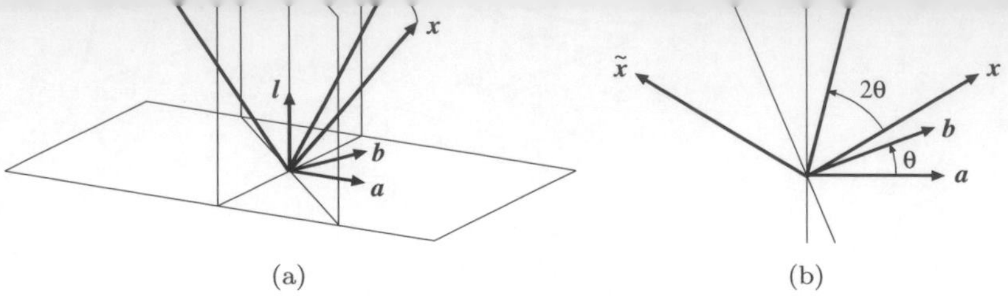

FIGURE 6.2 (a) If $\tilde{\boldsymbol{x}}$ is the reflection of vector $\boldsymbol{x}$ with respect to a plane orthogonal to $\boldsymbol{a}$, and $\boldsymbol{x}$ the reflection of $\tilde{\boldsymbol{x}}$ with respect to a plane orthogonal to $\boldsymbol{b}$, the vector $\boldsymbol{x}'$ is a rotation of $\boldsymbol{x}$ aroun the vector $\boldsymbol{l}$. (b) Top view. If $\boldsymbol{a}$ and $\boldsymbol{b}$ make an angle θ, the vector $\boldsymbol{x}'$ is a rotation of $\boldsymbol{x}$ by angle 2θ

where we let

$$\mathcal{R} = \boldsymbol{ba}. \tag{6.63}$$

Viewed as an operator that acts in the form of Eq. (6.62), this $\mathcal{R}$ is called a *rotor*. If $\boldsymbol{a}$ an $\boldsymbol{b}$ make an angle θ, the vector $\boldsymbol{x}'$ is a rotation of $\boldsymbol{x}$ around $\boldsymbol{l}$ by angle 2θ. This is easily see from Fig. 6.2(b), which is a top view of Fig. 6.2(a). If vector $\boldsymbol{x}$ makes an angle ϕ from $\boldsymbol{a}$, w see that $\tilde{\boldsymbol{x}}$ makes the angle $\pi - \phi$ from $\boldsymbol{a}$ and that $\boldsymbol{x}'$ makes the angle $2\theta + \phi$ from $\boldsymbol{a}$. Henc $\boldsymbol{x}'$ makes angle 2θ from $\boldsymbol{x}$. Since Eq. (6.62) contains both $\mathcal{R}$ and $\mathcal{R}^{-1}$, the result does no depend on the norms of $\boldsymbol{a}$ and $\boldsymbol{b}$, i.e., the rotation is defined only by the orientations of and $\boldsymbol{b}$.

If a rotation of $\boldsymbol{x}$ specified by rotor $\mathcal{R}$ is followed by another rotation specified by roto $\mathcal{R}'$, i.e., if $\boldsymbol{x}' = \mathcal{R}\boldsymbol{x}\mathcal{R}^{-1}$ is rotated by a rotation $\mathcal{R}'$, we obtain

$$\boldsymbol{x}'' = \mathcal{R}'\boldsymbol{x}'\mathcal{R}'^{-1} = \mathcal{R}'\mathcal{R}\boldsymbol{x}\mathcal{R}^{-1}\mathcal{R}'^{-1} = (\mathcal{R}'\mathcal{R})\boldsymbol{x}(\mathcal{R}'\mathcal{R})^{-1}. \tag{6.64}$$

Hence, the composition of the two rotations is specified by the rotor

$$\mathcal{R}'' = \mathcal{R}'\mathcal{R}. \tag{6.65}$$

Namely, the composition of rotations is given by *the geometric product of the respectiv rotors*.

6.7.2 Representation by surface element

If we want to describe a rotation of angle θ made by the vectors $\boldsymbol{a}$ and $\boldsymbol{b}$, we compute th vector $\boldsymbol{c}$ that bisects that angle and let $\mathcal{R} = \boldsymbol{ca}$. For this, we normalize $\boldsymbol{a}$ and $\boldsymbol{b}$ to uni vectors and compute $\boldsymbol{c}$ by the following equation (Fig. 6.3):

$$\boldsymbol{c} = \frac{(\boldsymbol{a}+\boldsymbol{b})/2}{\|(\boldsymbol{a}+\boldsymbol{b})/2\|} = \frac{\boldsymbol{a}+\boldsymbol{b}}{\sqrt{2(1+\langle \boldsymbol{a}, \boldsymbol{b}\rangle)}}. \tag{6.66}$$

Hence, the rotor $\mathcal{R} = \boldsymbol{ca}$ is given by

$$\mathcal{R} = \frac{1+\boldsymbol{ba}}{\sqrt{2(1+\langle \boldsymbol{a}, \boldsymbol{b}\rangle)}}. \tag{6.67}$$

However, this formula cannot be used if $\boldsymbol{a}$ and $\boldsymbol{b}$ are exactly in opposite directions. Also numerical computation becomes unstable as the angle they make approaches π. So we seel an expression that directly involves the angle of rotation.

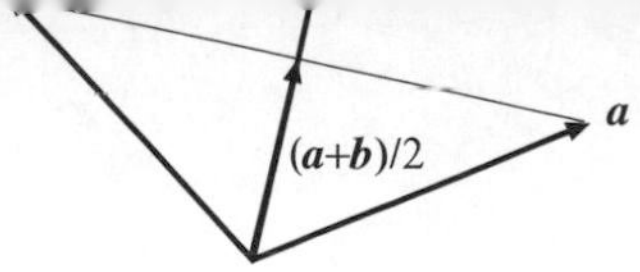

FIGURE 6.3 The unit vector $\boldsymbol{c}$ bisecting the angle between unit vectors $\boldsymbol{a}$ and $\boldsymbol{b}$.

According to the Grassmann algebra, the plane spanned by vectors $\boldsymbol{a}$ and $\boldsymbol{b}$ is specified by bivector $\boldsymbol{a} \wedge \boldsymbol{b}$. The side on which the rotation of $\boldsymbol{a}$ toward $\boldsymbol{b}$ is counterclockwise is regarded as the "front." If $\boldsymbol{a}$ and $\boldsymbol{b}$ make angle θ, the magnitude of $\boldsymbol{a} \wedge \boldsymbol{b}$ is $\|\boldsymbol{a}\|\|\boldsymbol{b}\| \sin\theta$, where θ is the oriented angle from $\boldsymbol{a}$ to $\boldsymbol{b}$. If we let

$$\mathcal{I} = \frac{\boldsymbol{a} \wedge \boldsymbol{b}}{\|\boldsymbol{a}\|\|\boldsymbol{b}\| \sin\theta}, \tag{6.68}$$

this is a unit (i.e., of norm 1) bivector. We call it the *surface element* of this plane. It specifies the orientation of this plane and is independent of the vectors $\boldsymbol{a}$ and $\boldsymbol{b}$. In other words, different vectors $\boldsymbol{a}'$ and $\boldsymbol{b}'$ on this plane define the same surface element $\mathcal{I}$ as long as they have the same relative orientation.

Let $\boldsymbol{a}$ and $\boldsymbol{b}$ be unit vectors that make angle $\Omega/2$. Then, Eq. (6.63) is written as

$$\mathcal{R} = \boldsymbol{ba} = \langle \boldsymbol{b}, \boldsymbol{a} \rangle + \boldsymbol{b} \wedge \boldsymbol{a} = \langle \boldsymbol{b}, \boldsymbol{a} \rangle - \boldsymbol{a} \wedge \boldsymbol{b} = \cos\frac{\Omega}{2} - \sin\frac{\Omega}{2}\frac{\boldsymbol{a} \wedge \boldsymbol{b}}{\sin\Omega/2} = \cos\frac{\Omega}{2} - \mathcal{I}\sin\frac{\Omega}{2}, \tag{6.69}$$

where the surface element $\mathcal{I}$ is defined by Eq. (6.68) (we let $\|\boldsymbol{a}\| = \|\boldsymbol{b}\| = 1$ and $\theta = \Omega/2$). The inverse $\mathcal{R}^{-1}$ is

$$\mathcal{R}^{-1} = (\boldsymbol{ba})^{-1} = \boldsymbol{ab} = \langle \boldsymbol{a}, \boldsymbol{b} \rangle + \boldsymbol{a} \wedge \boldsymbol{b} = \cos\frac{\Omega}{2} + \sin\frac{\Omega}{2}\frac{\boldsymbol{a} \wedge \boldsymbol{b}}{\sin\Omega/2} = \cos\frac{\Omega}{2} + \mathcal{I}\sin\frac{\Omega}{2}, \tag{6.70}$$

which is of course the sign reversal of Ω in Eq. (6.69).

6.7.3 Exponential expression of rotors

The important property of the surface element $\mathcal{I}$ of Eq. (6.68) is that

$$\mathcal{I}^2 = -1. \tag{6.71}$$

In other words, *the surface element $\mathcal{I}$ plays the same role as the imaginary unit i*. This can be shown as follows. From Eq. (6.29), the bivector $\boldsymbol{a} \wedge \boldsymbol{b}$ $(= -\boldsymbol{b} \wedge \boldsymbol{a})$ can be expressed in two ways:

$$\boldsymbol{a} \wedge \boldsymbol{b} = \boldsymbol{ab} - \langle \boldsymbol{a}, \boldsymbol{b} \rangle, \qquad \boldsymbol{a} \wedge \boldsymbol{b} = -\boldsymbol{b} \wedge \boldsymbol{a} = -\boldsymbol{ba} + \langle \boldsymbol{b}, \boldsymbol{a} \rangle. \tag{6.72}$$

Hence,

$$\begin{aligned}
\mathcal{I}^2 &= \frac{(\boldsymbol{ab} - \langle \boldsymbol{a}, \boldsymbol{b} \rangle)(\langle \boldsymbol{a}, \boldsymbol{b} \rangle - \boldsymbol{ba})}{\|\boldsymbol{a}\|^2\|\boldsymbol{b}\|^2 \sin^2\theta} = \frac{\langle \boldsymbol{a}, \boldsymbol{b} \rangle\boldsymbol{ab} - \boldsymbol{abba} - \langle \boldsymbol{a}, \boldsymbol{b} \rangle^2 + \langle \boldsymbol{a}, \boldsymbol{b} \rangle\boldsymbol{ba}}{\|\boldsymbol{a}\|^2\|\boldsymbol{b}\|^2 \sin^2\theta} \\
&= \frac{\langle \boldsymbol{a}, \boldsymbol{b} \rangle(\boldsymbol{ab} + \boldsymbol{ba}) - \boldsymbol{a}\|\boldsymbol{b}\|^2\boldsymbol{a} - \langle \boldsymbol{a}, \boldsymbol{b} \rangle^2}{\|\boldsymbol{a}\|^2\|\boldsymbol{b}\|^2 \sin^2\theta} = \frac{2\langle \boldsymbol{a}, \boldsymbol{b} \rangle^2 - \|\boldsymbol{a}\|^2\|\boldsymbol{b}\|^2 - \langle \boldsymbol{a}, \boldsymbol{b} \rangle^2}{\|\boldsymbol{a}\|^2\|\boldsymbol{b}\|^2 \sin^2\theta} \\
&= -\frac{\|\boldsymbol{a}\|^2\|\boldsymbol{b}\|^2 - \langle \boldsymbol{a}, \boldsymbol{b} \rangle^2}{\|\boldsymbol{a}\|^2\|\boldsymbol{b}\|^2 \sin^2\theta} = -\frac{\|\boldsymbol{a}\|^2\|\boldsymbol{b}\|^2(1 - \cos^2\theta)}{\|\boldsymbol{a}\|^2\|\boldsymbol{b}\|^2 \sin^2\theta} = -1.
\end{aligned} \tag{6.73}$$

$$\begin{aligned}\mathcal{R}\mathcal{R}^{-1} &= \Big(\cos\frac{\Omega}{2} - \mathcal{I}\sin\frac{\Omega}{2}\Big)\Big(\cos\frac{\Omega}{2} + \mathcal{I}\sin\frac{\Omega}{2}\Big)\\ &= \cos^2\frac{\Omega}{2} + \mathcal{I}\cos\frac{\Omega}{2}\sin\frac{\Omega}{2} - \mathcal{I}\sin\frac{\Omega}{2}\cos\frac{\Omega}{2} - \mathcal{I}^2\sin^2\frac{\Omega}{2} = \cos^2\frac{\Omega}{2} + \sin^2\frac{\Omega}{2} = 1.\end{aligned} \tag{6.74}$$

From Eq. (6.71), the rotor $\mathcal{R}$ of Eq. (6.69) can be written as

$$\mathcal{R} = \exp(-\frac{\Omega}{2}\mathcal{I}), \tag{6.75}$$

where the exponential function "exp" is defined via the Taylor expansion (↪ Eq. (4.38) in Chapter 4). In fact, noting that $\mathcal{I}^2 = -1$, we see that

$$\begin{aligned}\exp(-\frac{\Omega}{2}\mathcal{I}) &= 1 + (-\frac{\Omega}{2}\mathcal{I}) + \frac{1}{2!}(-\frac{\Omega}{2}\mathcal{I})^2 + \frac{1}{3!}(-\frac{\Omega}{2}\mathcal{I})^3 + \cdots\\ &= \Big(1 - \frac{1}{2!}(\frac{\Omega}{2})^2 + \frac{1}{4!}(\frac{\Omega}{2})^4 - \cdots\Big) - \mathcal{I}\Big(\frac{\Omega}{2} - \frac{1}{3!}(\frac{\Omega}{2})^3 + \cdots\Big) = \cos\frac{\Omega}{2} - \mathcal{I}\sin\frac{\Omega}{2}.\end{aligned} \tag{6.76}$$

6.8 VERSORS

Consider the product of k vectors $\boldsymbol{v}_1$, $\boldsymbol{v}_2$, ..., $\boldsymbol{v}_k$

$$\mathcal{V} = \boldsymbol{v}_k\boldsymbol{v}_{k-1}\cdots\boldsymbol{v}_1. \tag{6.77}$$

Its inverse is $\mathcal{V}^{-1} = \boldsymbol{v}_1^{-1}\boldsymbol{v}_2^{-1}\cdots\boldsymbol{v}_k^{-1}$. We multiply it by $(-1)^k$ and write

$$\mathcal{V}^\dagger \equiv (-1)^k\mathcal{V}^{-1} = (-1)^k\boldsymbol{v}_1^{-1}\boldsymbol{v}_2^{-1}\cdots\boldsymbol{v}_k^{-1}. \tag{6.78}$$

Consider a transformation of vector $\boldsymbol{x}$ in the form

$$\boldsymbol{x}' = \mathcal{V}\boldsymbol{x}\mathcal{V}^\dagger = (-1)^k\boldsymbol{v}_k\boldsymbol{v}_{k-1}\cdots\boldsymbol{v}_1\boldsymbol{x}\boldsymbol{v}_1^{-1}\boldsymbol{v}_2^{-1}\cdots\boldsymbol{v}_k^{-1}. \tag{6.79}$$

We regard Eq. (6.77) as an operator that transforms a vector to a vector, or the space as whole, in the form of Eq. (6.79) and call $\mathcal{V}$ a *versor* of *grade* k. We call it an *odd versor* if k is odd and an *even versor* if k is even. The transformation by a versor is independent of the norms of the constituent vectors $\boldsymbol{v}_i$, because the norms are canceled by v_i on the right and v_i^{-1} on the left. Hence, only the orientations of individual $\boldsymbol{v}_i$ matter. Transformation of the vector $\boldsymbol{x}'$ in Eq. (6.79) by another versor $\mathcal{V}'$ results in

$$\boldsymbol{x}'' = \mathcal{V}'\boldsymbol{x}'\mathcal{V}'^\dagger = \mathcal{V}'\mathcal{V}\boldsymbol{x}\mathcal{V}^\dagger\mathcal{V}'^\dagger = (\mathcal{V}'\mathcal{V})\boldsymbol{x}(\mathcal{V}'\mathcal{V})^\dagger. \tag{6.80}$$

In other words,

Proposition 6.10 (Composition of versors) *The composition $\mathcal{V}''$ of two versors $\mathcal{V}$ and $\mathcal{V}'$ is given by their geometric product:*

$$\mathcal{V}'' = \mathcal{V}'\mathcal{V}. \tag{6.81}$$

1 are reflectors, while versors of grade 2 are rotors. Evidently, Eq. (6.79) defines a linear mapping, and the norm is unchanged. A linear mapping that preserves the norm is called an *orthogonal transformation*, which is known to be either a rotation or a composition of reflections and rotations. Hence, we conclude that

Proposition 6.11 (Orthogonal transformation) *An orthogonal transformation of the space is specified by a versor. An even versor defines a rotation, and an odd versor defines a composition of rotations and reflections.*

The important property of a versor is that it acts not only on vectors but also on subspaces in the same way. Consider a plane defined by bivector $\boldsymbol{a} \wedge \boldsymbol{b}$, for example. If we rotate this plane around some axis by some angle, the resulting plane is spanned by the vectors $\boldsymbol{a}'$ and $\boldsymbol{b}'$ that are individually rotated. Hence, if the rotation of vectors is specified by rotor $\mathcal{R}$, the rotation of the plane is specified by

$$\begin{aligned}
\boldsymbol{a}' \wedge \boldsymbol{b}' &= (\mathcal{R}\boldsymbol{a}\mathcal{R}^\dagger) \wedge (\mathcal{R}\boldsymbol{b}\mathcal{R}^\dagger) = (\mathcal{R}\boldsymbol{a}\mathcal{R}^{-1}) \wedge (\mathcal{R}\boldsymbol{b}\mathcal{R}^{-1}) \\
&= \frac{(\mathcal{R}\boldsymbol{a}\mathcal{R}^{-1})(\mathcal{R}\boldsymbol{b}\mathcal{R}^{-1}) - (\mathcal{R}\boldsymbol{b}\mathcal{R}^{-1})(\mathcal{R}\boldsymbol{a}\mathcal{R}^{-1})}{2} = \frac{\mathcal{R}\boldsymbol{a}\boldsymbol{b}\mathcal{R}^{-1} - \mathcal{R}\boldsymbol{b}\boldsymbol{a}\mathcal{R}^{-1}}{2} \\
&= \mathcal{R}\Big(\frac{\boldsymbol{a}\boldsymbol{b} - \boldsymbol{b}\boldsymbol{a}}{2}\Big)\mathcal{R}^{-1} = \mathcal{R}(\boldsymbol{a} \wedge \boldsymbol{b})\mathcal{R}^\dagger,
\end{aligned} \tag{6.82}$$

since the rotor $\mathcal{R}$ is an even versor and hence $\mathcal{R}^\dagger = \mathcal{R}^{-1}$. The same holds for trivector $\boldsymbol{a} \wedge \boldsymbol{b} \wedge \boldsymbol{c}$.

Consider an odd versor, e.g., a reflector with respect to a plane with unit surface normal $\boldsymbol{n}$. If we apply this to a plane $\boldsymbol{a} \wedge \boldsymbol{b}$, we obtain a plane spanned by the vectors $\boldsymbol{a}'$ and $\boldsymbol{b}'$ that are individually reflected *with the surface orientation reversed*. Hence, the reflected plane is given by

$$\begin{aligned}
-\boldsymbol{a}' \wedge \boldsymbol{b}' &= -(\boldsymbol{n}\boldsymbol{a}\boldsymbol{n}^\dagger) \wedge (\boldsymbol{n}\boldsymbol{b}\boldsymbol{n}^\dagger) = -(\boldsymbol{n}\boldsymbol{a}\boldsymbol{n}^{-1}) \wedge (\boldsymbol{n}\boldsymbol{b}\boldsymbol{n}^{-1}) \\
&= -\frac{(\boldsymbol{n}\boldsymbol{a}\boldsymbol{n}^{-1})(\boldsymbol{n}\boldsymbol{b}\boldsymbol{n}^{-1}) - (\boldsymbol{n}\boldsymbol{b}\boldsymbol{n}^{-1})(\boldsymbol{n}\boldsymbol{a}\boldsymbol{n}^{-1})}{2} = -\frac{\boldsymbol{n}\boldsymbol{a}\boldsymbol{b}\boldsymbol{n}^{-1} - \boldsymbol{n}\boldsymbol{b}\boldsymbol{a}\boldsymbol{n}^{-1}}{2} \\
&= -\boldsymbol{n}\Big(\frac{\boldsymbol{a}\boldsymbol{b} - \boldsymbol{b}\boldsymbol{a}}{2}\Big)\boldsymbol{n}^{-1} = \boldsymbol{n}(\boldsymbol{a} \wedge \boldsymbol{b})\boldsymbol{n}^\dagger.
\end{aligned} \tag{6.83}$$

On the other hand, the reflection of $\boldsymbol{a} \wedge \boldsymbol{b} \wedge \boldsymbol{c}$ is a space spanned by their individual reflections $\boldsymbol{a}'$, $\boldsymbol{b}'$, and $\boldsymbol{c}'$ of $\boldsymbol{a}$, $\boldsymbol{b}$, and $\boldsymbol{c}$, respectively (note that $\boldsymbol{a}' \wedge \boldsymbol{b}' \wedge \boldsymbol{c}'$ and $\boldsymbol{a} \wedge \boldsymbol{b} \wedge \boldsymbol{c}$ have opposite orientations). Hence, the reflected space is given by

$$\begin{aligned}
\boldsymbol{a}' \wedge \boldsymbol{b}' \wedge \boldsymbol{c}' &= (\boldsymbol{n}\boldsymbol{a}\boldsymbol{n}^\dagger) \wedge (\boldsymbol{n}\boldsymbol{b}\boldsymbol{n}^\dagger) \wedge (\boldsymbol{n}\boldsymbol{c}\boldsymbol{n}^\dagger) = -(\boldsymbol{n}\boldsymbol{a}\boldsymbol{n}^{-1}) \wedge (\boldsymbol{n}\boldsymbol{b}\boldsymbol{n}^{-1}) \wedge (\boldsymbol{n}\boldsymbol{c}\boldsymbol{n}^{-1}) \\
&= -\boldsymbol{n}(\boldsymbol{a} \wedge \boldsymbol{b} \wedge \boldsymbol{c})\boldsymbol{n}^{-1} = \boldsymbol{n}(\boldsymbol{a} \wedge \boldsymbol{b} \wedge \boldsymbol{c})\boldsymbol{n}^\dagger,
\end{aligned} \tag{6.84}$$

where the third expression reduces to the fourth expression by a manipulation similar to that of Eq. (6.83) using Eq. (6.22) ($\hookrightarrow$ Exercise 6.6). Thus, we conclude for both even and odd versors that

Proposition 6.12 (Transformation of subspaces) *The transformation of the space by versor $\mathcal{V}$ induces the transformation of subspaces $\boldsymbol{a}$, $\boldsymbol{a} \wedge \boldsymbol{b}$, and $\boldsymbol{a} \wedge \boldsymbol{b} \wedge \boldsymbol{c}$ to $\mathcal{V}\boldsymbol{a}\mathcal{V}^\dagger$, $\mathcal{V}(\boldsymbol{a} \wedge \boldsymbol{b})\mathcal{V}^\dagger$, and $\mathcal{V}(\boldsymbol{a} \wedge \boldsymbol{b} \wedge \boldsymbol{c})\mathcal{V}^\dagger$, respectively.*

mathematician, who defined a general algebra that contains Hamilton's quaternion alg-bra (and hence complex numbers) and Grassmann's outer product algebra. In contrast, a already pointed out in previous sections, Gibbs simplified the Hamilton algebra and th Grassmann algebra to establish today's vector calculus. Although Gibbs' vector calculus limited to 3D only, it is sufficient to describe almost all problems of physics and engineerin so it has become a fundamental tool for geometry today. Because of the huge success o Gibbs' vector calculus, the Clifford algebra has been almost forgotten except among som mathematicians. It was *David Orlon Hestenes* (1933–), an American physicist, who cast new light on the Clifford algebra, which he called *geometric algebra*, and actively advocate its application to physics and engineering. This has had a big impact on such engineer-ing domains as control theory, e.g., robotic arm control, computer graphics, e.g., geometr modeling and rendering, including ray-tracing, and computer vision, e.g., description an analysis of camera imaging geometry.

In contrast to Gibbs' vector calculus, however, the Clifford algebra appears to be difficu to understand and inaccessible to many. This is mainly due to the number of operation involved: the inner product, the outer product, the contraction, the geometric product, an their combination lead to a multitude of formulas, which are, although smart in appearanc difficult to remember or to invoke intuitive meanings. As a result, textbooks of geometri algebra would look like a list of formulas.

To overcome this problem, Perwass [17] offers a software tool called `CLUCalc` and Dor et al. [6] offer their tool called `GAViewer`. Bayro-Corrochano [3] lists URLs of various too currently available on the Web, including his own. Users only need to input required dat and specify the geometric relationship that they want to compute, and these tools execut them inside the computer according to the operation rules of the Clifford algebra.

Note that the term "algebra" has two meanings. One is the study of operations o symbols or the name of that domain of mathematics, e.g., linear algebra. The other is set of elements that is closed under sums, scalar multiples, and products, e.g., commutativ algebra.

The "surface element" $\mathcal{I}$ introduced in Sec. 6.7.2 is simply the (2D) volume element o that plane when regarded as an entire space. As mentioned in the supplement to Chapter 5 the volume element is called the "pseudoscalar" by many authors [2, 3, 4, 5, 12, 16]. The also calle this $\mathcal{I}$ the "pseudoscalar," too, and make distinctions, whenever necessary, b saying "the pseudoscalar of the space" and "the pseudoscalar of the plane." The term "versor" in Sec. 6.8 was introduced by Hestenes and Sobczyk [12]. The dagger notation o Eq. (6.79) is this book's own usage.

6.10 EXERCISES

6.1. Show that Eqs. (6.5) and (6.6) are satisfied if i, j, and k are defined by Eq. (6.19).

6.2. Derive Eq. (6.38) using the definition of the outer product in Eq. (6.20) and th expression of the inner product in Eq. (6.28).

6.3. Show that the projection $\boldsymbol{x}_{\parallel}$, the rejection $\boldsymbol{x}_{\perp}$, and the reflection $\boldsymbol{x}_{\top}$ of vector $\boldsymbol{x}$ wit respect to plane $\boldsymbol{a} \wedge \boldsymbol{b}$ are given by

$$\boldsymbol{x}_{\parallel} = (\boldsymbol{x} \cdot \boldsymbol{a} \wedge \boldsymbol{b})(\boldsymbol{a} \wedge \boldsymbol{b})^{-1}, \qquad \boldsymbol{x}_{\perp} = \boldsymbol{x} \wedge \boldsymbol{a} \wedge \boldsymbol{b}(\boldsymbol{a} \wedge \boldsymbol{b})^{-1}, \qquad \boldsymbol{x}_{\top} = -(\boldsymbol{a} \wedge \boldsymbol{b})\boldsymbol{x}(\boldsymbol{a} \wedge \boldsymbol{b})^{-1}$$

Let $\boldsymbol{u}'$ and $\boldsymbol{v}'$ be the vectors obtained by applying the rotor of Eq. (6.69) to $\boldsymbol{u}$ and $\boldsymbol{v}$, respectively. Show that

$$\boldsymbol{u}' = \boldsymbol{u}\cos\Omega + \boldsymbol{v}\sin\Omega, \qquad \boldsymbol{v}' = -\boldsymbol{u}\sin\Omega + \boldsymbol{v}\cos\Omega,$$

when the rotation of $\boldsymbol{u}$ toward $\boldsymbol{v}$ has the same sense as specified by the surface element $\mathcal{I}$.

6.6. Show that the third expression in Eq. (6.84) reduces to the fourth expression by using Eq. (6.22).

Homogeneous Space and Grassmann–Cayley Algebra

In Chapter 5, we considered subspaces, i.e., lines and planes passing through the origin, as well as the origin itself and the entire space. In this chapter, we consider points not necessarily at the origin and lines and planes not necessarily passing through the origin. We first show that points, lines, and planes in 3D can be regarded as subspaces in 4D by adding an extra dimension. This enables us to deal with them by the Grassmann algebra in that 4D space. There, a position in 3D and a direction in 3D are represented differently; the latter is identified with a "point at infinity." There exist duality relations among points, lines, and planes, and exploiting the duality, we can describe the "join" (the line passing through two points and the plane passing through a point and a line or through three points) and the "meet" (the intersection of a line with a plane and of two or three planes) in a systematic manner.

7.1 HOMOGENEOUS SPACE

So far, we have considered a space generated by symbols e_1, e_2, and e_3, which we identify with the unit vectors along the x-, y-, and z-axes, respectively, of the 3D xyz space. We now introduce a new symbol e_0, which we interpret to be the unit vector along an axis orthogonal to the 3D xyz space and consider the formal sum

$$x_0e_0 + x_1e_1 + x_2e_2 + x_3e_3. \tag{7.1}$$

The 4D space consisting of all elements in this form is called the *homogeneous space*. We identify a point (x, y, z) in 3D with the element

$$p = e_0 + xe_1 + ye_2 + ze_3 \tag{7.2}$$

in this space; we call it simply "point p." Since the origin $(0, 0, 0)$ in 3D corresponds to point $p = e_0$, we identify the symbol e_0 with the origin in 3D and hereafter call it simply "the origin e_0." Note that this is different from the origin O of this 4D homogeneous space corresponding to Eq. (7.1) with $x_0 = x_1 = x_2 = x_3 = 0$. In other words, the origin e_0 of the 3D xyz space is not the origin O of the 4D homogeneous space.

Now, we assume the "homogeneity" of this space in the sense that, for an arbitrary $\alpha \neq 0$, the element

$$p' = \alpha e_0 + \alpha xe_1 + \alpha ye_2 + \alpha ze_3 \tag{7.3}$$

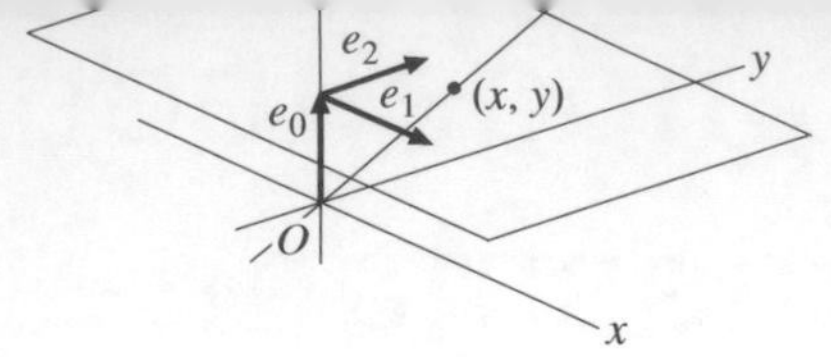

FIGURE 7.1 Interpretation of the 4D homogeneous space and the 3D xyz space. The symbols e e_1, e_2, and e_3 are thought of as the unit vectors along the four coordinate axes. Here, we omit th z-axes for the sake of illustrating the 4D space three dimensionally.

is regarded as representing the same point as the point p given by Eq. (7.2). The ter "homogeneous space" originates from this. It follows from this homogeneity that, for x $\neq 0$, Eq. (7.1) represents the point $(x_1/x_0, x_2/x_0, x_3/x_0)$ in 3D. We adopt the followin interpretation (Fig. 7.1). For an arbitrary real number α, Eq. (7.3) is interpreted to b representing a line in 4D passing through the origin O, i.e., a 1D subspace. However, we ar unable to perceive this 4D space entirely; all we can see is the 3D "cross section" of th space that passes through the origin e_0 and is parallel to the x-, y-, and z-axes, which w identify with the 3D xyz space with origin at e_0. Then, the point p given by Eq. (7.2) regarded as the "intersection" of this xyz space with the line given by Eq. (7.3).

7.2 POINTS AT INFINITY

If a point (x, y, z) in 3D is moved away from the origin e_0 α ($\neq 0$) times, it comes t $(\alpha x, \alpha y, \alpha z)$. This point is represented in the 4D homogeneous space by

$$p = e_0 + \alpha x e_1 + \alpha y e_2 + \alpha z e_3, \tag{7.4}$$

which is also represented by

$$p' = \frac{e_0}{\alpha} + x e_1 + y e_2 + z e_3, \tag{7.5}$$

due to the homogeneity of the space. In the limit of $\alpha \to \infty$, p' approaches

$$q = x e_1 + y e_2 + z e_3, \tag{7.6}$$

which we interpret to be the point infinitely far away along the line starting from the origi e_0 and passing through the point (x, y, z). We regard such a limit as if it is a point an call it a *point at infinity*. Note that Eq. (7.6) is also a limit of $\alpha \to -\infty$ in Eq. (7.5). Th means that if we move along the line passing through the origin e_0 and the point $(x, y, z$ indefinitely in either direction, we reach the same point at infinity. Hence, we are obliged t regard a line in 3D as if it is a closed circle of infinite radius, meeting at a point at infinit

In the following, we identify the position vector $\boldsymbol{x} = x e_1 + y e_2 + z e_3$ in 3D with th point p of Eq. (7.2) in the 4D homogeneous space and write simply $p = e_0 + \boldsymbol{x}$. On the othe hand, the direction vector $\boldsymbol{u} = u_1 e_1 + u_2 e_2 + u_3 e_3$ in 3D is regarded as an element of the 4D homogeneous space as it is. Due to the homogeneity of the space, it has the same meaning multiplied by a nonzero number. Hence, only the direction of $\boldsymbol{u}$ matters; its magnitude doe not have a meaning. At the same time, it is regarded as a point at infinity, as stated abov

as stated at the beginning of Chapter 2. In the 4D homogeneous space, in contrast, they are treated differently: a position vector is regarded as a point $p = e_0 + \boldsymbol{x}$, while a direction vector is identified with a point $\boldsymbol{u}$ at infinity.

Traditional World 7.1 (Projective geometry) Traditional *projective geometry* represents a point (x, y, z) in 3D by four numbers (X, Y, Z, W), called the *homogeneous coordinates*. If $W \neq 0$, a point with homogeneous coordinates (X, Y, Z, W) is identified with a point $(X/W, Y/W, Z/W)$ in 3D, and if $W = 0$, it is regarded as the point at infinity in the direction of (X, Y, Z). Hence, only the ratio of the homogeneous coordinates (X, Y, Z, W) has a meaning; it represents the same point as (cX, cY, cZ, cW) for any $c \neq 0$. As opposed to the homogeneous coordinates, the usual coordinates (x, y, z) in 3D are called *inhomogeneous coordinates*. The set of all points specified by homogeneous coordinates, namely, the 3D space augmented by points at infinity, is called the 3D *projective space*.

The same holds in other dimensions. For example, we can express a point in 2D by three homogeneous coordinates (X, Y, Z): if $Z \neq 0$, it is identified with a point $(X/Z, Y/Z)$, and if $Z = 0$, it is regarded as the point at infinity in the direction of (X, Y). The 2D plane augmented by points at infinity is called the 2D projective space. One of the merits of using homogeneous coordinates for geometric computation is that most of the transformations we frequently encounter can be performed as *linear* operations in homogeneous coordinates. For example, a similarity in 2D is written as

$$\begin{pmatrix} X' \\ Y' \\ Z' \end{pmatrix} \simeq \left(\begin{array}{c|c} s\boldsymbol{R} & \boldsymbol{t} \\ \hline 0 \;\; 0 & 1 \end{array}\right) \begin{pmatrix} X \\ Y \\ Z \end{pmatrix}, \tag{7.7}$$

where $\simeq$ denotes equality up to a nonzero scalar multiplier. Here, $\boldsymbol{R}$ is a rotation matrix, $\boldsymbol{t}$ a translation vector, and s a scale factor of expansion/contraction. Equation (7.7) describes a rigid motion for $s = 1$ and a pure rotation for $\boldsymbol{t} = \boldsymbol{0}$ in addition. It describes a general affine transformation if $s\boldsymbol{R}$ is replaced by a general nonsingular matrix $\boldsymbol{A}$. Equation (7.7) can be further generalized into the form

$$\begin{pmatrix} X' \\ Y' \\ Z' \end{pmatrix} \simeq \boldsymbol{H} \begin{pmatrix} X \\ Y \\ Z \end{pmatrix}, \tag{7.8}$$

for a general nonsingular matrix $\boldsymbol{H}$. This defines a mapping called (2D or planar) *projective transformation* or *homography*. This is the most general transformation that maps lines to lines. For example, a square is mapped to a general quadrilateral. The set of all projective transformations forms a group of transformations, which includes, as its subgroups, affine transformations, similarities, rigid motions, rotations, translations, scale changes, and the identity.

Traditional World 7.2 (Perspective projection) The geometric meaning of homogeneous coordinates is best understood if we consider the *perspective projection* of a 3D scene onto a 2D image. The camera imaging geometry can be idealized as shown in Fig. 7.2, where the lens center is at the origin O and the *optical axis* (the axis of symmetry passing through the lens center) is along the Z-axis. We identify a plane that passes through a point on

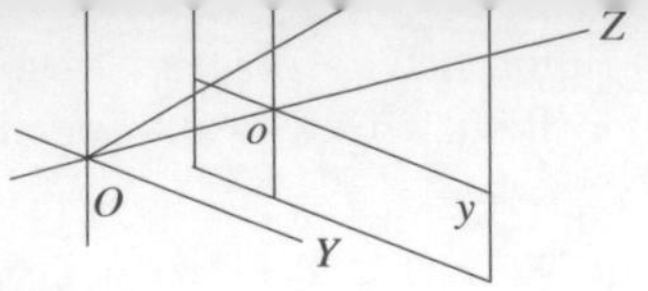

FIGURE 7.2 Perspective projection. A point (X, Y, Z) in the scene is mapped onto a point $(x,$ on the image plane.

the Z-axis and is parallel to the XY plane as the *image plane.* A point (X, Y, Z) in th scene is mapped to the intersection of the image plane with the line connecting this poi and the origin O. In a real camera, the image plane is behind the lens, but the geometr relationship is the same if it is placed in front of the lens. In such a modeling, the origin O called the *viewpoint* or the *center of projection* and the line passing through the viewpoi O and the point (X, Y, Z) in the scene is called the *line of sight* or simply the *ray*. Th distance between the viewpoint O and the image plane is often called the *focal length*. If w take this as the unit of length on the image plane, the relationship between the scene poi (X, Y, Z) and the image point (x, y) is written in the following form:

$$x = \frac{X}{Z}, \qquad y = \frac{Y}{Z}. \qquad (7.9$$

The image origin o, i.e., the origin of the image coordinate system, is at the intersectio of the image plane with the optical axis and is called the *principal point*. Thus, the imag origin o and the origin O of the outside scene are different points.

If we move the scene point (X, Y, Z) along its line of sight, its projection (x, y) remai the same. Thus, the 3D coordinates (X, Y, Z) play the role of the homogeneous coordinat of the image point (x, y). As the point (x, y) moves infinitely far away from the image origi o within the image plane, its line of sight becomes more and more parallel to the imag plane, and the Z-coordinate of points on the line of sight approaches 0. Hence, the scen point $(X, Y, 0)$ corresponds to the point at infinity in the direction of (x, y) on the imag plane.

7.3 PLÜCKER COORDINATES OF LINES

We now show the following facts about lines in 3D:

- The line L passing through two points p_1 and p_2 is represented by a bivector in th 4D homogeneous space in the form $L = p_1 \wedge p_2$.
- The equation of line L has the form $p \wedge L = 0$.
- A line is specified by its Plücker coordinates $\boldsymbol{m}$ and $\boldsymbol{n}$.
- The line that passes through a given point and extends in a given direction is inter preted to be the line connecting that point and the point at infinity in that directio
- The line passing through two given points at infinity is characterized only by it orientation and interpreted to be a line located infinitely far away.

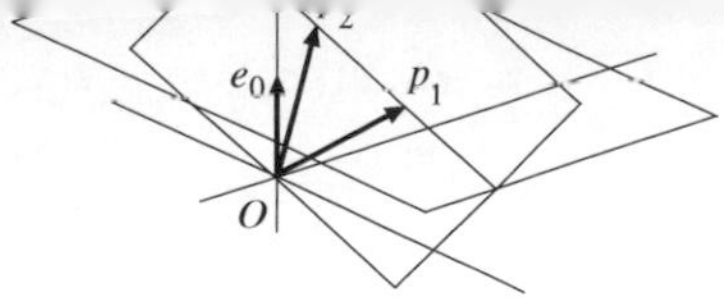

FIGURE 7.3 The line L passing through points p_1 and p_2 is regarded as the intersection of the plane $p_1 \wedge p_2$ spanned by 4D vectors p_1 and p_2 in the 4D homogeneous space with the 3D space passing through the origin e_0.

7.3.1 Representation of a line

Consider a line L passing through two pints at $\boldsymbol{x}_1$ and $\boldsymbol{x}_2$ (the position vectors in 3D). In the 4D homogeneous space, these points are represented by $p_1 = e_0 + \boldsymbol{x}_1$ and $p_2 = e_0 + \boldsymbol{x}_2$, and the line L is represented by their bivector

$$L = p_1 \wedge p_2. \tag{7.10}$$

This is based on the following interpretation. From the definition of the outer product, the bivector $p_1 \wedge p_2$ defines a plane, i.e., 2D subspace, spanned by the 4D vectors p_1 and p_2. However, we are unable to perceive the entire 4D space; all we can see is its 3D cross section containing the origin e_0, with which the 2D subspace $p_1 \wedge p_2$ intersects along the line L (Fig. 7.3). Since L is a bivector, it has the following expression with respect to the basis $\{e_0, e_1, e_2, e_3\}$:

$$L = m_1 e_0 \wedge e_1 + m_2 e_0 \wedge e_2 + m_3 e_0 \wedge e_3 + n_1 e_2 \wedge e_3 + n_2 e_3 \wedge e_1 + n_3 e_1 \wedge e_2. \tag{7.11}$$

The coefficients m_i and n_i, $i = 1, 2, 3$, are called the *Plücker coordinates* of this line. Due to the homogeneity of the space, p_1 and p_2 represent the same points if they are multiplied by any nonzero number. It follows that the bivector L also represents the same line if it is multiplied by any nonzero number. Hence, only the ratios among m_i and n_i have a geometric meaning. We express this fact by saying that *the Plücker coordinates are homogeneous coordinates of a line*.

Let $\boldsymbol{m} = m_1 e_1 + m_2 e_2 + m_3 e_3$ and $\boldsymbol{n} = n_1 e_1 + n_2 e_2 + n_3 e_3$ by regarding the Plücker coordinates m_i and n_i as vectors. Then, Eq. (7.11) is written as

$$L = e_0 \wedge \boldsymbol{m} - \boldsymbol{n}^*, \tag{7.12}$$

where we let $\boldsymbol{n}^* = -\boldsymbol{n} \cdot I$, a bivector dual to $\boldsymbol{n}$ in 3D with $I = e_1 \wedge e_2 \wedge e_3$ the volume element in 3D (↪ Eq. (5.55) in Chapter 5). With respect to the basis, $\boldsymbol{n}^*$ has the following expression (↪ Eq. (5.67) in Chapter 5):

$$\boldsymbol{n}^* = -n_1 e_2 \wedge e_3 - n_2 e_3 \wedge e_1 - n_3 e_1 \wedge e_2. \tag{7.13}$$

7.3.2 Equation of a line

A point $p = e_0 + \boldsymbol{x}$ is on the line $L = p_1 \wedge p_2$ if and only if $p \wedge p_1 \wedge p_2 = 0$, namely,

$$p \wedge L = 0. \tag{7.14}$$

$$= -e_0 \wedge (\boldsymbol{x} \wedge \boldsymbol{m} + \boldsymbol{n}^*) + \boldsymbol{x} \wedge \boldsymbol{n}^*, \quad (7.1$$

so the equation $p \wedge L = 0$ is equivalent to

$$\boldsymbol{x} \wedge \boldsymbol{m} = -\boldsymbol{n}^*, \qquad \boldsymbol{x} \wedge \boldsymbol{n}^* = 0. \quad (7.1$$

The dual of the former is $(\boldsymbol{x} \wedge \boldsymbol{m})^* = \boldsymbol{n}$, but the expression $(\boldsymbol{x} \wedge \boldsymbol{m})^*$ is nothing but th vector product $\boldsymbol{x} \times \boldsymbol{m}$ (↪ Eq. (5.73) in Chapter 5). The latter is equivalent to $\boldsymbol{x} \cdot \boldsymbol{n} = 0$ (↪ Eq. (5.79) in Chapter 5). Hence, in terms of the notation of Chapter 2, we can equivalent write Eq. (7.16) as follows (↪ Eqs. (2.60) and (2.65) in Chapter 2):

$$\boldsymbol{x} \times \boldsymbol{m} = \boldsymbol{n}, \qquad \langle \boldsymbol{x}, \boldsymbol{n} \rangle = 0. \quad (7.1$$

This implies that $\boldsymbol{m}$ is the direction vector of the line L and $\boldsymbol{n}$ is the surface normal to th supporting plane of the line L.

7.3.3 Computation of a line

We now consider how to compute the Plücker coordinates $\boldsymbol{m}$ and $\boldsymbol{n}$ of the line L passin through points p_1 and p_2. The bivector expression of L is

$$\begin{aligned} L = p_1 \wedge p_2 &= (e_0 + \boldsymbol{x}_1) \wedge (e_0 + \boldsymbol{x}_2) = e_0 \wedge \boldsymbol{x}_2 + \boldsymbol{x}_1 \wedge e_0 + \boldsymbol{x}_1 \wedge \boldsymbol{x}_2 \\ &= e_0 \wedge (\boldsymbol{x}_2 - \boldsymbol{x}_1) + \boldsymbol{x}_1 \wedge \boldsymbol{x}_2. \end{aligned} \quad (7.1$$

Comparing this with Eq. (7.12), we find that $\boldsymbol{m} = \boldsymbol{x}_2 - \boldsymbol{x}_1$ and $-\boldsymbol{n}^* = \boldsymbol{x}_1 \wedge \boldsymbol{x}_2$. The du of the latter is $\boldsymbol{n} = \boldsymbol{x}_1 \times \boldsymbol{x}_2$. Hence, we obtain the following Plücker coordinates:

$$\boldsymbol{m} = \boldsymbol{x}_2 - \boldsymbol{x}_1, \qquad \boldsymbol{n} = \boldsymbol{x}_1 \times \boldsymbol{x}_2. \quad (7.1$$

Since the Plücker coordinates are homogeneous coordinates, this result is equivalent Eq. (2.68) in Chapter 2.

Next, let L be the line that passes through point $p = e_0 + \boldsymbol{x}$ and has direction $\boldsymbol{u}$. Th line can be regarded as passing through the finite point p and the point $\boldsymbol{u}$ at infinity. Henc can write

$$L = p \wedge \boldsymbol{u} = (e_0 + \boldsymbol{x}) \wedge \boldsymbol{u} = e_0 \wedge \boldsymbol{u} + \boldsymbol{x} \wedge \boldsymbol{u}. \quad (7.2$$

Comparing this with Eq. (7.12), we see that $\boldsymbol{m} = \boldsymbol{u}$ and $-\boldsymbol{n}^* = \boldsymbol{x} \wedge \boldsymbol{u}$. The dual of th latter is $\boldsymbol{n} = \boldsymbol{x} \times \boldsymbol{u}$. Hence, we obtain the following Plücker coordinates:

$$\boldsymbol{m} = \boldsymbol{u}, \qquad \boldsymbol{n} = \boldsymbol{x} \times \boldsymbol{u}. \quad (7.2$$

This coincides with Eq. (7.19) if we let $\boldsymbol{u} = \boldsymbol{x}_2 - \boldsymbol{x}_1$.

Finally, consider the line L_∞ passing through two points $\boldsymbol{u}_1$ and $\boldsymbol{u}_2$ at infinity:

$$L_\infty = \boldsymbol{u}_1 \wedge \boldsymbol{u}_2. \quad (7.2$$

Such a line is called a *line at infinity*. Comparing Eq. (7.22) with Eq. (7.12), we see tha $\boldsymbol{m} = 0$ and $-\boldsymbol{n}^* = \boldsymbol{u}_1 \wedge \boldsymbol{u}_2$. The dual of the latter is $\boldsymbol{n} = \boldsymbol{u}_1 \times \boldsymbol{u}_2$. Hence, the Plücke coordinates are

$$\boldsymbol{m} = 0, \qquad \boldsymbol{n} = \boldsymbol{u}_1 \times \boldsymbol{u}_2. \quad (7.2$$

L_∞ passing through points $\boldsymbol{u}_1$ and $\boldsymbol{u}_2$ at infinity is interpreted to be the "boundary" of the plane spanned by the directions $\boldsymbol{u}_1$ and $\boldsymbol{u}_2$, surrounding it like a "circle" of infinite radius. The above results are summarized as follows:

Proposition 7.1 (Description of lines)

1. *The line passing through two points p_1 and p_2 is represented by $L = p_1 \wedge p_2$, where one or both of p_1 and p_2 can be at infinity.*
2. *A point at infinity can be identified with the direction toward it.*
3. *The equation of the line L has the form $p \wedge L = 0$.*

7.4 PLÜCKER COORDINATES OF PLANES

In this section, we show the following facts about planes in 3D:

- The plane Π passing through three points p_1, p_2, and p_3 is represented by a trivector in the 4D homogeneous space in the form $\Pi = p_1 \wedge p_2 \wedge p_3$.
- The equation of the plane Π has the form $p \wedge \Pi = 0$.
- A plane is specified by its Plücker coordinates $\boldsymbol{n}$ and h.
- The plane that passes through two given points and contains a given direction is interpreted to be the plane connecting those two points and the point at infinity corresponding to that direction.
- The plane that passes through a given point and contains two given directions is interpreted to be the plane connecting that point and the two points at infinity corresponding to those two directions.
- The plane passing through three points at infinity has no direction and is interpreted to be a plane located infinitely far away.

7.4.1 Representation of a plane

Consider a plane Π passing through three points at $\boldsymbol{x}_1$, $\boldsymbol{x}_2$, and $\boldsymbol{x}_3$ (the position vectors in 3D). In the 4D homogeneous space, these points are represented by $p_1 = e_0 + \boldsymbol{x}_1$, $p_2 = e_0 + \boldsymbol{x}_2$, and $p_3 = e_0 + \boldsymbol{x}_3$, and the plane Π is represented by their trivector

$$L = p_1 \wedge p_2 \wedge p_3. \tag{7.24}$$

This is based on the following interpretation. From the definition of the outer product, the trivector $p_1 \wedge p_2 \wedge p_3$ defines a space, i.e., a 3D subspace, spanned by the 4D vectors p_1, p_2, and p_3. However, we are unable to perceive the entire 4D space; all we can see is its 3D cross section containing the origin e_0, with which the 3D subspace $p_1 \wedge p_2 \wedge p_3$ intersects along the plane Π. Since Π is a trivector, it has the following expression with respect to the basis $\{e_0, e_1, e_2, e_3\}$:

$$\Pi = n_1 e_0 \wedge e_2 \wedge e_3 + n_2 e_0 \wedge e_3 \wedge e_1 + n_3 e_0 \wedge e_1 \wedge e_2 + h e_1 \wedge e_2 \wedge e_3. \tag{7.25}$$

by any nonzero number. Hence, only the ratios among n_i and h have a geometric meani
The Plücker coordinates of a plane are *homogeneous coordinates* in this sense.

Let $\boldsymbol{n} = n_1 e_1 + n_2 e_2 + n_3 e_3$ by regarding the Plücker coordinates n_i as a vector. The Eq. (7.25) is written as

$$\Pi = -e_0 \wedge \boldsymbol{n}^* + hI, \tag{7.26}$$

where we let $\boldsymbol{n}^* = -\boldsymbol{n} \cdot I$, a bivector dual to $\boldsymbol{n}$ in 3D with $I = e_1 \wedge e_2 \wedge e_3$ the volun element in 3D.

7.4.2 Equation of a plane

A point $p = e_0 + \boldsymbol{x}$ is on the plane $\Pi = p_1 \wedge p_2 \wedge p_3$ if and only if $p \wedge p_1 \wedge p_2 \wedge p_3 =$ namely,

$$p \wedge \Pi = 0. \tag{7.27}$$

Hence, this can be regarded as the "equation" of the plane Π. From Eq. (7.26), we have

$$\begin{aligned} p \wedge \Pi &= (e_0 + \boldsymbol{x}) \wedge (-e_0 \wedge \boldsymbol{n}^* + hI) = -\boldsymbol{x} \wedge e_0 \wedge \boldsymbol{n}^* + he_0 \wedge I \\ &= e_0 \wedge \boldsymbol{x} \wedge \boldsymbol{n}^* + he_0 \wedge I = e_0 \wedge (\boldsymbol{x} \wedge \boldsymbol{n}^* + hI) = (h - \langle \boldsymbol{n}, \boldsymbol{x} \rangle) e_0 \wedge I, \end{aligned} \tag{7.28}$$

where we have noted that $\boldsymbol{x} \wedge \boldsymbol{n}^* = -\langle \boldsymbol{n}, \boldsymbol{x} \rangle I$ from Eq. (7.13). Hence, the equation $p \wedge \Pi$ 0 is equivalent to

$$\langle \boldsymbol{n}, \boldsymbol{x} \rangle = h. \tag{7.29}$$

As we saw in Chapter 2, the vector $\boldsymbol{n}$ is the surface normal to this plane and $h/\|\boldsymbol{n}\|$ is th distance of this plane from the origin e_0.

7.4.3 Computation of a plane

We now consider how to compute the Plücker coordinates $\boldsymbol{n}$ and h of the plane Π passin through points p_1, p_2, and p_3. The trivector expression of Π is

$$\begin{aligned} \Pi &= p_1 \wedge p_2 \wedge p_3 = (e_0 + \boldsymbol{x}_1) \wedge (e_0 + \boldsymbol{x}_2) \wedge (e_0 + \boldsymbol{x}_3) \\ &= e_0 \wedge \boldsymbol{x}_2 \wedge \boldsymbol{x}_3 + \boldsymbol{x}_1 \wedge e_0 \wedge \boldsymbol{x}_3 + \boldsymbol{x}_1 \wedge \boldsymbol{x}_2 \wedge e_0 + \boldsymbol{x}_1 \wedge \boldsymbol{x}_2 \wedge \boldsymbol{x}_3 \\ &= e_0 \wedge (\boldsymbol{x}_2 \wedge \boldsymbol{x}_3 + \boldsymbol{x}_3 \wedge \boldsymbol{x}_1 + \boldsymbol{x}_1 \wedge \boldsymbol{x}_2) + \boldsymbol{x}_1 \wedge \boldsymbol{x}_2 \wedge \boldsymbol{x}_3. \end{aligned} \tag{7.30}$$

Comparing this with Eq. (7.26), we find that $-\boldsymbol{n}^* = \boldsymbol{x}_2 \wedge \boldsymbol{x}_3 + \boldsymbol{x}_3 \wedge \boldsymbol{x}_1 + \boldsymbol{x}_1 \wedge \boldsymbol{x}_2$ and $hI =$ $\boldsymbol{x}_1 \wedge \boldsymbol{x}_2 \wedge \boldsymbol{x}_3$. Replacing both sides with their duals and recalling $I^* = 1$ and $(\boldsymbol{x}_1 \wedge \boldsymbol{x}_2 \wedge \boldsymbol{x}_3)$ $= |\boldsymbol{x}_1, \boldsymbol{x}_2, \boldsymbol{x}_3|$, we obtain the following Plücker coordinates ($\hookrightarrow$ Eqs. (5.64) and (5.73) o Chapter 5):

$$\boldsymbol{n} = \boldsymbol{x}_2 \times \boldsymbol{x}_3 + \boldsymbol{x}_3 \times \boldsymbol{x}_1 + \boldsymbol{x}_1 \times \boldsymbol{x}_2, \qquad h = |\boldsymbol{x}_1, \boldsymbol{x}_2, \boldsymbol{x}_3|. \tag{7.31}$$

This is equivalent to Eq. (2.53) in Chapter 2 except for scale normalization.

The plane Π passing through points p_1, p_2, and p_3 can be regarded as the plane passin through point p_1 and the line L connecting points p_2 and p_3. If we write p_1 as $p = e_0 +$ and let $L = e_0 \wedge \boldsymbol{m} - \boldsymbol{n}_L^*$, we can write the trivector Π as follows:

$$\begin{aligned} \Pi &= (e_0 + \boldsymbol{p}) \wedge (e_0 \wedge \boldsymbol{m} - \boldsymbol{n}_L^*) = -e_0 \wedge \boldsymbol{n}_L^* + \boldsymbol{p} \wedge e_0 \wedge \boldsymbol{m} - \boldsymbol{p} \wedge \boldsymbol{n}_L^* \\ &= -e_0 \wedge (\boldsymbol{n}_L^* + \boldsymbol{p} \wedge \boldsymbol{m}) - \boldsymbol{p} \wedge \boldsymbol{n}_L^* = -e_0 \wedge (\boldsymbol{n}_L - (\boldsymbol{p} \wedge \boldsymbol{m})^*)^* - \boldsymbol{p} \wedge \boldsymbol{n}_L^* \\ &= -e_0 \wedge (\boldsymbol{n}_L - \boldsymbol{p} \times \boldsymbol{m})^* - \boldsymbol{p} \wedge \boldsymbol{n}_L^* = -e_0 \wedge (\boldsymbol{n}_L - \boldsymbol{p} \times \boldsymbol{m})^* + \langle \boldsymbol{p}, \boldsymbol{n}_L \rangle I, \end{aligned} \tag{7.32}$$

$$\boldsymbol{n} = \boldsymbol{n}_L - \boldsymbol{p} \times \boldsymbol{m}, \qquad h = \langle \boldsymbol{p}, \boldsymbol{n}_L \rangle. \tag{7.33}$$

This coincides with Eq. (2.83) in Chapter 2 up to scale normalization.

Next, let Π be the plane that passes through points $p_1 = e_0 + \boldsymbol{x}_1$ and $p_2 = e_0 + \boldsymbol{x}_2$ at $\boldsymbol{x}_1$ and $\boldsymbol{x}_2$, respectively, and contains a direction vector $\boldsymbol{u}$. This plane can be regarded as passing through the finite points p_1 and p_2 and the point $\boldsymbol{u}$ at infinity. Hence,

$$\begin{aligned}\Pi &= p_1 \wedge p_2 \wedge \boldsymbol{u} = (e_0 + \boldsymbol{x}_1) \wedge (e_0 + \boldsymbol{x}_2) \wedge \boldsymbol{u} = e_0 \wedge \boldsymbol{x}_2 \wedge \boldsymbol{u} + \boldsymbol{x}_1 \wedge e_0 \wedge \boldsymbol{u}\boldsymbol{x}_1 \wedge \boldsymbol{x}_2 \wedge \boldsymbol{u} \\ &= e_0 \wedge (\boldsymbol{x}_2 - \boldsymbol{x}_1) \wedge \boldsymbol{u} + \boldsymbol{x}_1 \wedge \boldsymbol{x}_2 \wedge \boldsymbol{u}.\end{aligned} \tag{7.34}$$

Comparing this with Eq. (7.26), we see that $-\boldsymbol{n}^* = (\boldsymbol{x}_2 - \boldsymbol{x}_1) \wedge \boldsymbol{u}$ and $hI = \boldsymbol{x}_1 \wedge \boldsymbol{x}_2 \wedge \boldsymbol{u}$. Replacing both sides with their duals, we obtain the Plücker coordinates:

$$\boldsymbol{n} = (\boldsymbol{x}_2 - \boldsymbol{x}_1) \times \boldsymbol{u}, \qquad h = |\boldsymbol{x}_1, \boldsymbol{x}_2, \boldsymbol{u}|, \tag{7.35}$$

which is equivalent to the plane in Fig. 2.10 in Chapter 2 if $\boldsymbol{u}$ is replaced by $\boldsymbol{x}_3 - \boldsymbol{x}_1$. If we write the line passing through p_1 and p_2 as $L = e_0 \wedge \boldsymbol{m} - \boldsymbol{n}_L^*$, the above result can be rewritten in the form

$$\begin{aligned}\Pi &= (e_0 \wedge \boldsymbol{m} - \boldsymbol{n}_L^*) \wedge \boldsymbol{u} = e_0 \wedge \boldsymbol{m} \wedge \boldsymbol{u} - \boldsymbol{n}_L^* \wedge \boldsymbol{u} \\ &= -e_0 \wedge (\boldsymbol{m} \wedge \boldsymbol{u})^{**} + \langle \boldsymbol{n}_L, \boldsymbol{u} \rangle I = -e_0 \wedge (\boldsymbol{m} \times \boldsymbol{u})^* + \langle \boldsymbol{n}_L, \boldsymbol{u} \rangle I,\end{aligned} \tag{7.36}$$

where we have used $\boldsymbol{n}_L^* \wedge \boldsymbol{u} = -\langle \boldsymbol{n}_L, \boldsymbol{u} \rangle I$ obtained from Eq. (7.13). Comparing Eq. (7.36) with Eq. (7.26), we obtain the following expression of the Plücker coordinates:

$$\boldsymbol{n} = \boldsymbol{m} \times \boldsymbol{u}, \qquad h = \langle \boldsymbol{n}_L, \boldsymbol{u} \rangle. \tag{7.37}$$

This coincides with the result of Exercise 2.14 in Chapter 2 up to scale normalization.

Now, let Π be the plane that passes through a point $p = e_0 + \boldsymbol{x}$ at $\boldsymbol{x}$ and contains direction vectors $\boldsymbol{u}$ and $\boldsymbol{v}$. This plane can be regarded as passing through the finite point p and the two points $\boldsymbol{u}$ and $\boldsymbol{v}$ at infinity. Hence, we can write

$$\Pi = p \wedge \boldsymbol{u} \wedge \boldsymbol{v} = (e_0 + \boldsymbol{x}) \wedge \boldsymbol{u} \wedge \boldsymbol{v} = e_0 \wedge \boldsymbol{u} \wedge \boldsymbol{v} + \boldsymbol{x} \wedge \boldsymbol{u} \wedge \boldsymbol{v}. \tag{7.38}$$

Comparing this with Eq. (7.26), we see that $-\boldsymbol{n}^* = \boldsymbol{u} \wedge \boldsymbol{v}$ and $hI = \boldsymbol{x} \wedge \boldsymbol{u} \wedge \boldsymbol{v}$. Replacing both sides with their duals, we obtain the following Plücker coordinates:

$$\boldsymbol{n} = \boldsymbol{u} \times \boldsymbol{v}, \qquad h = |\boldsymbol{x}, \boldsymbol{u}, \boldsymbol{v}|. \tag{7.39}$$

This coincides with the result of Exercise 2.15 in Chapter 2 up to scale normalization. In terms of the L_∞ in Eq. (7.22), Eq. (7.38) can also be written as $\Pi = p \wedge L_\infty$.

Finally, consider the plane passing through three points $\boldsymbol{u}$, $\boldsymbol{v}$, and $\boldsymbol{w}$ at infinity:

$$\Pi_\infty = \boldsymbol{u} \wedge \boldsymbol{v} \wedge \boldsymbol{w}. \tag{7.40}$$

Such a plane is called a *plane at infinity*. Comparing Eq. (7.40) with Eq. (7.26), we see that $-\boldsymbol{n}^* = 0$ and $hI = \boldsymbol{u} \wedge \boldsymbol{v} \wedge \boldsymbol{w}$. Replacing these with their duals, we obtain the following Plücker coordinates:

$$\boldsymbol{n} = 0, \qquad h = |\boldsymbol{u}, \boldsymbol{v}, \boldsymbol{w}|. \tag{7.41}$$

not defined. Due to the homogeneity of the Plücker coordinates, only the ratio between
and h has a meaning, so if $\boldsymbol{n} = 0$, the absolute value h is irrelevant as long as it is nonzer
Hence, we can let $h = 1$ and write Eq. (7.40) as

$$\Pi_\infty = I \ \ (= e_1 \wedge e_2 \wedge e_3), \tag{7.4}$$

without losing generality. Thus, the existence of the plane Π_∞ at infinity is *unique* irre spective of the points $\boldsymbol{u}$, $\boldsymbol{v}$, and $\boldsymbol{w}$ at infinity that define Π. We can interpret it as th "boundary" of the entire 3D space, surrounding it like a "sphere" of infinite radius. Th fact that Π_∞ does not contain any finite points is easily seen if we note that no finite $\boldsymbol{x}$ ca satisfy Eq. (7.29) for $\boldsymbol{n} = 0$ and $h \neq 0$. The above results are summarized as follows:

Proposition 7.2 (Description of planes)

1. *The plane passing through three points p_1, p_2, and p_3 is represented by $\Pi = p_1 \wedge p_2 \wedge p_3$ where any or all of p_1, p_2, and p_3 can be at infinity.*
2. *The plane passing through point p and line L is represented by $\Pi = p \wedge L$, where can be a point at infinity and L can be a line at infinity.*
3. *The equation of the plane Π has the form $p \wedge \Pi = 0$.*

7.5 DUAL REPRESENTATION

As pointed out in Chapter 5, a subspace can also be specified by its orthogonal complemen In the 4D homogeneous space, a line is represented by a bivector, i.e., a 2D subspace, s its orthogonal complement is also 2D, which represents a line in 3D. Similarly, a plane represented by a trivector in the 4D homogeneous space, i.e., a 3D subspace, so its orthogona complement is 1D, which represents a point in 3D. In describing such duality relationship the *volume element*

$$I_4 = e_0 \wedge e_1 \wedge e_2 \wedge e_3 \tag{7.43}$$

of the 4D homogeneous space plays a fundamental role. We define the dual of a k-vecto $(\cdots)$, $k = 0, 1, 2, 3, 4$, by

$$(\cdots)^* = (\cdots) \cdot I_4, \tag{7.44}$$

which is a $(4-k)$-vector. Note that we defined the dual in 3D by $(\cdots)^* = -(\cdots) \cdot I$ (
Eq. (5.54) in Chapter 5). In the general nD, the dual is defined by $(-1)^{n(n-1)/2}(\cdots) \cdot I_n$ and Eq. (7.44) is the expression for $n = 4$. From this definition, we obtained the followin results:

Proposition 7.3 (Duality of basis) *The outer products of the basis elements e_0, e_1, e_2 and e_3 have the following dual expressions:*

$$1^* = e_0 \wedge e_1 \wedge e_2 \wedge e_3, \qquad e_0^* = e_1 \wedge e_2 \wedge e_3, \tag{7.45}$$

$$e_1^* = -e_0 \wedge e_2 \wedge e_3, \qquad e_2^* = -e_0 \wedge e_3 \wedge e_1, \qquad e_3^* = -e_0 \wedge e_1 \wedge e_2, \tag{7.46}$$

$$(e_0 \wedge e_1)^* = -e_2 \wedge e_3, \quad (e_0 \wedge e_2)^* = -e_3 \wedge e_1, \quad (e_0 \wedge e_3)^* = -e_1 \wedge e_2, \tag{7.47}$$

$$(e_2 \wedge e_3)^* = -e_0 \wedge e_1, \quad (e_3 \wedge e_1)^* = -e_0 \wedge e_2, \quad (e_1 \wedge e_2)^* = -e_0 \wedge e_3, \tag{7.48}$$

$$(e_0 \wedge e_2 \wedge e_3)^* = -e_1, \quad (e_0 \wedge e_3 \wedge e_1)^* = -e_2, \quad (e_0 \wedge e_1 \wedge e_2)^* = -e_3, \tag{7.49}$$

$$(e_1 \wedge e_2 \wedge e_3)^* = e_0, \qquad (e_0 \wedge e_1 \wedge e_2 \wedge e_3)^* = 1. \tag{7.50}$$

planes, and points in terms of their Plücker coordinates.

7.5.1 Dual representation of lines

From Eqs. (7.47) and (7.48), the direct representation of line L in Eq. (7.11) has its dual

$$L^* = -m_1 e_2 \wedge e_3 - m_2 e_3 \wedge e_1 - m_3 e_1 \wedge e_2 - n_1 e_0 \wedge e_1 - n_2 e_0 \wedge e_2 - n_3 e_0 \wedge e_3, \tag{7.51}$$

which can be rewritten as

$$L^* = -e_0 \wedge \boldsymbol{n} + \boldsymbol{m}^*, \tag{7.52}$$

where $\boldsymbol{m}^*$ is the dual of $\boldsymbol{m}$ in 3D.

As stated in Chapter 5 ($\hookrightarrow$ Eq. (5.79) in Chapter 5), the direct expression $p \wedge L = 0$ for the equation of line L should be equivalently written in its dual expression as $p \cdot L^* = 0$. This is confirmed as follows. From Eq. (7.52), we see that

$$\begin{aligned} p \cdot L^* &= (e_0 + \boldsymbol{x}) \cdot (-e_0 \wedge \boldsymbol{n} + \boldsymbol{m}^*) = -e_0 \cdot e_0 \wedge \boldsymbol{n} + e_0 \cdot \boldsymbol{m}^* - \boldsymbol{x} \cdot e_0 \wedge \boldsymbol{n} + \boldsymbol{x} \cdot \boldsymbol{m}^* \\ &= -e_0 \cdot e_0 \wedge \boldsymbol{n} - \boldsymbol{n} + e_0(\boldsymbol{x} \cdot \boldsymbol{n}) + \boldsymbol{x} \cdot \boldsymbol{m}^* = -\boldsymbol{n} + e_0(\boldsymbol{x} \cdot \boldsymbol{n}) - \boldsymbol{x} \cdot (\boldsymbol{m} \cdot I) \\ &= \langle \boldsymbol{n}, \boldsymbol{x} \rangle e_0 + (\boldsymbol{x} \cdot \boldsymbol{m})^* - \boldsymbol{n} = \langle \boldsymbol{n}, \boldsymbol{x} \rangle e_0 + \boldsymbol{x} \times \boldsymbol{m} - \boldsymbol{n}, \end{aligned} \tag{7.53}$$

where we have noted that $\boldsymbol{n}$ and $\boldsymbol{m}^*$ do not contain e_0 so contraction of them by e_0 vanishes, that $\boldsymbol{x} \wedge \boldsymbol{m} \cdot I = \boldsymbol{x} \cdot (\boldsymbol{m} \cdot I)$ from the rule of contraction, and that $-\boldsymbol{x} \wedge \boldsymbol{m} \cdot I = (\boldsymbol{x} \wedge \boldsymbol{m})^*$ $= \boldsymbol{x} \times \boldsymbol{m}$ holds. Thus, $p \cdot L^* = 0$ is equivalent to Eq. (7.17).

7.5.2 Dual representation of planes

From Eqs. (7.49) and (7.50), the direct representation of plane Π in Eq. (7.26) has its dual

$$\Pi^* = -n_1 e_1 - n_2 e_2 - n_3 e_3 + h e_0, \tag{7.54}$$

which can be rewritten as

$$\Pi^* = h e_0 - \boldsymbol{n}. \tag{7.55}$$

As in the case of lines, the direct expression $p \wedge \Pi = 0$ for the equation of plane Π should be equivalently written in its dual expression as $p \cdot \Pi^* = 0$. In fact, from Eq. (7.54) we see that

$$p \cdot \Pi^* = (e_0 + \boldsymbol{x}) \cdot (h e_0 - \boldsymbol{n}) = h - \boldsymbol{x} \cdot \boldsymbol{n} = h - \langle \boldsymbol{n}, \boldsymbol{x} \rangle, \tag{7.56}$$

where we have noted that $\boldsymbol{x}$ and $\boldsymbol{n}$ do not contain e_0, so contraction of them by e_0 vanishes. Thus, $p \cdot \Pi^* = 0$ is equivalent to Eq. (7.29).

7.5.3 Dual representation of points

Let us write a point at $\boldsymbol{y} = y_1 e_1 + y_2 e_2 + y_3 e_3$ in 3D as

$$q = e_0 + \boldsymbol{y}. \tag{7.57}$$

Then, point $p = e_0 + \boldsymbol{x}$ coincides with point q if and only if $p \wedge q = 0$. In fact, from

$$p \wedge q = (e_0 + \boldsymbol{x}) \wedge (e_0 + \boldsymbol{y}) = e_0 \wedge \boldsymbol{y} + \boldsymbol{x} \wedge e_0 + \boldsymbol{x} \wedge \boldsymbol{y} = e_0 \wedge (\boldsymbol{y} - \boldsymbol{x}) + \boldsymbol{x} \wedge \boldsymbol{y}, \tag{7.58}$$

		direct	dual
points	representation	$q = e_0 + \boldsymbol{y}$	$q^* = e_0 \wedge \boldsymbol{y}^* + I$
	equation	$p \wedge q = 0$	$p \cdot q^* = 0$
lines	representation	$L = e_0 \wedge \boldsymbol{m} - \boldsymbol{n}^*$	$L^* = -e_0 \wedge \boldsymbol{n} + \boldsymbol{m}^*$
	equation	$p \wedge L = 0$	$p \cdot L^* = 0$
planes	representation	$\Pi = -e_0 \wedge \boldsymbol{n}^* + hI$	$\Pi^* = he_0 - \boldsymbol{n}$
	equation	$p \wedge \Pi = 0$	$p \cdot \Pi^* = 0$

the condition $p \wedge q = 0$ is equivalent to

$$\boldsymbol{x} = \boldsymbol{y}, \qquad \boldsymbol{x} \wedge \boldsymbol{y} = 0. \tag{7.59}$$

In other words, $p \wedge q = 0$ is the "equation" of point q. From Eqs. (7.45) and (7.46), the du of Eq. (7.57) is

$$q^* = e_1 \wedge e_2 \wedge e_3 - e_0 \wedge (y_1 e_2 \wedge e_3 + y_2 e_2 \wedge e_3 + y_3 e_2 \wedge e_3), \tag{7.60}$$

which can be written as

$$q^* = e_0 \wedge \boldsymbol{y}^* + I, \tag{7.61}$$

where $\boldsymbol{y}^*$ is the dual of $\boldsymbol{y}$ in 3D and I is the volume element in 3D. Hence, $p \cdot q^*$ is

$$\begin{aligned} p \cdot q^* &= (e_0 + \boldsymbol{x}) \cdot (e_0 \wedge \boldsymbol{y}^* + I) = e_0 \cdot e_0 \wedge \boldsymbol{y}^* + \boldsymbol{x} \cdot e_0 \wedge \boldsymbol{y}^* + \boldsymbol{x} \cdot I \\ &= \boldsymbol{y}^* - e_0 \wedge (\boldsymbol{x} \cdot \boldsymbol{y}^*) - \boldsymbol{x}^* = e_0 \wedge (\boldsymbol{x} \wedge \boldsymbol{y})^* + (\boldsymbol{x} - \boldsymbol{y})^*, \end{aligned} \tag{7.62}$$

where we have noted that $\boldsymbol{y}^*$ does not contain e_0, so contraction of it by e_0 vanishes an that $\boldsymbol{x} \cdot \boldsymbol{y}^* = (\boldsymbol{x} \wedge \boldsymbol{y})^*$ holds (↪ Eq. (5.78) in Chapter 5). Thus, $p \cdot q^* = 0$ is equivalent t Eq. (7.59). Namely, $p \cdot q^* = 0$ is the equation of point q. Table 7.1 summarizes the resul we have observed so far.

7.6 DUALITY THEOREM

We now show the following:

- To points, lines, and planes, their "dual planes," "dual lines," and "dual points correspond.
- For the "join" and " meet" involving points, lines, and planes, a duality holds in th sense that "the dual of join is the meet of the duals" and "the dual of meet is the joi of the duals."

7.6.1 Dual points, dual lines, and dual planes

A line is represented by a bivector L, so its dual representation L^* is a bivector. Hence, represents some line, which we call the *dual line* of L. Comparing Eq. (7.12) with Eq. (7.52 we see that the direction $\boldsymbol{m}$ of line L and the surface normal $\boldsymbol{n}$ to its supporting plane ar related to those of the dual line L^* by interchange of $\boldsymbol{m}$ and $\boldsymbol{n}$ and their sign reversa Since the supporting point of line L is at distance $h = \|\boldsymbol{n}\|/\|\boldsymbol{m}\|$ from the origin e_0 (↪ Eq. (2.61) in Chapter 2), the supporting point of the dual line L^* is at distance $1/h$ fro e_0 (Fig. 7.4(a)).

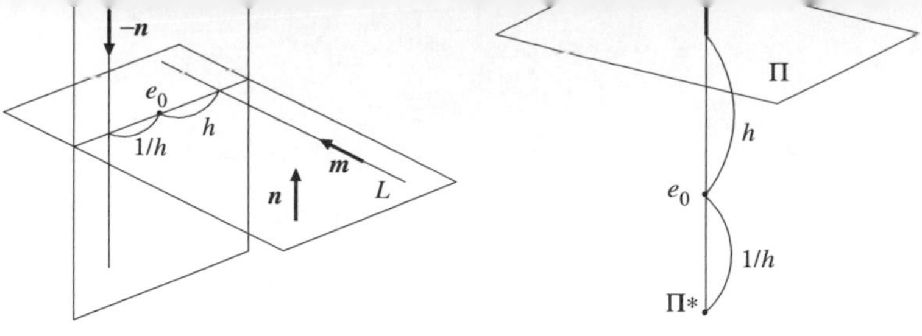

(a) (b)

FIGURE 7.4 (a) Line L and its dual line L^*. (b) Plane Π and its dual point Π^*.

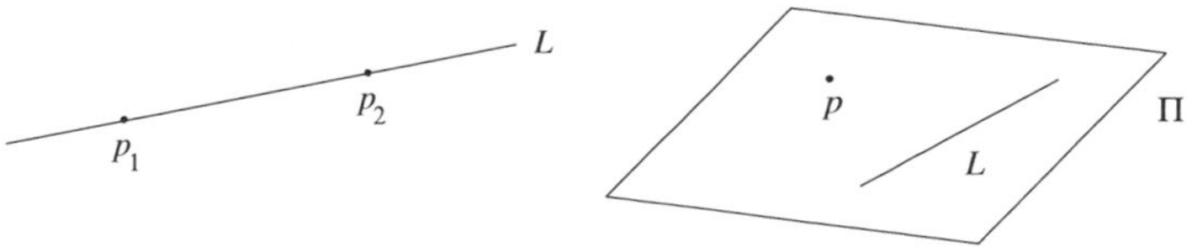

(a) (b)

FIGURE 7.5 (a) Two points p_1 and p_2 and their join $L = p_1 \cup p_2$ $(= p_1 \wedge p_2)$. (b) A point p, a line L, and their join $\Pi = p \cup L$ $(= p \wedge L)$.

A plane is represented by a trivector Π, so its dual representation Π^* is a vector. Hence, it represents some point, which we call the *dual point* of Π. If Eq. (7.26) is interpreted to be a point, it is at $-\boldsymbol{n}/h$ in 3D. Namely, it is located at distance $1/h$ from the origin e_0 along the surface normal to Π in the opposite direction (Fig. 7.4(b)).

A point $q = e_0 + \boldsymbol{y}$ has its dual q^* that has the form of Eq. (7.61), which is a trivector. Hence, it represents some plane, which we call the *dual plane* of q. Due to the homogeneity, we can write q^* as $e_0 \wedge (\boldsymbol{y}/\|\boldsymbol{y}\|)^* + (1/\|\boldsymbol{y}\|)I$. Comparing this with Eq. (7.26), we see that q^* represents a plane at distance $1/\|\boldsymbol{y}\|$ from the origin e_0 in the opposite direction of $\boldsymbol{y}$. This is the same relationship as that between a plane Π and its dual point Π^*.

7.6.2 Join and meet

We now introduce new terminologies to describe the duality theorems. The line L that passes through points p_1 and p_2 is called their *join* (Fig. 7.5(a)) and is written as

$$L = p_1 \cup p_2. \tag{7.63}$$

In terms of the 4D vectors p_1 and p_2, this is given by the bivector

$$L = p_1 \wedge p_2. \tag{7.64}$$

The plane Π that passes through a point p and a line L is called their *join* (Fig. 7.5(b)) and is written as

$$\Pi = p \cup L. \tag{7.65}$$

In terms of the 4D vector p and the bivector L, this is given by the trivector

$$\Pi = p \wedge L. \tag{7.66}$$

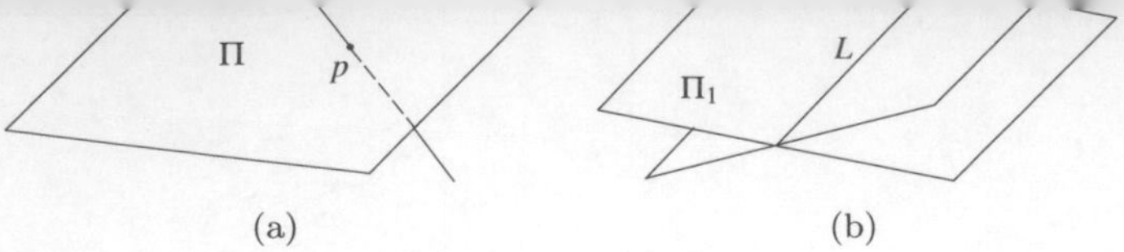

FIGURE 7.6 (a) A line L, a plane Π, and their meet $p = L \cap \Pi$. (b) Two planes Π_1 and Π_2, and their meet $L = \Pi_1 \cap \Pi_2$.

The intersection point p of a line L and a plane Π is called their *meet* (Fig. 7.6(a)) and is written as

$$p = L \cap \Pi. \tag{7.67}$$

It is not straightforward to express p in terms of the bivector L and the trivector Π. The intersection line L of two planes Π_1 and Π_2 is called their *meet* (Fig. 7.6(b)) and is written as

$$L = \Pi_1 \cap \Pi_2. \tag{7.68}$$

It is not straightforward to express L in terms of the trivectors Π_1 and Π_2. However, the meet of a line with a plane and the meet of two planes are easily computed by using the duality theorems in the following subsections.

7.6.3 Join of two points and meet of a plane with a line

In terms of the terminologies introduced above, the following duality theorem holds:

Proposition 7.4 (Duality theorem 1) *The join of point p and line L has its dual point given by the meet of the dual plane p^* with the dual line L^*:*

$$(p \cup L)^* = p^* \cap L^*. \tag{7.69}$$

The meet of line L with plane Π has its dual plane given by the join of the dual line L^* and the dual point Π^*:

$$(L \cap \Pi)^* = L^* \cup \Pi^*. \tag{7.70}$$

As we saw earlier, dual of dual is the original expression. Hence, Eqs. (7.69) and (7.70) state the same thing. So let us consider Eq. (7.70). It can be confirmed as follows. From Table 7.1, the dual of line $L = e_0 \wedge \boldsymbol{m} - \boldsymbol{n}_L^*$ and the dual of plane $\Pi = -e_0 \wedge \boldsymbol{n}_\Pi + hI$ are, respectively, line $L^* = -e_0 \wedge \boldsymbol{n}_L + \boldsymbol{m}^*$ and point $\Pi^* = he_0 - \boldsymbol{n}_\Pi$. Their join is

$$\begin{aligned} L^* \cup \Pi^* &= L^* \wedge \Pi^* = (-e_0 \wedge \boldsymbol{n}_L + \boldsymbol{m}^*) \wedge (he_0 - \boldsymbol{n}_\Pi) \\ &= e_0 \wedge \boldsymbol{n}_L \wedge \boldsymbol{n}_\Pi + \boldsymbol{m}^* \wedge he_0 + \boldsymbol{m}^* \wedge \boldsymbol{n}_\Pi = e_0 \wedge (\boldsymbol{n}_L \wedge \boldsymbol{n}_\Pi - h\boldsymbol{m}^*) - \boldsymbol{m}^* \wedge \boldsymbol{n}_\Pi \\ &= e_0 \wedge (-(\boldsymbol{n}_L \wedge \boldsymbol{n}_\Pi)^* + h\boldsymbol{m})^* + \langle \boldsymbol{m}, \boldsymbol{n}_\Pi \rangle I \\ &= e_0 \wedge (\boldsymbol{n}_\Pi \times \boldsymbol{n}_L + h\boldsymbol{m})^* + \langle \boldsymbol{m}, \boldsymbol{n}_\Pi \rangle I. \end{aligned} \tag{7.71}$$

From Table 7.1, its dual is

$$(L^* \cup \Pi^*)^* = \langle \boldsymbol{m}, \boldsymbol{n}_\Pi \rangle e_0 + \boldsymbol{n}_\Pi \times \boldsymbol{n}_L + h\boldsymbol{m}. \tag{7.72}$$

$$p = \frac{\qquad}{\langle m, n_{\Pi}\rangle} \qquad (7.73)$$

in 3D. As shown in Eq. (2.91) in Chapter 2, this is exactly the meet $L \cap \Pi$ of line L with plane Π, which confirms Eq. (7.70).

7.6.4 Join of two points and meet of two planes

The following duality theorem also holds.

Proposition 7.5 (Duality theorem 2) *The join of two points p_1 and p_2 has its dual line given by the meet of the dual planes p_1^* and p_2^*:*

$$(p_1 \cup p_2)^* = p_1^* \cap p_2^*. \qquad (7.74)$$

The meet L of planes Π_1 and Π_2 has its dual given by the join of the dual points Π_1^* and Π_2^*:

$$(\Pi_1 \cap \Pi_2)^* = \Pi_1^* \cup \Pi_2^*. \qquad (7.75)$$

Equations (7.74) and (7.75) state the same thing. Equation (7.75) can be confirmed as follows. According to Table 7.1, the dual of planes $\Pi_i = -e_0 \wedge \boldsymbol{n}_i^* + h_i I$, $i = 1, 2$, are points $\Pi_i^* = h_i e_0 - \boldsymbol{n}_i$, $i = 1, 2$. Their join is

$$\begin{aligned}
\Pi_1^* \cup \Pi_2^* &= \Pi_1^* \wedge \Pi_2^* = (h_1 e_0 - \boldsymbol{n}_1) \wedge (h_2 e_0 - \boldsymbol{n}_2) \\
&= -h_1 e_0 \wedge \boldsymbol{n}_2 - \boldsymbol{n}_1 \wedge h_2 e_0 + \boldsymbol{n}_1 \wedge \boldsymbol{n}_2 = e_0 \wedge (h_2 \boldsymbol{n}_1 - h_1 \boldsymbol{n}_2) + \boldsymbol{n}_1 \wedge \boldsymbol{n}_2 \\
&= e_0 \wedge (h_2 \boldsymbol{n}_1 - h_1 \boldsymbol{n}_2) - (\boldsymbol{n}_1 \wedge \boldsymbol{n}_2)^{**} = e_0 \wedge (h_2 \boldsymbol{n}_1 - h_1 \boldsymbol{n}_2) - (\boldsymbol{n}_1 \times \boldsymbol{n}_2)^*. \qquad (7.76)
\end{aligned}$$

From Table 7.1, its dual is

$$(\Pi_1^* \cup \Pi_2^*)^* = -e_0 \wedge (\boldsymbol{n}_1 \times \boldsymbol{n}_2) + (h_2 \boldsymbol{n}_1 - h_1 \boldsymbol{n}_2)^*. \qquad (7.77)$$

This can be identified with a line $e_0 \wedge \boldsymbol{m} - \boldsymbol{n}_L^*$. Due to the homogeneity, the Plücker coordinates are, up to a nonzero constant,

$$\boldsymbol{m} = \boldsymbol{n}_1 \times \boldsymbol{n}_2, \qquad \boldsymbol{n}_L = h_2 \boldsymbol{n}_1 - h_1 \boldsymbol{n}_2. \qquad (7.78)$$

This coincides with Eq. (2.96) in Chapter 2 up to scale normalization, i.e., the meet $\Pi_1 \cap \Pi_1$ of planes Π_1 and Π_2. Hence, Eq. (7.75) holds.

Since the joins of points, lines, and planes can be directly computed using the outer product operation $\wedge$, the meets of points, lines, and planes can be computed by first computing the joins of their duals and then computing their duals, using Propositions 7.4 and 7.5. Possible combinations are shown in Table 7.2.

7.6.5 Join of three points and meet of three planes

Propositions 7.4 and 7.5 can be combined to extend to three objects. For three points p_1, p_2, and p_3, the plane Π that passes through them is called their *join* (Fig. 7.7(a)) and is written as

$$\Pi = p_1 \cup p_2 \cup p_3. \qquad (7.79)$$

	point p_2	line L_2	plane Π_2
point p_1	$p_1 \cup p_2 = p_1 \wedge p_2$	$p_1 \cup L_2 = p_1 \wedge L_2$	—
line L_1	$L_1 \cup p_2 = L_1 \wedge p_2$	—	$L_1 \cap \Pi_2 = (L_1^* \wedge \Pi_2^*)^*$
plane Π_1	—	$\Pi_1 \cap L_2 = (\Pi_1^* \wedge L_2^*)^*$	$\Pi_1 \cap \Pi_2 = (\Pi_1^* \wedge \Pi_2^*)^*$

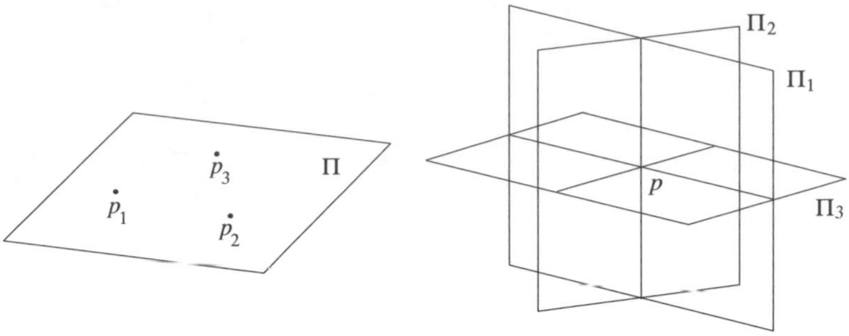

(a) (b)

FIGURE 7.7 (a) Three points p_1, p_2, and p_3 and their join $\Pi = p_1 \cup p_2 \cup p_3$ $(= p_1 \wedge p_2 \wedge p_3)$. (b) Three planes Π_1, Π_2, and Π_3 and their meet $\boldsymbol{p} = \Pi_1 \cap \Pi_2 \cap \Pi_3$.

For three planes Π_1, Π_2, and Π_3, their intersection p is called their *meet* (Fig. 7.7(b)) ar is written as

$$p = \Pi_1 \cap \Pi_2 \cap \Pi_3. \tag{7.80}$$

Combining Propositions 7.4 and 7.5, we can obtain the following duality theorem for thr points and three planes ($\hookrightarrow$ Exercise 7.6):

Proposition 7.6 (Duality theorem 3) *The join Π of three points p_i, $i = 1, 2, 3$ has i dual point Π^* given by the meet of the respective dual planes p_i^*, $i = 1, 2, 3$:*

$$(p_1 \cup p_2 \cup p_3)^* = p_1^* \cap p_2^* \cap p_3^*. \tag{7.81}$$

The meet p of three planes Π_i, $i = 1, 2, 3$, has its dual plane p^* given by the join of the respective dual points Π_i^*, $i = 1, 2, 3$:

$$(\Pi_1 \cap \Pi_2 \cap \Pi_3)^* = \Pi_1^* \cup \Pi_2^* \cup \Pi_3^* \tag{7.82}$$

As an application, let us compute the intersection of three planes $\langle \boldsymbol{n}_i, \boldsymbol{x} \rangle = h_i$, $i = 1,$ 3. From Table 7.1, each plane is represented by the trivector $\Pi_i = -e_0 \wedge \boldsymbol{n}_i^* + h_i I$, whos dual point is $\Pi_i^* = h_i e_0 - \boldsymbol{n}_i$. The join of the three dual points is

$$\begin{aligned} \Pi_1^* \cup \Pi_2^* \cup \Pi_3^* &= \Pi_1^* \wedge \Pi_2^* \wedge \Pi_3^* = (h_1 e_0 - \boldsymbol{n}_1) \wedge (h_2 e_0 - \boldsymbol{n}_2) \wedge (h_3 e_0 - \boldsymbol{n}_3) \\ &= e_0 \wedge (h_1 \boldsymbol{n}_2 \wedge \boldsymbol{n}_3 + h_2 \boldsymbol{n}_3 \wedge \boldsymbol{n}_1 + h_3 \boldsymbol{n}_1 \wedge \boldsymbol{n}_2) - \boldsymbol{n}_1 \wedge \boldsymbol{n}_2 \wedge \boldsymbol{n}_3 \\ &= -e_0 \wedge (h_1 (\boldsymbol{n}_2 \wedge \boldsymbol{n}_3)^* + h_2 (\boldsymbol{n}_3 \wedge \boldsymbol{n}_1)^* + h_3 (\boldsymbol{n}_1 \wedge \boldsymbol{n}_2)^*)^* - |\boldsymbol{n}_1, \boldsymbol{n}_2, \boldsymbol{n}_3| I \\ &= -e_0 \wedge (h_1 \boldsymbol{n}_2 \times \boldsymbol{n}_3 + h_2 \boldsymbol{n}_3 \times \boldsymbol{n}_1 + h_3 \boldsymbol{n}_1 \times \boldsymbol{n}_2)^* - |\boldsymbol{n}_1, \boldsymbol{n}_2, \boldsymbol{n}_3| I. \end{aligned} \tag{7.83}$$

From Table 7.1, its dual is

$$\Pi_1 \cap \Pi_2 \cap \Pi_3 = (\Pi_1^* \cup \Pi_2^* \cup \Pi_3^*)^* = -|\boldsymbol{n}_1, \boldsymbol{n}_2, \boldsymbol{n}_3| e_0 - (h_1 \boldsymbol{n}_2 \times \boldsymbol{n}_3 + h_2 \boldsymbol{n}_3 \times \boldsymbol{n}_1 + h_3 \boldsymbol{n}_1 \times \boldsymbol{n}_2 \tag{7.84}$$

Proposition 7.7 (Intersection of three planes) *Three planes* $\langle \boldsymbol{n}_i, \boldsymbol{x} \rangle = h_i$, $i = 1, 2, 3$, *have their intersection*

$$\boldsymbol{p} = \frac{h_1 \boldsymbol{n}_2 \times \boldsymbol{n}_3 + h_2 \boldsymbol{n}_3 \times \boldsymbol{n}_1 + h_3 \boldsymbol{n}_1 \times \boldsymbol{n}_2}{|\boldsymbol{n}_1, \boldsymbol{n}_2, \boldsymbol{n}_3|}. \tag{7.85}$$

In fact, Eq. (7.85) satisfies $\langle \boldsymbol{n}_i, \boldsymbol{p} \rangle = h_i$, $i = 1, 2, 3$, so this point is on the three planes.

Traditional World 7.3 (Cramer's formula) Since planes are described by linear equations in coordinates, computing their intersection lines and intersection points reduces to solving a set of simultaneous linear equations, for which various numerical schemes are known, including *Gaussian elimination*, the *LU decomposition*, and *Gauss–Seidel iterations*. However, we can also express the solution in an analytical form in terms of determinants known as *Cramer's formula*. Using this, we can express the solution of linear equations

$$\begin{aligned} a_{11}x_1 + a_{12}x_2 + \cdots + a_{1n}x_n &= b_1, \\ a_{21}x_1 + a_{22}x_2 + \cdots + a_{2n}x_n &= b_2, \\ &\vdots \\ a_{n1}x_1 + a_{n2}x_2 + \cdots + a_{nn}x_n &= b_n. \end{aligned} \tag{7.86}$$

as

$$x_i = \left| \begin{array}{ccccc} a_{11} & \cdots & \overset{(i)}{b_1} & \cdots & a_{1n} \\ a_{21} & \cdots & b_2 & \cdots & a_{2n} \\ \vdots & \cdots & \vdots & \cdots & \vdots \\ a_{n1} & \cdots & b_n & \cdots & a_{nn} \end{array} \right| \Bigg/ \left| \begin{array}{cccc} a_{11} & a_{12} & \cdots & a_{1n} \\ a_{21} & a_{22} & \cdots & a_{2n} \\ \vdots & \vdots & \ddots & \vdots \\ a_{n1} & a_{n2} & \cdots & a_{nn} \end{array} \right|, \tag{7.87}$$

where the numerator is obtained from the denominator by replacing the ith column by b_1, b_2, ..., b_n. Cramer's formula is very convenient for theoretical analysis but is not suitable for numerical computation, because it takes a long computation time for a large n. For three planes, the equations have the form

$$\begin{aligned} n_1 x + n_2 y + n_3 z &= h, \\ n_1' x + n_2' y + n_3' z &= h', \\ n_1'' x + n_2'' y + n_3'' z &= h''. \end{aligned} \tag{7.88}$$

Hence, if we let

$$\Delta = \left| \begin{array}{ccc} n_1 & n_2 & n_3 \\ n_1' & n_2' & n_3' \\ n_1'' & n_2'' & n_3'' \end{array} \right|, \tag{7.89}$$

Cramer's formula gives the solution in the form

$$x = \frac{1}{\Delta} \left| \begin{array}{ccc} h & n_2 & n_3 \\ h' & n_2' & n_3' \\ h'' & n_2'' & n_3'' \end{array} \right| = \frac{h(n_2' n_3'' - n_3' n_2'') + h'(n_2'' n_3 - n_3'' n_2) + h''(n_2 n_3' - n_3 n_2')}{\Delta},$$

$$y = \frac{1}{\Delta} \left| \begin{array}{ccc} n_1 & h & n_3 \\ n_1' & h' & n_3' \\ n_1'' & h'' & n_3'' \end{array} \right| = \frac{h(n_3' n_1'' - n_1' n_3'') + h'(n_3'' n_1 - n_1'' n_3) + h''(n_3 n_1' - n_3 n_3')}{\Delta},$$

(7.9

In the notation of the standard vector calculus, it is written as

$$\begin{pmatrix} x \\ y \\ z \end{pmatrix} = \frac{1}{\Delta}\left(h \begin{pmatrix} n_1' \\ n_2' \\ n_3' \end{pmatrix} \times \begin{pmatrix} n_1'' \\ n_2'' \\ n_3'' \end{pmatrix} + h' \begin{pmatrix} n_1'' \\ n_2'' \\ n_3'' \end{pmatrix} \times \begin{pmatrix} n_1 \\ n_2 \\ n_3 \end{pmatrix} + h'' \begin{pmatrix} n_1 \\ n_2 \\ n_3 \end{pmatrix} \times \begin{pmatrix} n_1' \\ n_2' \\ n_3' \end{pmatrix} \right), \quad (7.9$$

from which Eq. (7.85) is obtained.

7.7 SUPPLEMENTAL NOTE

This chapter presents a formulation for representing points that are not necessarily a the origin and lines and planes that do not necessarily pass through the origin in 3D b regarding them as subspaces in 4D and applying the Grassmann algebra in that spac Regarding points, lines, and planes in 3D as subspaces in 4D is the basic principle *projective geometry*, and one of its most fundamental characteristics is the *duality* betwee joins and meets. Combination of this projective geometric structure with the Grassman algebra was named *Grassmann–Cayley algebra* by 20th century mathematicians in honor the English mathematician *Arthur Cayley* (1821–1895), who studied the algebraic structu of projective geometry. It is also called simply *Cayley algebra* or *double algebra*.

What we call the "homogeneous space" in this chapter is commonly known as the *pro jective space*, but strictly speaking it is the "3D projective space," the usual 3D spa augmented by adding to it points and lines at infinity and the plane at infinity. This spa is realized as a set of subspaces in 4D. In general, the nD projective space $\mathbb{P}^n$ is realize as a set of subspaces of an $(n+1)$D space $\mathbb{R}^{n+1}$. In this chapter, we use the term "4 homogeneous space" rather than "3D projective space" to emphasize the fact that we a working in 4D.

Historically, projective geometry developed mainly as plane geometry. It is an extensio of plane Euclidean geometry, to which are added points at infinity, also called *ideal point* and the line at infinity, also called the *ideal line*. Then, geometric relationships among point lines, and quadratic curves or *conics* (ellipses, parabolas, hyperbolas, and their degeneracie are described in terms of *cross-ratios* (also called *anharmonic ratios*) (↪ Exercise 5.3). Fo this, Semple and Kneebone [19] is a well-known textbook. Study of intersections betwee lines, planes, and polynomial surfaces in general dimensions from the viewpoint of projectiv geometry is called *algebraic geometry*, for which Semple and Roth [20] is a classic textboo

Applying projective geometry to computer vision problems, Kanatani [14] presented unified computational scheme for computing the relationships between points, lines, plane and quadratic surfaces in 3D and their 2D images taken by a perspective camera. Projectiv geometry also plays an important role in analyzing the relationships among multiple image of the same scene taken by different cameras. For this, Hartley and Zisserman [10] is a wel known textbook. Faugeras and Luong [7] formulated this "multiview geometry," as it called now, in terms of the Grassmann–Cayley algebra.

As we see from the description in this chapter, the outer product operation $\wedge$ and th join relation $\cup$ have essentially the same meaning. The Grassmann–Cayley algebra regar the meet relation $\cap$ as its counterpart and gives a unifying framework in which these tw operations have equal footing. However, there are some notational variations. In man textbooks [2, 3, 4, 12, 16], the outer product symbol $\wedge$ is used for joins, and the symb

Faugeras and Luong [7] use ∇ for outer products and joins and $\triangle$ for meets. Grassmann himself defined the meet operation $\cap$ as the dual of the outer product operation $\wedge$, but later mathematicians defined the meet operator independently of the outer product through a process called *shuffle*, and the meet operation is termed the *shuffle product*. It is shown that this is an antisymmetric operation that satisfies associativity, defining an algebra in its own right. Thus, the Grassmann–Cayley algebra has two algebraic structures simultaneously: one based on the outer product (or join), the other based on the shuffle product (or meet). Hence, the name "double algebra," and the two are shown to be dual to each other.

This double structure is defined in terms of determinant operations of linear algebra, viewing vectors as arrays of numbers. There, what we call a k-vector $\boldsymbol{a}_1 \wedge \cdots \wedge \boldsymbol{a}_k$ in this chapter is called an *extensor* of *step k*. However, describing this formulation involves considerable complications, so we did not go into the details of the double structure, although it is the core of the Grassmann–Cayley algebra. Here, we merely described it as having a dual structure from the viewpoint of the Grassmann algebra.

Some readers might feel uneasy about the projective geometric formulation of this chapter. As we saw, vectors are classified into positions, i.e., expressions that contain the symbol e_0, and into directions, i.e., expressions that do not contain the symbol e_0, and we understand that the orientation is the only attribute of direction; its sign and magnitude do not have any meaning. However, although it is understandable that $\boldsymbol{u}$ and $2\boldsymbol{u}$ represent the same orientation, it is intuitively difficult to accept that $\boldsymbol{u}$ and $-\boldsymbol{u}$ represent the "same" orientation. But this is a consequence of the fact that if we move along a line in the projective space in either direction, we arrive at the same point at infinity, which both $\boldsymbol{u}$ and $-\boldsymbol{u}$ represent. Thus, we cannot speak of "opposite orientations." This seeming anomaly is resolved by the *oriented projective geometry* of Stolfi [23]. Here, a line is interpreted actually as a superposition of two lines, say, the plus and the minus lines, like an electric cord. If the orientation of this line is specified by $\boldsymbol{u}$, we imagine that we arrive at $+\infty$ if we move on along the positive line in the direction of $\boldsymbol{u}$ and at $-\infty$ if we move in the direction of $-\boldsymbol{u}$. Similarly, we imagine that we arrive at $+\infty'$ if we move on along the negative line in the direction of $\boldsymbol{u}$ and at $-\infty'$ if we move in the direction of $-\boldsymbol{u}$. Then, we identify $+\infty$ with $-\infty'$ and $-\infty'$ with $+\infty'$.

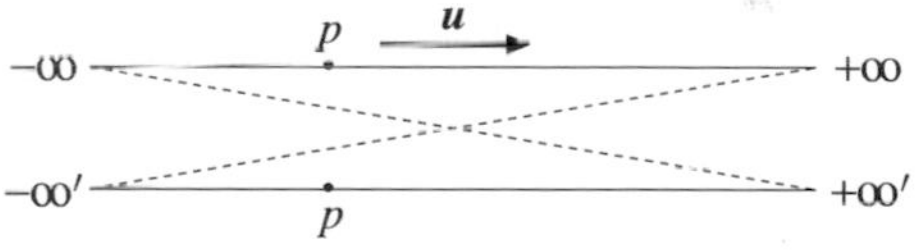

If we move along this line from a point p in the direction of $\boldsymbol{u}$ indefinitely, we arrive at $+\infty$ and jump to the other end $-\infty'$ of the minus line. If we keep moving on along it, we pass through the point p and reach $+\infty'$. Then, we jump to $-\infty$ of the plus line. Thus, we return to the starting point p after traversing this line twice. In other words, although it appears that we go back to the same point p after one traversal, we are actually on the "reverse side" of the line that happens to coincide with the original point on the front side. This is a reasonable interpretation of projective geometry from the viewpoint of topology. It follows that the "opposite direction" makes sense if $\boldsymbol{u}$ is associated with $+\infty$ and $-\boldsymbol{u}$ with $-\infty$.

Still, $\boldsymbol{u}$ and $2\boldsymbol{u}$ represent the "same direction" even in the framework of oriented projective geometry. This means that we cannot describe the magnitude of displacement. This is an inevitable consequence of identifying a direction with a point at infinity. This difficulty can be resolved only by introducing a formulation that distinguishes directions and

7.8 EXERCISES

7.1. Let $p_1 = e_0 + \boldsymbol{x}_1$, $p_2 = e_0 + \boldsymbol{x}_2$, and $p_3 = e_0 + \boldsymbol{x}_3$ be the representations of thre points at positions $\boldsymbol{x}_1$, $\boldsymbol{x}_2$, and $\boldsymbol{x}_3$ in 3D. Show that the three points $\boldsymbol{x}_1$, $\boldsymbol{x}_2$, and $\boldsymbol{x}$ are collinear if and only if

$$p_1 \wedge p_2 \wedge p_3 = 0.$$

7.2. Let $p_1 = e_0 + \boldsymbol{x}_1$, $p_2 = e_0 + \boldsymbol{x}_2$, $p_3 = e_0 + \boldsymbol{x}_3$, and $p_4 = e_0 + \boldsymbol{x}_4$ be the representatio of four points at positions $\boldsymbol{x}_1$, $\boldsymbol{x}_2$, $\boldsymbol{x}_3$, and $\boldsymbol{x}_4$ in 3D. Show that the four points $\boldsymbol{x}$ $\boldsymbol{x}_2$, $\boldsymbol{x}_3$, and $\boldsymbol{x}_4$ are coplanar if and only if

$$p_1 \wedge p_2 \wedge p_3 \wedge p_4 = 0.$$

7.3. Show that in the 4D homogeneous space, a bivector

$$L = m_1 e_0 \wedge e_1 + m_2 e_0 \wedge e_2 + m_3 e_0 \wedge e_3 + n_1 e_2 \wedge e_3 + n_2 e_3 \wedge e_1 + n_3 e_1 \wedge e_2$$

can be *factorized* ($\hookrightarrow$ Supplemental note to Chapter 5 and Exercise 5.1), i.e., it expressed in the form

$$L = x \wedge y$$

for some elements x and y, if and only if

$$m_1 n_1 + m_2 n_2 + m_3 n_3 = 0.$$

This condition for factorization is called the *Plücker condition.*

7.4. Show that in the 4D homogeneous space, a trivector

$$\Pi = n_1 e_0 \wedge e_2 \wedge e_3 + n_2 e_0 \wedge e_3 \wedge e_1 + n_3 e_0 \wedge e_1 \wedge e_2 + h e_1 \wedge e_2 \wedge e_3$$

is always factorized, i.e., it is expressed in the form

$$\Pi = x \wedge y \wedge z$$

for some elements x, y, and z. In other words, show that the Plücker condition doe not exist for trivectors.

7.5. Consider the following two bivectors in the 4D homogeneous space that represent line in 3D:

$$L = m_1 e_0 \wedge e_1 + m_2 e_0 \wedge e_2 + m_3 e_0 \wedge e_3 + n_1 e_2 \wedge e_3 + n_2 e_3 \wedge e_1 + n_3 e_1 \wedge e_2,$$
$$L' = m_1' e_0 \wedge e_1 + m_2' e_0 \wedge e_2 + m_3' e_0 \wedge e_3 + n_1' e_2 \wedge e_3 + n_2' e_3 \wedge e_1 + n_3' e_1 \wedge e_2.$$

(1) Show that the outer product of L and L' is expressed in the following form:

$$L \wedge L' = (m_1 n_1' + m_2 n_2' + m_3 n_3' + n_1 m_1' + n_2 m_2' + n_3 m_3') e_0 \wedge e_1 \wedge e_2 \wedge e_3.$$

Also, show that in terms of 3D vectors $\boldsymbol{m} = m_1e_1 + m_2e_2 + m_3e_3$, $\boldsymbol{n} = n_1e_1 + n_2e_2 + n_3e_3$, $\boldsymbol{m}' = m_1'e_1 + m_2'e_2 + m_3'e_3$, and $\boldsymbol{n}' = n_1'e_1 + n_2'e_2 + n_3'e_3$, this condition is written as follows ($\hookrightarrow$ Eq. (2.71) in Chapter 2):

$$\langle \boldsymbol{m}, \boldsymbol{n}' \rangle + \langle \boldsymbol{n}, \boldsymbol{m}' \rangle = 0.$$

7.6. (1) Show that Eq. (7.81) holds.

(2) Show that Eq. (7.82) holds.

Conformal Space and Conformal Geometry: Geometric Algebra

In the preceding chapter, we considered a 4D space obtained by adding the origin e_0 to the 3D space. In this chapter, we consider a 5D space, called the "conformal space," obtained by further adding the point at infinity e_∞. Basic geometric elements of this space are spheres and circles: a point is regarded as a sphere of radius 0, a plane as a sphere of radius ∞ passing through e_∞, and a line as a circle of radius ∞ passing through e_∞. In this space, translation, being interpreted to be rotation around an axis placed infinitely far apart, is treated equivalently as rotation. The "conformal mappings" that map a sphere to a sphere are generated from translation, rotation, reflection, inversion, and dilation, which are described in terms of the geometric product of the Clifford algebra. The content of this chapter is the core of what is now known as "geometric algebra."

8.1 CONFORMAL SPACE AND INNER PRODUCT

In this book, we identify the symbols e_1, e_2, and e_3 with the orthonormal basis of the 3D Euclidean space and represent a vector in 3D in the form $\boldsymbol{a} = a_1e_1 + a_2e_2 + a_3e_3$. In the preceding chapter, we introduced a new symbol e_0; identifying it with the origin of the 3D space, we considered a 4D space spanned by $\{e_0, e_1, e_2, e_3\}$. In this chapter, we introduce yet another new symbol e_∞ and consider a 5D space spanned by $\{e_0, e_1, e_2, e_3, e_\infty\}$. Thus, an element of this space has the form

$$x = x_0e_0 + x_1e_1 + x_2e_2 + x_3e_3 + x_\infty e_\infty. \tag{8.1}$$

In the preceding chapter, we regarded e_0 as a unit vector orthogonal to the basis $\{e_1, e_2, e_3\}$ of the 3D space. In this chapter, we define the inner products among the basis elements as follows:

$$\langle e_0, e_0\rangle = 0, \qquad \langle e_0, e_\infty\rangle = -1, \qquad \langle e_\infty, e_\infty\rangle = 0,$$
$$\langle e_0, e_i\rangle = 0, \qquad \langle e_i, e_j\rangle = \delta_{ij}, \qquad i, j = 1, 2, 3. \tag{8.2}$$

In other words, the symbols e_0 and e_∞ are *orthogonal to all basis elements including themselves* except that $\langle e_0, e_\infty\rangle = -1$. It follows that if we let $y = y_0e_0+y_1e_1+y_2e_2+y_3e_3+y_\infty e_\infty$,

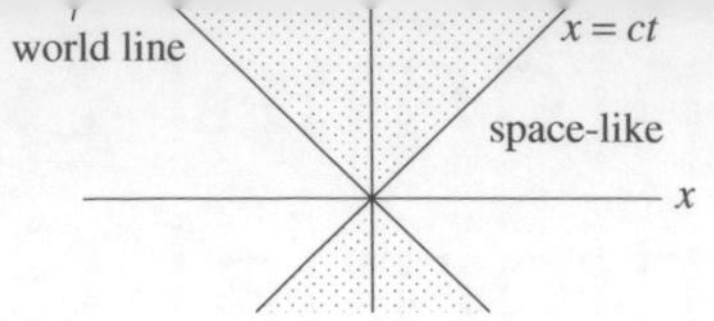

FIGURE 8.1 The xt plane of the 4D $xyzt$ space-time. It is divided by the world line $x = ct$ into t space-like region with positive Minkowski norm and the time-like region with negative Minkows norm.

the inner product of x and y has the form

$$\langle x, y\rangle = x_1y_1 + x_2y_2 + x_3y_3 - x_0y_\infty - x_\infty y_0. \qquad (8.$$

Letting $x = y$ in particular, we see that the square norm is

$$\|x\|^2 = x_1^2 + x_2^2 + x_3^2 - 2x_0x_\infty. \qquad (8.$$

In the following, norms always appear in the form of squares, so we call $\|x\|^2$ simply t "norm," rather than the squared norm.

Note that $\|x\|^2$ for $x \neq 0$ need not be positive in this space; there exist elements $x \neq$ for which $\|x\|^2 = 0$. In fact, we see that $\|x\|^2 = 0$ for $x = e_0$ and $x = e_\infty$. There are al elements x for which $\|x\|^2 < 0$. We call the 5D space with this inner product the *conform space* (↪ Exercise 8.2(1)).

Traditional World 8.1 (Non-Euclidean space) In Sec. 2.3 of Chapter 2, the inn product was defined by positivity, symmetry, and linearity. As a result, the norm is positi or zero. Such an inner product is called a *Euclidean metric*, and a space equipped with Euclidean metric is said to be a *Euclidean space*. If we drop the positivity, the resultir inner product is called a *non-Euclidean metric*, and a space equipped with a such a metr is said to be a *non-Euclidean space*. The conformal space of this chapter is a non-Euclidea space.

Mathematically, if the metric tensor, i.e., the $n \times n$ matrix (g_{ij}) whose (i, j) element $g_{ij} = \langle e_i, e_j\rangle$, has p positive eigenvalues and q negative eigenvalues, we say that the metr has *signature* (p, q). The space with a metric with signature (p, q) is denoted by $\mathbb{R}^{p,q}$. Th means that we can define a set of n orthonormal basis vectors, p of which have positi norms and q of which have negative norms. By definition, the space is Euclidean if (p, q) $(n, 0)$ and non-Euclidean otherwise.

A well-known non-Euclidean space is the 4D $xyzt$ space-time of Einstein's special theor of relativity, in which a space-time point (x, y, z, t) has the norm

$$x^2 + y^2 + z^2 - c^2t^2. \qquad (8.$$

Its signature is $(3, 1)$. A metric with this signature ($(n - 1, 1)$ in general nD) is calle the *Minkowski metric*, and the resulting norm the *Minkowski norm*. In the special theor of relativity, a space-time point is said to be *space-like* if its Minkowski norm is positiv and *time-like* if its Minkowski norm is negative (Fig. 8.1). The direction along which th Minkowski norm is 0 is called a *world line*. The set of all world lines forms a cone aroun the time axis with the apex at the origin.

$$\|\boldsymbol{x}\|^2 = x_1'^2 + \cdots + x_p'^2 - x_{p+1}'^2 - \cdots - x_{p+q}'^2. \tag{8.6}$$

This mapping is not unique, but p and q are always the same whatever mapping is used. This fact is known as *Sylvester's law of inertia*. The metric defined by the inner product of Eq. (8.3) has signature $(4,1)$ and hence is a Minkowski metric. This can be confirmed by rewriting Eq. (8.4) in the form

$$\|x\|^2 = x_1^2 + x_2^2 + x_3^2 + \left(\frac{x_0}{2} - x_\infty\right)^2 - \left(\frac{x_0}{2} + x_\infty\right)^2. \tag{8.7}$$

Thus, this conformal space is the non-Euclidean space $\mathbb{R}^{4,1}$ ($\hookrightarrow$ Exercise 8.2(2), (3)).

8.2 REPRESENTATION OF POINTS, PLANES, AND SPHERES

Points, planes, and spheres in 3D can be represented by elements of the 5D conformal space. By "represented," we mean that their equations have the form $p \cdot (\cdots) = 0$. It is what we called the *dual representation* in Chapters 5 and 7.

8.2.1 Representation of points

In this conformal space, points are represented rather differently than in previous chapters: a point p at $\boldsymbol{x} = xe_1 + ye_2 + ze_3$ in 3D is represented by the element

$$p = e_0 + \boldsymbol{x} + \frac{1}{2}\|\boldsymbol{x}\|^2 e_\infty \tag{8.8}$$

of the 5D conformal space. From Eq. (8.4), its norm is

$$\|p\|^2 = 0. \tag{8.9}$$

In other words, *all 3D points have norm 0* in this space ($\hookrightarrow$ Exercise 8.1). From Eq. (8.3), the inner product with another point $q = e_0 + \boldsymbol{y} + \|\boldsymbol{y}\|^2 e_\infty/2$ is given by

$$\langle p, q\rangle = \langle \boldsymbol{x}, \boldsymbol{y}\rangle - \frac{1}{2}\|\boldsymbol{x}\|^2 - \frac{1}{2}\|\boldsymbol{y}\|^2 = -\frac{1}{2}\langle \boldsymbol{x} - \boldsymbol{y}, \boldsymbol{x} - \boldsymbol{y}\rangle = -\frac{1}{2}\|\boldsymbol{x} - \boldsymbol{y}\|^2. \tag{8.10}$$

Hence, the square distance between 3D positions $\boldsymbol{x}$ and $\boldsymbol{y}$ is expressed in the form

$$\|\boldsymbol{x} - \boldsymbol{y}\|^2 = -2\langle p, q\rangle. \tag{8.11}$$

If we let $\boldsymbol{x} = 0$ in Eq. (8.8), we have $p = e_0$. Hence, *the symbol e_0 is identified with the origin in 3D* as in the preceding chapter. We also regard this space as homogeneous in the same sense as the 4D space in the preceding chapter. Namely, for any scalar $\alpha \neq 0$, we regard elements x and αx as representing the same geometric object. Hence, Eq. (8.8) and

$$\frac{p}{\|\boldsymbol{x}\|^2/2} = \frac{e_0}{\|\boldsymbol{x}\|^2/2} + \frac{\boldsymbol{x}}{\|\boldsymbol{x}\|^2/2} + e_\infty \tag{8.12}$$

represent the same 3D point. Since this converges to e_∞ in the limit $\|\boldsymbol{x}\|^2 \to \infty$, *the symbol e_∞ is identified with a point at infinity in 3D*. In the preceding chapter, we associated different 3D directions with different points at infinity. In this chapter, we adopt a different interpretation: in whatever direction we move, we reach in the end *a unique point e_∞ at infinity*. Hereafter, we call e_∞ simply "the infinity".

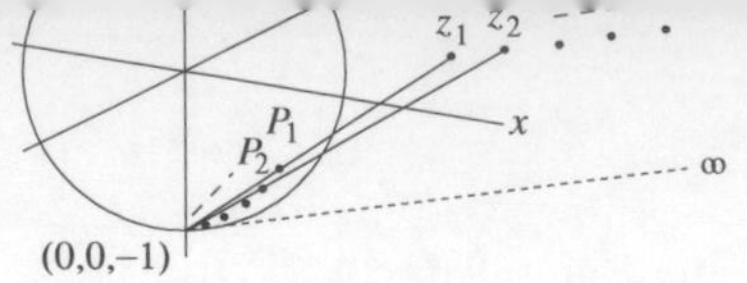

FIGURE 8.2 For a divergent sequence z_1, z_2, ... in the complex plane, the point sequence P_1, P ... on the unit sphere obtained by stereographic projection converges to the south pole $(0, 0, 1$ In the complex plane, this is interpreted to be an accumulation in the neighborhood of the poi ∞ at infinity. The complex plane identified with the unit sphere in this way is called the Riemar surface.

Traditional World 8.2 (One-point compactification) A typical example of regardi infinity as a unique point is complex analysis. If a point (x, y) is associated with the compl number $z = x + iy$, the xy plane is called the *complex plane*. If the limit $\lim_{z\to 0} 1/z$ is al regarded as a complex number, denoted by ∞, the set of all complex numbers including c has the same topology with a sphere, i.e., there exists a one-to-one correspondence betwee the complex plane and a sphere by a continuous mapping. Such a mapping is given by tl stereographic projection shown in Fig. 4.2 in Chapter 4. By this mapping, the "south pol $(0, 0, -1)$ of the sphere corresponds to $z = \infty$ (Fig. 8.2).

Spaces that have the same topology as a finite closed space like a sphere are said to l *compact*. To be precise, a space is compact if any infinite sequence in it has an accumulati point, i.e., a point any neighborhood of which contains an infinite number of points. divergent sequence in the complex plane does not have accumulation points, but if c is added to the complex plane, a divergent sequence is regarded as accumulating in tl neighborhood of ∞. The process of making a space compact by adding one point like th is called *one-point compactification*. As a result, the complex plane can be identified with sphere, called the *Riemann sphere*. The symbol e_∞ of the conformal space has a meani similar to ∞ of the complex plane, and the entire space can be regarded as topological closed.

8.2.2 Representation of planes

For a 3D vector $\boldsymbol{n} = n_1 e_1 + n_2 e_2 + n_3 e_3$, the element

$$\pi = \boldsymbol{n} + h e_\infty \tag{8.13}$$

of the conformal space represents the plane $\langle \boldsymbol{n}, \boldsymbol{x} \rangle = h$ in 3D (Fig. 8.3(a)). By "represents we mean that the equation of the plane is written in terms of the point p in Eq. (8.8) i the form

$$p \cdot \pi = 0, \tag{8.14}$$

where the dot $\cdot$ indicates the contraction operation introduced in Chapter 5; contractic between two elements of the conformal space is simply their inner product. It is easy to se that Eq. (8.14) is indeed the equation of the plane. In fact,

$$p \cdot \pi = \langle e_0 + \boldsymbol{x} + \frac{1}{2}\|\boldsymbol{x}\|^2 e_\infty, \boldsymbol{n} + h e_\infty \rangle = \langle \boldsymbol{x}, \boldsymbol{n} \rangle + \langle e_0, h e_\infty \rangle = \langle \boldsymbol{x}, \boldsymbol{n} \rangle - h. \tag{8.15}$$

This is a consequence of the fact that e_0 and e_∞ are orthogonal to all basis elemen including themselves except for $\langle e_0, e_\infty \rangle = -1$.

(a) (b)

FIGURE 8.3 (a) The plane π with unit surface normal $\boldsymbol{n}$ at distance h from the origin e_0. Its equation is written as $p \cdot \pi = 0$. (b) The sphere σ of center c and radius r. Its equation is written as $p \cdot \sigma = 0$.

In Chapters 5 and 7, we introduced two types of representation for a geometric object: if its equation has the form $p \wedge (\cdots) = 0$, then $(\cdots)$ is called its "direct representation"; if its equation is $p \cdot (\cdots) = 0$, then $(\cdots)$ is its "dual representation." Thus, Eq. (8.13) is the dual representation of the plane $\langle \boldsymbol{n}, \boldsymbol{x} \rangle = h$. Since this space is homogeneous, any scalar multiple $\alpha\pi$ of Eq. (8.13) for $\alpha \neq 0$ also represents the same plane. Letting $h = 0$ reduces Eq. (8.13) to $\pi = \boldsymbol{n}$. Hence, we can view a 3D vector in the form $\boldsymbol{n} = n_1 e_1 + n_2 e_2 + n_3 e_3$ as the dual representation of the orientation of the plane *irrespective of its position*.

Instead of the surface normal $\boldsymbol{n}$ and the distance h from the origin e_0, we can define a plane by specifying two points p_1 and p_2: the plane is defined to be their orthogonal bisector. This can be done by letting

$$\pi = p_1 - p_2. \tag{8.16}$$

For $p_i = e_0 + \boldsymbol{x}_i + \|\boldsymbol{x}_i\|^2 e_\infty/2$, $i = 1, 2$, and p of Eq. (8.8), we see that Eq. (8.10) implies

$$p \cdot \pi = \langle p, p_1 \rangle - \langle p, p_2 \rangle = -\frac{1}{2}\|\boldsymbol{x} - \boldsymbol{x}_1\|^2 + \frac{1}{2}\|\boldsymbol{x} - \boldsymbol{x}_2\|^2. \tag{8.17}$$

Hence, Eq. (8.14) means $\|\boldsymbol{x} - \boldsymbol{x}_1\|^2 = \|\boldsymbol{x} - \boldsymbol{x}_2\|^2$, i.e., the orthogonal bisector of $\boldsymbol{x}_1$ and $\boldsymbol{x}_2$, so Eq. (8.16) is its dual representation.

Using the same logic, we see that Eq. (8.8) itself is the dual representation. In fact, for a point $q = e_0 + \boldsymbol{y} + \|\boldsymbol{y}\|^2 e_\infty/2$ at $\boldsymbol{y}$, Eq. (8.10) implies $p \cdot q = -\|\boldsymbol{x} - \boldsymbol{y}\|^2/2$, so $p \cdot q = 0$ means $\boldsymbol{x} = \boldsymbol{y}$, which is the "equation of the point $\boldsymbol{y}$."

8.2.3 Representation of spheres

In the following, we call the element $c = e_0 + \boldsymbol{c} + \|\boldsymbol{c}\|^2 e_\infty/2$ of the conformal space that represents a 3D position $\boldsymbol{c} = c_1 e_1 + c_2 e_2 + c_3 e_3$ simply "point c" and identify it with the 3D position $\boldsymbol{c}$. Now, the expression

$$\sigma = c - \frac{r^2}{2} e_\infty \tag{8.18}$$

represents a sphere of radius r centered at c (Fig. 8.3(b)). This is confirmed by computing $p \cdot \sigma$:

$$p \cdot \sigma = \langle p, c - \frac{r^2}{2} e_\infty \rangle = \langle p, c \rangle - \frac{r^2}{2}\langle p, e_\infty \rangle = -\frac{1}{2}\|\boldsymbol{x} - \boldsymbol{c}\|^2 - \frac{r^2}{2}. \tag{8.19}$$

Note that $\langle p, e_\infty \rangle = \langle e_0, e_\infty \rangle = -1$. Thus,

$$p \cdot \sigma = 0 \tag{8.20}$$

which is the point c itself, *a point in 3D is interpreted to be a sphere of radius 0*.

8.3 GRASSMANN ALGEBRA IN CONFORMAL SPACE

The outer product operation $\wedge$ can be defined by the antisymmetry $e_i \wedge e_j = -e_i \wedge e_j$, $i, j =$ 1, 2, 3, ∞, independent of the inner product. Hence, we can consider the Grassmann algebra of Chapter 5 in this space and define the "direct representation" of geometric objects in the sense that their equations have the form $p \wedge (\cdots) = 0$. In the following, we derive direct representations of lines, planes, spheres, circles, and point pairs. It turns out that circles and spheres are the most fundamental objects and that lines and planes are interpreted to be circles and spheres of infinite radius passing through infinity.

8.3.1 Direct representations of lines

In the 4D homogeneous space of the preceding chapter, a line L passing through two points p_1 and p_2 was represented by $p_1 \wedge p_2$. In the 5D conformal space, we now show that it becomes

$$L = p_1 \wedge p_2 \wedge e_\infty. \tag{8.21}$$

As in Chapters 5 and 7, we mean by this that its equation is written in terms of point p of Eq. (8.8) in the form

$$p \wedge L = 0. \tag{8.22}$$

This is shown as follows. Letting $p_i = e_0 + \boldsymbol{x}_i + \|\boldsymbol{x}_i\|^2 e_\infty/2$, $i = 1, 2$, we see that

$$\begin{aligned} p \wedge L &= (e_0 + \boldsymbol{x} + \frac{1}{2}\|\boldsymbol{x}\|^2 e_\infty) \wedge (e_0 + \boldsymbol{x}_1 + \frac{1}{2}\|\boldsymbol{x}_1\|^2 e_\infty) \wedge (e_0 + \boldsymbol{x}_2 + \frac{1}{2}\|\boldsymbol{x}_2\|^2 e_\infty) \wedge e_\infty \\ &= (e_0 + \boldsymbol{x}) \wedge (e_0 + \boldsymbol{x}_1) \wedge (e_0 + \boldsymbol{x}_2) \wedge e_\infty. \end{aligned} \tag{8.23}$$

Note that the last term $\wedge e_\infty$ of L annihilates the symbol e_∞ in the expressions of p_1 and p_2. Since e_∞ is orthogonal to e_1, e_2, and e_3, it is linearly independent of the first three factors. Hence, $p \wedge L = 0$ implies

$$(e_0 + \boldsymbol{x}) \wedge (e_0 + \boldsymbol{x}_1) \wedge (e_0 + \boldsymbol{x}_2) = 0 \tag{8.24}$$

in the 4D homogeneous space considered in the preceding chapter, which confirms that Eq. (8.21) is the direct representation of line L.

We see that, thanks to the the expression $(\cdots) \wedge e_\infty$ in Eq. (8.21), all the results in the 4D homogeneous space of the preceding chapter hold *as if the symbol e_∞ did not exist*. The existence of the extra term $\wedge e_\infty$ in Eq. (8.21) is a natural consequence of our interpretation that *a line passes through the infinity e_∞*. In fact, Eq. (8.22) is satisfied for $p = p_1$, p_2, e_∞ by the antisymmetry of the outer product (Fig. 8.4). By a similar logic, the line passing through p and extending in the direction $\boldsymbol{u}$ is represented by

$$L = p \wedge \boldsymbol{u} \wedge e_\infty. \tag{8.25}$$

If we let $q = e_0 + \boldsymbol{y} + \|\boldsymbol{y}\|^2 e_\infty/2$, the equality $p \wedge q = 0$ implies that p is a scalar multiple of q. Hence, $p \wedge q = 0$ is the "equation of point q." This means that Eq. (8.8) is the direct representation of point p. At the same time, it is its dual representation, as pointed out at the end of Sec. 8.2.2.

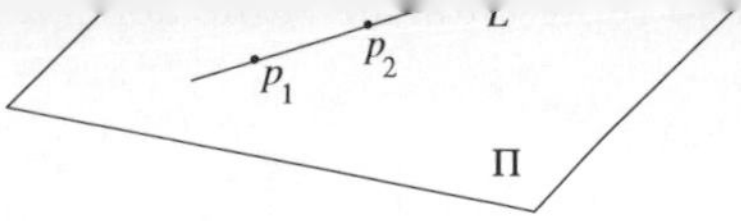

FIGURE 8.4 The line L passing through two points p_1 and p_2 and the plane Π passing through three points p_1, p_2, and p_3. Both pass through the infinity e_∞, so they have respective representations $L = p_1 \wedge p_2 \wedge e_\infty$ and $\Pi = p_1 \wedge p_2 \wedge p_3 \wedge e_\infty$, and their equations have the form $p \wedge L = 0$ and $p \wedge \Pi = 0$, respectively.

8.3.2 Direct representation of planes

By the same logic as above, the direct representation of the plane Π passing through three points p_1, p_2, and p_3 is

$$\Pi = p_1 \wedge p_2 \wedge p_3 \wedge e_\infty, \tag{8.26}$$

and its equation is written in terms of the point p in Eq. (8.8) as

$$p \wedge \Pi = 0. \tag{8.27}$$

The last term $\wedge e_\infty$ indicates that *a plane passes through the infinity* e_∞, and Eq. (8.27) is satisfied for $p = p_1$, p_2, p_3, e_∞ (Fig. 8.4). Thanks to the last term $\wedge e_\infty$ of Eq. (8.26), all the results in the preceding chapter hold for planes, as pointed out earlier. For example, the plane that passes through points p_1 and p_2 and contains the direction $\boldsymbol{u}$ is represented by

$$\Pi = p_1 \wedge p_2 \wedge \boldsymbol{u} \wedge e_\infty. \tag{8.28}$$

This can also be written as $\Pi = -L \wedge \boldsymbol{u}$ in terms of the line L of Eq. (8.21). Similarly, the plane that passes through point p and contains directions $\boldsymbol{u}$ and $\boldsymbol{v}$ has the representation

$$\Pi = p \wedge \boldsymbol{u} \wedge \boldsymbol{v} \wedge e_\infty, \tag{8.29}$$

which can also be written as $\Pi = -L \wedge \boldsymbol{v}$ in terms of the line L of Eq. (8.25).

8.3.3 Direct representation of spheres

For $p_i = e_0 + \boldsymbol{x}_i + \|\boldsymbol{x}_i\|^2 e_\infty/2$, $i = 1, 2, 3, 4$, representing four positions $\boldsymbol{x}_i$ in 3D, the expression

$$\Sigma = p_1 \wedge p_2 \wedge p_3 \wedge p_4 \tag{8.30}$$

represents the sphere that passes through them (Fig. 8.5(a)), and

$$p \wedge \Sigma = 0 \tag{8.31}$$

is its equation. This can be shown as follows. Expanding the left side, we see that

$$\begin{aligned} p \wedge \Sigma &= (e_0 + \boldsymbol{x} + \frac{1}{2}\|\boldsymbol{x}\|^2 e_\infty) \wedge (e_0 + \boldsymbol{x}_1 + \frac{1}{2}\|\boldsymbol{x}_1\|^2 e_\infty) \wedge (e_0 + \boldsymbol{x}_2 + \frac{1}{2}\|\boldsymbol{x}_1\|^2 e_\infty) \\ &\quad \wedge (e_0 + \boldsymbol{x}_3 + \frac{1}{2}\|\boldsymbol{x}_3\|^2 e_\infty) \wedge (e_0 + \boldsymbol{x}_4 + \frac{1}{2}\|\boldsymbol{x}_4\|^2 e_\infty) \\ &= (\cdots) e_0 \wedge e_1 \wedge e_2 \wedge e_3 \wedge e_\infty, \end{aligned} \tag{8.32}$$

(a) (b)

FIGURE 8.5 (a) The sphere Σ passing through four points p_1, p_2, p_3, and p_4 has the representatic $\Sigma = p_1 \wedge p_2 \wedge p_3 \wedge p_4$. Its equation is $p \wedge \Sigma = 0$. (b) The circle S passing through three points p p_2, and p_3 has the representation $S = p_1 \wedge p_2 \wedge p_3$. Its equation is $p \wedge S = 0$.

where $(\cdots)$ is a linear expression in $\boldsymbol{x}$ and $\|\boldsymbol{x}\|^2$. If we let this expression be 0, we obtai the equation of a sphere, and Eq. (8.31) is automatically satisfied if p coincides with ar of p_i due the properties of the outer product. Thus, Σ represents the sphere that pass through the four points p_i, $i = 1, 2, 3$. Note that if we let $p_4 = e_\infty$, Eq. (8.30) reduces Eq. (8.26). This fact provides the interpretation that *a plane is a sphere of infinite radi passing through the infinity* e_∞.

8.3.4 Direct representation of circles and point pairs

For $p_i = e_0 + \boldsymbol{x}_i + \|\boldsymbol{x}_i\|^2 e_\infty/2$, $i = 1, 2, 3$, representing three positions $\boldsymbol{x}_i$ in 3D, th expression

$$S = p_1 \wedge p_2 \wedge p_3 \qquad (8.33)$$

represents the circle that passes through them (Fig. 8.5(b)), and

$$p \wedge S = 0 \qquad (8.34)$$

is its equation. The derivation is a little complicated, but this can be intuitively understoo as follows:

It is clear that the object S defined by Eq. (8.34) passes through the three points p p_2, p_3 from the properties of the outer product. As in the case of Eq. (8.32), expansion the left sides leads to the form

$$\begin{aligned} p \wedge S = &(\cdots)e_1 \wedge e_2 \wedge e_3 \wedge e_\infty + (\cdots)e_0 \wedge e_2 \wedge e_3 \wedge e_\infty + (\cdots)e_0 \wedge e_3 \wedge e_1 \wedge e_\infty \\ &+(\cdots)e_0 \wedge e_1 \wedge e_2 \wedge e_\infty + (\cdots)e_0 \wedge e_1 \wedge e_2 \wedge e_3, \end{aligned} \qquad (8.35)$$

where $(\cdots)$ are all linear in $\boldsymbol{x}$ and $\|\boldsymbol{x}\|^2$. The equation $p \wedge S = 0$ means $p \wedge p_1 \wedge p_2 \wedge p_3 =$ 0, implying no sphere that passes through the four points p, p_1, p_2, and p_3 exists. Henc point p should be on the plane passing through p_1, p_2, and p_3. Thus, S is on a plane an specified by linear equations in $\boldsymbol{x}$ and $\|\boldsymbol{x}\|^2$. This implies that S is an intersection betwee a sphere and plane, i.e., a circle. If we let $p_3 = e_\infty$, Eq. (8.33) reduces to Eq. (8.21). Th fact provides the interpretation that *a line is a circle of infinite radius passing through th infinity* e_∞.

Now, consider the two-point version of Eq. (8.33), i.e.,

$$p_1 \wedge p_2. \qquad (8.36)$$

This is a point pair, which can be regarded as a low-dimensional sphere (Fig. 8.6(a)). Th is understood by the reasoning that a sphere (2D sphere) is "the set of points equidistan

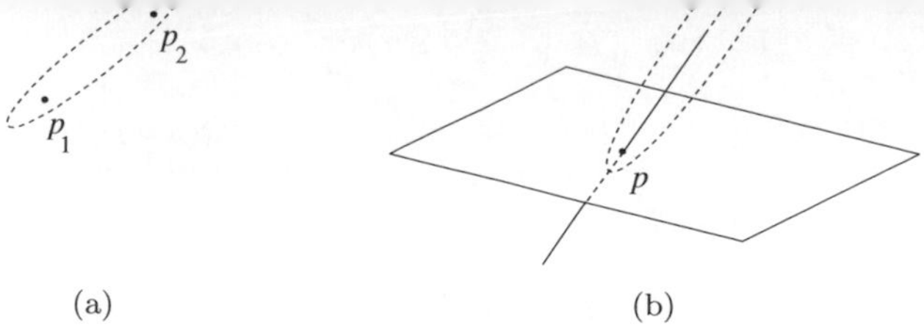

FIGURE 8.6 (a) A point pair $\{p_1, p_2\}$ is regarded as a low-dimensional sphere and hence is represented by $p_1 \wedge p_2$. (b) A flat point p such as the intersection of a plane and a line is a point pair with the infinity e_∞ and hence is represented by $p \wedge e_\infty$.

from a point in 3D" and a circle (1D sphere) is "the set of points equidistant from a point in 2D." Hence, a point pair, i.e.,"the set of points equidistant from a point in 1D" is a 0D sphere. If one of the two points is replaced by e_∞, the point pair $p \wedge e_\infty$ represents one point, but this is geometrically distinguished from an isolated point p (= a sphere of radius 0) given by Eq. (8.8) and called a *flat point*. A flat point $p \wedge e_\infty$ appears as an intersection of a plane and a line and an intersection of three planes. This is because lines and planes all pass through the infinity e_∞, so their intersections should necessarily contain e_∞ (Fig. 8.6(b)).

8.4 DUAL REPRESENTATION

The volume element of the 5D conformal space is

$$I_5 = e_0 \wedge e_1 \wedge e_2 \wedge e_3 \wedge e_\infty, \tag{8.37}$$

and dual expression is given by

$$(\cdots)^* = -(\cdots) \cdot I_5. \tag{8.38}$$

From the rule of the contraction ($\hookrightarrow$ Sec. 5.3 of Chapter 5), we observe the following ($\hookrightarrow$ Eq. (5.78) in Chapter 5):

$$(p \wedge (\cdots))^* = p \cdot (\cdots)^*. \tag{8.39}$$

Hence, if $p \wedge (\cdots) = 0$, then $p \cdot (\cdots)^* = 0$, and if $p \cdot (\cdots) = 0$, then $p \wedge (\cdots)^* = 0$ ($\hookrightarrow$ Eq. (5.79) in Chapter 5).

In the following, we consider the explicit form of the dual representations for planes, lines, circles, point pairs, and flat points. We will see that the outer product $\wedge$ means "join" for direct representations and "meet" for dual representations. To distinguish them, we use uppercase letters for direct representations and lowercase letters for dual representation.

8.4.1 Dual representation for planes

Consider the plane Π passing through three points $p_i = e_0 + \boldsymbol{x}_i + \|\boldsymbol{x}_i\|^2 e_\infty/2$, $i = 1, 2, 3$:

$$\begin{aligned}\Pi &= p_1 \wedge p_2 \wedge p_3 \wedge e_\infty \\ &= (e_0 + \boldsymbol{x}_1 + \frac{1}{2}\|\boldsymbol{x}_1\|^2 e_\infty) \wedge (e_0 + \boldsymbol{x}_2 + \frac{1}{2}\|\boldsymbol{x}_2\|^2 e_\infty) \wedge (e_0 + \boldsymbol{x}_3 + \frac{1}{2}\|\boldsymbol{x}_3\|^2 e_\infty) \wedge e_\infty \\ &= (e_0 + \boldsymbol{x}_1) \wedge (e_0 + \boldsymbol{x}_2) \wedge (e_0 + \boldsymbol{x}_3) \wedge e_\infty.\end{aligned} \tag{8.40}$$

$$\Pi = (n_1 e_0 \wedge e_2 \wedge e_3 + n_2 e_0 \wedge e_3 \wedge e_1 + n_3 e_0 \wedge e_1 \wedge e_2 + h e_1 \wedge e_2 \wedge e_3) \wedge e_\infty. \tag{8.41}$$

Recalling that contraction is defined by consecutive inner products "from inside" with alternating signs ($\hookrightarrow$ Sec. 5.3 in Chapter 5), we see that the dual of $e_0 \wedge e_2 \wedge e_3 \wedge e_\infty$ is, from its definition,

$$\begin{aligned}
&(e_0 \wedge e_2 \wedge e_3 \wedge e_\infty)^* = -e_0 \wedge e_2 \wedge e_3 \wedge e_\infty \cdot e_0 \wedge e_1 \wedge e_2 \wedge e_3 \wedge e_\infty \\
&= -e_0 \wedge e_2 \wedge e_3 \cdot (e_\infty \cdot e_0 \wedge e_1 \wedge e_2 \wedge e_3 \wedge e_\infty) \\
&= -e_0 \wedge e_2 \wedge e_3 \cdot \langle e_\infty, e_0 \rangle \wedge e_1 \wedge e_2 \wedge e_3 \wedge e_\infty \\
&= e_0 \wedge e_2 \wedge e_3 \cdot e_1 \wedge e_2 \wedge e_3 \wedge e_\infty = e_0 \wedge e_2 \cdot (e_3 \cdot e_1 \wedge e_2 \wedge e_3 \wedge e_\infty) \\
&= e_0 \wedge e_2 \cdot e_1 \wedge e_2 \wedge \langle e_3, e_3 \rangle e_\infty = e_0 \wedge e_2 \cdot e_1 \wedge e_2 \wedge e_\infty = e_0 \cdot (e_2 \cdot e_1 \wedge e_2 \wedge e_\infty) \\
&= -e_0 \cdot e_1 \wedge \langle e_2, e_2 \rangle e_\infty = -e_0 \cdot e_1 \wedge e_\infty = e_1 \langle e_0, e_\infty \rangle = -e_1.
\end{aligned} \tag{8.42}$$

Similarly,

$$(e_0 \wedge e_3 \wedge e_1 \wedge e_\infty)^* = -e_2, \qquad (e_0 \wedge e_1 \wedge e_2 \wedge e_\infty)^* = -e_3. \tag{8.43}$$

The dual of $e_1 \wedge e_2 \wedge e_3 \wedge e_\infty$ is obtained also by consecutive contraction "from inside" in the form

$$\begin{aligned}
&(e_1 \wedge e_2 \wedge e_3 \wedge e_\infty)^* = -e_1 \wedge e_2 \wedge e_3 \wedge e_\infty \cdot e_0 \wedge e_1 \wedge e_2 \wedge e_3 \wedge e_\infty \\
&= -e_1 \wedge e_2 \wedge e_3 \cdot (e_\infty \cdot e_0 \wedge e_1 \wedge e_2 \wedge e_3 \wedge e_\infty) \\
&= -e_1 \wedge e_2 \wedge e_3 \cdot (\langle e_\infty, e_0 \rangle e_1 \wedge e_2 \wedge e_3 \wedge e_\infty) = e_1 \wedge e_2 \wedge e_3 \cdot e_1 \wedge e_2 \wedge e_3 \wedge e_\infty \\
&= e_1 \wedge e_2 \cdot (e_3 \cdot e_1 \wedge e_2 \wedge e_3 \wedge e_\infty) = e_1 \wedge e_2 \cdot e_1 \wedge e_2 \wedge \langle e_3, e_3 \rangle e_\infty \\
&= e_1 \cdot (e_2 \cdot e_1 \wedge e_2 e_\infty) - e_1 \cdot e_1 \wedge \langle e_2, e_2 \rangle e_\infty = -e_1 \cdot e_1 \wedge e_\infty = -\langle e_1, e_1 \rangle e_\infty \\
&= -e_\infty.
\end{aligned} \tag{8.44}$$

Hence, the dual of Eq. (8.41) is

$$\Pi^* = -n_1 e_1 - n_2 e_2 - n_3 e_3 - h e_\infty = -(\boldsymbol{n} + h e_\infty), \tag{8.45}$$

which agrees with the dual representation π in Eq. (8.13) except for the sign. Since the conformal space is homogeneous, sign change or scalar multiplication does not affect the representation. By similar calculations, it is shown that the dual Σ^* of the sphere Σ in Eq. (8.30) agrees with the dual sphere representation σ in Eq. (8.18) up to sign and scalar multiplication.

8.4.2 Dual representation for lines

Instead of specifying two points as in Eq. (8.21), a line can be defined as the intersection of two planes. Consider two planes Π_1 and Π_2. As pointed out in Sec. 7.6.3 in Chapter 7, their intersection $\Pi_1 \cap \Pi_2$ has its dual $\pi_1 \cup \pi_2$ ($= \pi_1 \wedge \pi_2$), where $\pi_i = \Pi_i^*$, $i = 1, 2$. Hence

$$l = \pi_1 \wedge \pi_2 \tag{8.46}$$

should be the dual representation of the intersection of the two planes. We now confirm this. Using Eq. (8.13) as the dual representation of the plane, let $\pi_i = \boldsymbol{n}_i + h_i e_\infty$, $i = 1, 2$

$$\begin{aligned}
p\cdot l &= p\cdot \pi_1\wedge\pi_1 = \langle p,\pi_1\rangle\pi_2 - \langle p,\pi_2\rangle\pi_1\\
&= (\langle \boldsymbol{n}_1,\boldsymbol{x}\rangle - h_1)(\boldsymbol{n}_2 + h_2 e_\infty) - (\langle \boldsymbol{n}_2,\boldsymbol{x}\rangle - h_2)(\boldsymbol{n}_1 + h_1 e_\infty)\\
&= (\langle \boldsymbol{n}_2,\boldsymbol{x}\rangle - h_2)\boldsymbol{n}_1 + (\langle \boldsymbol{n}_1,\boldsymbol{x}\rangle - h_1)\boldsymbol{n}_2 + \langle h_2\boldsymbol{n}_1 - h_1\boldsymbol{n}_2,\boldsymbol{x}\rangle e_\infty,
\end{aligned}\tag{8.47}$$

where we noted that $\langle p,\pi_i\rangle = \langle \boldsymbol{n}_i,\boldsymbol{x}\rangle - h_i$ from Eq. (8.15). Since the intersection exists when the surface normals $\boldsymbol{n}_1$ and $\boldsymbol{n}_2$ are linearly independent, the above expression vanishes only when the coefficients of $\boldsymbol{n}_1$, $\boldsymbol{n}_2$, and e_∞ are all 0. Hence, $p\cdot l = 0$ implies $\langle \boldsymbol{n}_i,\boldsymbol{x}\rangle = h_i$, $i=1, 2$, i.e., $\boldsymbol{x}$ satisfies the equation of the two planes, meaning that it is on their intersection. Also, the coefficient of e_∞ in Eq. (8.47) should be 0. Since the supporting plane of the intersection line has surface normal $\boldsymbol{n} = h_2\boldsymbol{n}_1 - h_1\boldsymbol{n}_2$ (↪ Eq. (2.96) in Chapter 2), we have $\langle \boldsymbol{n},\boldsymbol{x}\rangle = 0$ (↪ Eq. (2.62) in Chapter 2). Thus, $p\cdot l = 0$ is the equation of the intersection line, and Eq. (8.46) is its dual representation.

We can also use Eq. (8.16) to define dual representations of planes. If we let $\pi_1 = p_1 - p_2$ and $\pi_2 = p_2 - p_3$, Eq. (8.46) gives the dual representation of their intersection, which is the line perpendicular to the triangle $\triangle p_1p_2p_3$ passing through its circumcenter.

8.4.3 Dual representation of circles, point pairs, and flat points

A circle can be defined as the intersection of two spheres. We can write the dual representations of the two spheres as $\sigma_1 = c_i - r_i^2 e_\infty/2$, $i = 1, 2$, in the form of Eq. (8.18). Then, we obtain the dual representation of the intersection circle,

$$s = \sigma_1\wedge\sigma_2,\tag{8.48}$$

in the same way we define a line as the intersection of two planes. To confirm this, we compute $p\cdot s$. From Eq. (8.19), we see that

$$\begin{aligned}
p\cdot s &= p\cdot \sigma\wedge\sigma_2 = \langle p,\sigma_1\rangle\sigma_2 - \langle p,\sigma_2\rangle\sigma_1\\
&= \Big(-\frac{1}{2}\|\boldsymbol{x} - \boldsymbol{c}_1\|^2 - \frac{r_1^2}{2}\Big)(c_2 - \frac{r_2^2}{2}e_\infty) - \Big(-\frac{1}{2}\|\boldsymbol{x} - \boldsymbol{c}_2\|^2 - \frac{r_2^2}{2}\Big)(c_1 - \frac{r_1^2}{2}e_\infty).
\end{aligned}\tag{8.49}$$

Since the centers c_1 and c_2 of the two spheres are distinct, they are linearly independent, i.e., one is not a scalar multiple of the other. Hence, their coefficients are both 0, and we obtain $\|\boldsymbol{x} - \boldsymbol{c}_i\|^2 = r_i^2$, $i = 1, 2$, which means that $\boldsymbol{x}$ is on both spheres, defining a circle. Hence, Eq. (8.48) is its dual representation. Instead of regarding a circle as the intersection of two spheres, we can regard it as the intersection of a sphere σ and a plane π (= a sphere of infinite radius) to define it by $\sigma\wedge\pi$.

Similarly, a point pair is regarded as the intersection of a circle S and a sphere σ or a plane π (= a sphere of infinite radius), and hence its dual representation is given by $s\wedge\sigma$ or $s\wedge\pi$, where $s = S^*$ is the dual representation of the circle S. Using the same logic, we can regard a flat point whose direct representation is $p\wedge e_\infty$ as the intersection of a plane Π and a line L and hence its dual representation is given by $\pi\wedge l$, where π (= Π^*) and l (= L^*) are the dual representations of the plane Π and the line L, respectively. A flat point can also be regarded as the intersection of three planes, so we can also write its dual representation in the form $\pi_1\wedge\pi_2\wedge\pi_3$, where π_i (= Π_i^*), $i = 1, 2, 3$, are the dual representations of the three planes Π_i.

Table 8.1 summarizes the direct and dual representations in the conformal space described above. Due to the duality theorem in Sec. 7.6 in Chapter 7, the outer product $\wedge$

dual representation of a circle.

object	direct representation	dual representation
(isolated) point	$p = e_0 + \boldsymbol{x} + \|\boldsymbol{x}\|^2 e_\infty/2$	$p = e_0 + \boldsymbol{x} + \|\boldsymbol{x}\|^2 e_\infty/2$
line	$p_1 \wedge p_2 \wedge e_\infty$	$\pi_1 \wedge \pi_2$
	$p \wedge \boldsymbol{u} \wedge e_\infty$	
plane	$p_1 \wedge p_2 \wedge p_3 \wedge e_\infty$	$\boldsymbol{n} + he_\infty$
	$p_1 \wedge p_2 \wedge \boldsymbol{u} \wedge e_\infty$	$p_1 - p_2$
	$p \wedge \boldsymbol{u}_1 \wedge \boldsymbol{u}_2 \wedge e_\infty$	
sphere	$p_1 \wedge p_2 \wedge p_3 \wedge p_4$	$c - r^2 e_\infty/2$
circle	$p_1 \wedge p_2 \wedge p_3$	$\sigma_1 \wedge \sigma_2$
		$\sigma \wedge \pi$
point pair	$p_1 \wedge p_2$	$s \wedge \sigma$
		$s \wedge \pi$
flat point	$p \wedge e_\infty$	$\pi \wedge l$
		$\pi_1 \wedge \pi_2 \wedge \pi_3$
equation	$p \wedge (\cdots) = 0$	$p \cdot (\cdots) = 0$

indicates *join for direct representations* and *meet for dual representations*:

$$\text{(direct representation)} \wedge \text{(direct representation)} = \text{(direct representation of their join)},$$

$$\text{(dual representation)} \wedge \text{(dual representation)} = \text{(dual representation of their meet)}.$$

8.5 CLIFFORD ALGEBRA IN THE CONFORMAL SPACE

As stated in Chapter 6, the Clifford algebra unifies the inner product and the outer produc Namely, the geometric product (the Clifford product) is defined in such a way that *i symmetrization gives the inner product and its antisymmetrization gives the outer produc* This enables us to describe various transformations in 3D in terms of the geometric produc In the following, we introduce the geometric product to the 5D conformal space and sho how translations, rotations, and rigid motions as their compositions in 3D are described i terms of the geometric product.

8.5.1 Inner, outer, and geometric products

In order that the inner product given in Sec. 8.1 results, we define the geometric produ for the basis elements by the following rule (↪ Exercise 8.2(4)). First, we let

$$e_0^2 = e_\infty^2 = 0, \qquad e_0 e_\infty + e_\infty e_0 = -2, \tag{8.5C}$$

and for $i, j = 1, 2, 3$ we let

$$e_i e_0 + e_0 e_i = e_i e_\infty + e_\infty e_i = 0, \qquad e_i^2 = 1, \qquad e_i e_j + e_j e_i = 0. \tag{8.51}$$

We require that the associativity is satisfied and that the product is linearly distribute for linear combinations of the basis elements. Consequently, if we write Eq. (8.1) as $x =$ $x_0 e_0 + \boldsymbol{x} + x_\infty e_\infty$, where $\boldsymbol{x} = x_1 e_1 + x_2 e_2 + x_3 e_3$, and similarly write $y = y_0 e_0 + \boldsymbol{y} + y_\infty e_\infty$

$$\begin{aligned}
&= x_0 y_0 e_0^2 + x_0 e_0 \boldsymbol{y} + x_0 y_\infty e_0 e_\infty + y_0 \boldsymbol{x} e_0 + \boldsymbol{x}\boldsymbol{y} \\
&\quad + y_\infty \boldsymbol{x} e_\infty + x_\infty y_0 e_\infty e_0 + x_\infty e_\infty \boldsymbol{y} + x_\infty y_\infty e_\infty^2 \\
&= (y_0 \boldsymbol{x} - x_0 \boldsymbol{y}) e_0 + \boldsymbol{x}\boldsymbol{y} + x_0 y_\infty e_0 e_\infty + x_\infty y_0 e_\infty e_0 + (y_\infty \boldsymbol{x} - x_\infty \boldsymbol{y}) e_\infty, \qquad (8.52)
\end{aligned}$$

where we have noted that e_0 and e_∞ are both anticommutative with e_i, $i = 1, 2, 3$, and hence anticommutative with $\boldsymbol{x}$ and $\boldsymbol{y}$ as well. From this, we see that symmetrization of the geometric product is

$$\begin{aligned}
\frac{1}{2}(xy + xy) &= \frac{1}{2}\Big(\boldsymbol{x}\boldsymbol{y} + \boldsymbol{y}\boldsymbol{x} + x_0 y_\infty (e_0 e_\infty + e_\infty e_0) + x_\infty y_0 (e_0 e_\infty + e_\infty e_0)\Big) \\
&= \langle \boldsymbol{x}, \boldsymbol{y} \rangle - x_0 y_\infty - x_\infty y_0, \qquad (8.53)
\end{aligned}$$

which agrees with Eq. (8.3). Hence,

$$\langle x, y \rangle = \frac{1}{2}(xy + yx). \qquad (8.54)$$

In particular, $\|x\|^2 = \langle x, x \rangle = x^2$. The outer product is, on the other hand, defined by antisymmetrization as in Sec. 6:

$$\begin{aligned}
x \wedge y &\equiv \frac{1}{2}(xy - yx), \\
x \wedge y \wedge z &\equiv \frac{1}{6}(xyz + yzx + zxy - zyx - yxz - xzy), \\
x \wedge y \wedge z \wedge w &\equiv \frac{1}{24}(xyzw - yxzw + yzxw - yzwx + \cdots), \\
x \wedge y \wedge z \wedge w \wedge u &\equiv \frac{1}{120}(xyzwu - yxzwu + yzxwu - \cdots). \qquad (8.55)
\end{aligned}$$

The right sides are the sum of all permutations, each with its permutation signature, divided by the number of permutations. Outer products that involve more than six basis elements are defined to be 0. It is then shown that all the properties of the inner and outer products are satisfied. In particular, all the results for terms that involve e_1, e_2, and e_3, but not e_0 or e_∞, are identical to those in Chapter 6.

From Eq. (8.54) and the definition of the outer product $x \wedge y$, we obtain the same relationship that we saw in Chapter 6:

$$xy = \langle x, y \rangle + x \wedge y. \qquad (8.56)$$

Hence, for the element p that represents a position in the form of Eq. (8.8), we see from Eq. (8.9) that

$$p^2 = \langle p, p \rangle + p \wedge p = \|p\|^2 = 0. \qquad (8.57)$$

From Eq. (8.50), this also holds if p is e_0 or e_∞. Hence, *the square of all points (including infinity) is 0*.

8.5.2 Translator

We now show that the operation of translating a point by the vector $\boldsymbol{t} = t_1 e_1 + t_2 e_2 + t_3 e_3$ is given by

$$\mathcal{T}_{\boldsymbol{t}} = 1 - \frac{1}{2}\boldsymbol{t} e_\infty. \qquad (8.58)$$

where $\mathcal{T}_t^{-1}$ is the inverse of $\mathcal{T}_t$, i.e., the translator for $-\boldsymbol{t}$:

$$\mathcal{T}_t^{-1} = 1 + \frac{1}{2}\boldsymbol{t}e_\infty \quad (= \mathcal{T}_{-t}). \qquad (8.6$$

This is indeed the inverse of Eq. (8.58):

$$\mathcal{T}_t\mathcal{T}_t^{-1} = \left(1 - \frac{1}{2}\boldsymbol{t}e_\infty\right)\left(1 + \frac{1}{2}\boldsymbol{t}e_\infty\right) = 1 + \frac{1}{2}\boldsymbol{t}e_\infty - \frac{1}{2}\boldsymbol{t}e_\infty + \frac{1}{4}\boldsymbol{t}e_\infty\boldsymbol{t}e_\infty = 1. \qquad (8.6$$

The last term vanishes because e_∞ is anticommutative with e_i, $i = 1, 2, 3$, by Eq. (8.5 ($e_ie_\infty = -e_\infty e_i$), hence $\boldsymbol{t}e_\infty\boldsymbol{t}e_\infty = -\boldsymbol{t}^2e_\infty^2$, and $e_\infty^2 = 0$ from Eq. (8.50).

In order to see how the point p of Eq. (8.8) is moved by the translator of Eq. (8.59), suffices to check how the basis elements e_i, $i = 0, 1, 2, 3, \infty$, i.e., the origin e_0, a vector $= a_1e_1 + a_2e_2 + a_3e_3$, and the infinity e_∞, are mapped. We see the following:

origin e_0 The origin e_0 is translated by $\boldsymbol{t}$ in the form

$$\begin{aligned}\mathcal{T}_te_0\mathcal{T}_t^{-1} &= \left(1 - \frac{1}{2}\boldsymbol{t}e_\infty\right)e_0\left(1 + \frac{1}{2}\boldsymbol{t}e_\infty\right) = e_0 + \frac{1}{2}e_0\boldsymbol{t}e_\infty - \frac{1}{2}\boldsymbol{t}e_\infty e_0 - \frac{1}{4}\boldsymbol{t}e_\infty e_0\boldsymbol{t}e_\infty \\ &= e_0 + \frac{1}{2}(e_0\boldsymbol{t}e_\infty - \boldsymbol{t}e_\infty e_0) - \frac{1}{4}\boldsymbol{t}e_\infty e_0\boldsymbol{t}e_\infty = e_0 + \boldsymbol{t} + \frac{1}{2}\|\boldsymbol{t}\|^2e_\infty, \qquad (8.6\end{aligned}$$

where we use the following rules:

$$e_0\boldsymbol{t}e_\infty - \boldsymbol{t}e_\infty e_0 = -\boldsymbol{t}e_0e_\infty - \boldsymbol{t}e_\infty e_0 = -\boldsymbol{t}(e_0e_\infty + e_\infty e_0) = 2\boldsymbol{t}, \qquad (8.6$$

$$\begin{aligned}\boldsymbol{t}e_\infty e_0\boldsymbol{t}e_\infty &= -\boldsymbol{t}e_\infty\boldsymbol{t}e_0e_\infty = \boldsymbol{t}^2e_\infty e_0e_\infty = \|\boldsymbol{t}\|^2e_\infty e_0e_\infty \\ &= \|\boldsymbol{t}\|^2(-2 - e_0e_\infty)e_\infty = \|\boldsymbol{t}\|^2(-2e_\infty - e_0e_\infty^2) = -2\|\boldsymbol{t}\|^2e_\infty. \qquad (8.6\end{aligned}$$

Equation (8.62) states that the origin e_0 is translated by $\boldsymbol{t}$ to the point $t = e_0 + \boldsymbol{t}$ $\|\boldsymbol{t}\|^2e_\infty/2$.

vector $\boldsymbol{a}$ Vector $\boldsymbol{a}$ is translated by $\boldsymbol{t}$ in the form

$$\begin{aligned}\mathcal{T}_t\boldsymbol{a}\mathcal{T}_t^{-1} &= \left(1 - \frac{1}{2}\boldsymbol{t}e_\infty\right)\boldsymbol{a}\left(1 + \frac{1}{2}\boldsymbol{t}e_\infty\right) = \boldsymbol{a} + \frac{1}{2}\boldsymbol{a}\boldsymbol{t}e_\infty - \frac{1}{2}\boldsymbol{t}e_\infty\boldsymbol{a} - \frac{1}{4}\boldsymbol{t}e_\infty\boldsymbol{a}\boldsymbol{t}e_\infty \\ &= \boldsymbol{a} + \frac{1}{2}(\boldsymbol{a}\boldsymbol{t}e_\infty - \boldsymbol{t}e_\infty\boldsymbol{a}) - \frac{1}{4}\boldsymbol{t}e_\infty\boldsymbol{a}\boldsymbol{t}e_\infty = \boldsymbol{a} + \langle\boldsymbol{a}, \boldsymbol{t}\rangle e_\infty, \qquad (8.65\end{aligned}$$

where we have used the following rules:

$$\boldsymbol{a}\boldsymbol{t}e_\infty - \boldsymbol{t}e_\infty\boldsymbol{a} = \boldsymbol{a}\boldsymbol{t}e_\infty + \boldsymbol{t}\boldsymbol{a}e_\infty = (\boldsymbol{a}\boldsymbol{t} + \boldsymbol{t}\boldsymbol{a})e_\infty = 2\langle\boldsymbol{a}, \boldsymbol{t}\rangle e_\infty, \qquad (8.66$$

$$\boldsymbol{t}e_\infty\boldsymbol{a}\boldsymbol{t}e_\infty = -\boldsymbol{t}\boldsymbol{a}e_\infty\boldsymbol{t}e_\infty = \boldsymbol{t}\boldsymbol{a}\boldsymbol{t}e_\infty^2 = 0. \qquad (8.67$$

infinity e_∞ The infinity e_∞ is translated by $\boldsymbol{t}$ in the form

$$\mathcal{T}_te_\infty\mathcal{T}_t^{-1} = \left(1 - \frac{1}{2}\boldsymbol{t}e_\infty\right)e_\infty\left(1 + \frac{1}{2}\boldsymbol{t}e_\infty\right) = e_\infty + \frac{1}{2}e_\infty\boldsymbol{t}e_\infty - \frac{1}{2}\boldsymbol{t}e_\infty^2 - \frac{1}{4}\boldsymbol{t}e_\infty^2\boldsymbol{t}e_\infty = e_\infty \qquad (8.68$$

where we note the relationship $e_\infty\boldsymbol{t}e_\infty = -\boldsymbol{t}e_\infty^2 = 0$. The above result shows that th infinity e_∞ after the translation is still at the infinity e_∞.

$$
\begin{aligned}
\mathcal{T}_t p \mathcal{T}_t^{-1} &= \mathcal{T}_t\Big(e_0 + \boldsymbol{x} + \frac{1}{2}\|\boldsymbol{x}\|^2 e_\infty\Big)\mathcal{T}_t^{-1} = \mathcal{T}_t e_0 \mathcal{T}_t^{-1} + \mathcal{T}_t \boldsymbol{x} \mathcal{T}_t^{-1} + \frac{1}{2}\|\boldsymbol{x}\|^2 \mathcal{T}_t e_\infty \mathcal{T}_t^{-1} \\
&= e_0 + \boldsymbol{t} + \frac{1}{2}\|\boldsymbol{t}\|^2 e_\infty + \boldsymbol{x} + \langle \boldsymbol{x}, \boldsymbol{t}\rangle e_\infty + \frac{1}{2}\|\boldsymbol{x}\|^2 e_\infty \\
&= e_0 + \boldsymbol{x} + \boldsymbol{t} + \frac{1}{2}(\|\boldsymbol{x}\|^2 + 2\langle \boldsymbol{x}, \boldsymbol{t}\rangle + \|\boldsymbol{t}\|^2) e_\infty = e_0 + (\boldsymbol{x} + \boldsymbol{t}) + \frac{1}{2}\|\boldsymbol{x} + \boldsymbol{t}\|^2 e_\infty .
\end{aligned} \tag{8.69}
$$

Thus, it is translated to the position $\boldsymbol{x} + \boldsymbol{t}$.

Next, consider the composition of translations. If a point is translated by $\boldsymbol{t}_1$ and then by $\boldsymbol{t}_2$, the net result is

$$
\mathcal{T}_{\boldsymbol{t}_2}(\mathcal{T}_{\boldsymbol{t}_1}(\cdots)\mathcal{T}_{\boldsymbol{t}_1}^{-1})\mathcal{T}_{\boldsymbol{t}_2}^{-1} = (\mathcal{T}_{\boldsymbol{t}_2}\mathcal{T}_{\boldsymbol{t}_1})(\cdots)(\mathcal{T}_{\boldsymbol{t}_2}\mathcal{T}_{\boldsymbol{t}_1})^{-1}, \tag{8.70}
$$

which should be translation by $\boldsymbol{t}_1 + \boldsymbol{t}_2$. Hence, we obtain the following identity:

$$
\mathcal{T}_{\boldsymbol{t}_1 + \boldsymbol{t}_2} = \mathcal{T}_{\boldsymbol{t}_2}\mathcal{T}_{\boldsymbol{t}_1}. \tag{8.71}
$$

If $\boldsymbol{t}_1$ and $\boldsymbol{t}_2$ are interchanged, the right side is $\mathcal{T}_{\boldsymbol{t}_1}\mathcal{T}_{\boldsymbol{t}_2}$. Hence, $\mathcal{T}_{\boldsymbol{t}_2}\mathcal{T}_{\boldsymbol{t}_1} = \mathcal{T}_{\boldsymbol{t}_1}\mathcal{T}_{\boldsymbol{t}_2}$, i.e., translators are commutative with each other. Formally, we may write the translator in the form

$$
\mathcal{T}_{\boldsymbol{t}} = \exp(-\frac{1}{2}\boldsymbol{t}e_\infty), \tag{8.72}
$$

where the exponential function "exp" is defined via Taylor expansion (↪ Eqs. (4.38) in Chapter 4, Eq. (6.75) in Chapter 6). In fact, $(\boldsymbol{t}e_\infty)^2 = \boldsymbol{t}e_\infty \boldsymbol{t}e_\infty = -\boldsymbol{t}^2 e_\infty^2 = 0$ and similarly $(\boldsymbol{t}e_\infty)^k = 0$, $k = 2, 3, \ldots$, so

$$
\exp(-\frac{1}{2}\boldsymbol{t}e_\infty) = 1 - \frac{1}{2}\boldsymbol{t}e_\infty + \frac{1}{2!}(\frac{1}{2}\boldsymbol{t}e_\infty)^2 - \frac{1}{3!}(\frac{1}{2}\boldsymbol{t}e_\infty)^3 + \cdots = 1 - \frac{1}{2}\boldsymbol{t}e_\infty. \tag{8.73}
$$

8.5.3 Rotor and motor

We saw in Chapter 6 that a rotation around the origin is specified by two vectors $\boldsymbol{a}$ and $\boldsymbol{b}$ in the form

$$
\mathcal{R} = \boldsymbol{b}\boldsymbol{a}, \tag{8.74}
$$

which acts in the form $\mathcal{R}(\cdots)\mathcal{R}^{-1}$ (↪ Eq. (6.63) in Chapter 6). If the plane of rotation is specified by the surface element $\mathcal{I}$, the rotor of angle Ω was given as follows (↪ Eq. (6.69) in Chapter 6):

$$
\mathcal{R} = \cos\frac{\Omega}{2} - \mathcal{I}\sin\frac{\Omega}{2}. \tag{8.75}
$$

These results hold in the conformal space as well, because rotation around the origin does not affect the origin e_0 and the infinity e_∞. In fact, we see from the operation rules of the geometric product that both e_0 and e_∞ are anticommutative with all 3D vectors and hence

$$
\mathcal{R}e_0\mathcal{R}^{-1} = \boldsymbol{b}\boldsymbol{a}e_0\boldsymbol{a}^{-1}\boldsymbol{b}^{-1} = -\boldsymbol{b}e_0\boldsymbol{a}\boldsymbol{a}^{-1}\boldsymbol{b}^{-1} = e_0\boldsymbol{b}\boldsymbol{a}\boldsymbol{a}^{-1}\boldsymbol{b}^{-1} = e_0, \tag{8.76}
$$

$$
\mathcal{R}e_\infty\mathcal{R}^{-1} = \boldsymbol{b}\boldsymbol{a}e_\infty\boldsymbol{a}^{-1}\boldsymbol{b}^{-1} = -\boldsymbol{b}e_\infty\boldsymbol{a}\boldsymbol{a}^{-1}\boldsymbol{b}^{-1} = e_\infty\boldsymbol{b}\boldsymbol{a}\boldsymbol{a}^{-1}\boldsymbol{b}^{-1} = e_\infty. \tag{8.77}
$$

$$= e_0 + \mathcal{R}\boldsymbol{x}\mathcal{R}^{-1} + \frac{1}{2}\|\boldsymbol{x}\|^2 e_\infty = e_0 + \mathcal{R}\boldsymbol{x}\mathcal{R}^{-1} + \frac{1}{2}\|\mathcal{R}\boldsymbol{x}\mathcal{R}^{-1}\|^2 e_\infty, \quad (8.7$$

which is the point in the rotated position $\mathcal{R}\boldsymbol{x}\mathcal{R}^{-1}$. Note that the norm is preserved b rotation, i.e., $\|\mathcal{R}\boldsymbol{x}\mathcal{R}^{-1}\|^2 = \|\boldsymbol{x}\|^2$.

A rigid motion in 3D is a composition of a rotation around the origin and translatio Hence, if we let

$$\mathcal{M} = \mathcal{T}_{\boldsymbol{t}}\mathcal{R}, \quad (8.7$$

a rigid motion is computed in the form $\mathcal{M}(\cdots)\mathcal{M}^{-1}$. We call such a composition of tran lators and rotors a *motor* (↪ Exercise 8.4).

8.6 CONFORMAL GEOMETRY

Conformal geometry is the study of *conformal transformations*, for which two definitio exist. In a broad sense, they are transformations that preserve angles made by tangen to curves and surfaces at their intersections; in a narrow sense, they are transformatio defined throughout the space, including infinity, that map spheres to spheres. To speci cally mean the latter, they are referred to as *spherical conformal transformations* or *Möbi transformations*. Here, we consider conformal transformations in the latter sense that ma spheres to spheres and preserve the angles made by their tangent planes at their interse tions; a plane is regarded as a sphere of infinite radius. Since the intersection of two spher is a circle, circles are mapped to circles, and the angles made by their tangent lines at th intersections are preserved; a line is regarded as a circle of infinite radius.

In the following, we show that conformal transformations are defined by *versors* d scribed in Sec. 6.8 in Chapter 6 and derive specific versor forms that define reflectio inversion, and dilation. It is shown that reflection and inversion are the basic transform tions and that translation and rotation are obtained by a composition of reflections whi dilation is obtained by a composition of inversions. This gives the interpretation that translation can be viewed as a rotation around an axis located infinitely far away. Finall we summarize the properties of versors.

8.6.1 Conformal transformations and versors

The set of all conformal transformations constitutes a group of transformations, whic includes, among others, the following familiar subgroups:

- similarities
- rigid motions
- rotations
- reflections
- dilations
- translations
- the identity

These transformations do not move infinity. Compositions of rigid motions and reflectio are called *isometries* or *Euclid transformations*. They themselves constitute a closed sub group consisting of conformal transformations that preserve length; they include transla tions, rotations, and the identity as its subgroups.

and circles in terms of the outer product. Clifford algebra allows us to define conformal transformations in terms of the geometric product in the form of versors. A *versor* has the following form (↪ Eq. (6.77) in Chapter 6):

$$\mathcal{V} = v_k v_{k-1} \cdots v_1. \tag{8.80}$$

Here, each v_i has the form $a_0 e_0 + a_1 e_1 + a_2 e_2 + a_3 e_3 + a_\infty e_\infty$. The number k is called the *grade* of the versor. The versor is *odd* if k is odd, and *even* if k is even. We denote the inverse $\mathcal{V}^{-1} = v_1^{-1} v_2^{-1} \cdots v_k^{-1}$ multiplied by the sign $(-1)^k$ by

$$\mathcal{V}^\dagger \equiv (-1)^k \mathcal{V}^{-1} = (-1)^k v_1^{-1} v_2^{-1} \cdots v_k^{-1} \tag{8.81}$$

(↪ Eq. (6.78) in Chapter 6). Versors operate on elements of the conformal space in the following form (↪ Eq. (6.79) in Chapter 6):

$$\mathcal{V}(\cdots)\mathcal{V}^\dagger. \tag{8.82}$$

Since the norm of each v_i on the left is canceled by the norm of the v_i^{-1} on the right, the magnitude v_i does not affect the operation of the versor. If Eq. (8.82) is further transformed by another versor $\mathcal{V}'$, the net effect is

$$\mathcal{V}'\mathcal{V}(\cdots)\mathcal{V}^\dagger\mathcal{V}'^\dagger = (\mathcal{V}'\mathcal{V})(\cdots)(\mathcal{V}'\mathcal{V})^\dagger. \tag{8.83}$$

Hence, the composition of versors is given by the geometric product in the form $\mathcal{V}'' = \mathcal{V}'\mathcal{V}$ (↪ Proposition 6.10 in Chapter 6).

8.6.2 Reflectors

The most basic conformal transformation is reflection. We now show that reflection with respect to a plane π having unit surface normal $\boldsymbol{n}$ located at distance h (positive in the direction of $\boldsymbol{n}$) from the origin e_0 is specified by $\pi = \boldsymbol{n} + he_\infty$, which is the dual representation of that plane (↪ Eq. (8.13)) and is at the same time a versor of grade 1. This versor π is called the *reflector*, and its inverse π^{-1} is π itself. In fact,

$$\pi^2 = (\boldsymbol{n} + he_\infty)(\boldsymbol{n} + he_\infty) = \boldsymbol{n}^2 + h\boldsymbol{n}e_\infty + he_\infty\boldsymbol{n} + h^2 e_\infty^2 = 1 + h\boldsymbol{n}e_\infty - h\boldsymbol{n}e_\infty = 1, \tag{8.84}$$

which corresponds to the interpretation that reflection twice equals the identity.

To see how a given point is reflected, it is sufficient to know how the origin e_0, a vector $\boldsymbol{a}$, and the infinity e_∞ are reflected.

origin e_0 The origin e_0 is reflected to

$$\begin{aligned}
\pi e_0 \pi^\dagger &= -(\boldsymbol{n} + he_\infty)e_0(\boldsymbol{n} + he_\infty) = -\boldsymbol{n}e_0\boldsymbol{n} - h\boldsymbol{n}e_0e_\infty - he_\infty e_0\boldsymbol{n} - h^2 e_\infty e_0 e_\infty \\
&= \boldsymbol{n}^2 e_0 - h\boldsymbol{n}e_0e_\infty - h\boldsymbol{n}e_\infty e_0 - h^2(-2 - e_0e_\infty)e_\infty \\
&= e_0 - h\boldsymbol{n}(e_0e_\infty + e_\infty e_0) + 2h^2 e_\infty + e_0 e_\infty^2 e_\infty = e_0 + 2h\boldsymbol{n} + 2h^2 e_\infty \\
&= e_0 + 2h\boldsymbol{n} + \frac{1}{2}\|2h\boldsymbol{n}\|^2 e_\infty,
\end{aligned} \tag{8.85}$$

which represents a point at $2h\boldsymbol{n}$.

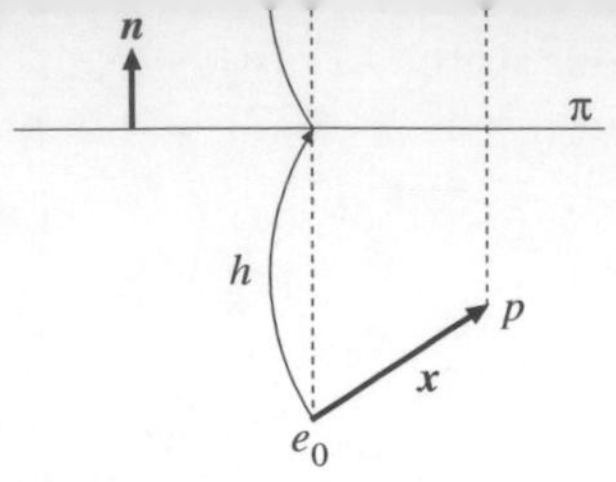

FIGURE 8.7 A point p at $\boldsymbol{x}$ is reflected with respect to the plane $\langle \boldsymbol{n}, \boldsymbol{x} \rangle = h$ to a point $p' = \pi p$ at $2h\boldsymbol{n} + \boldsymbol{x}'$, where $\boldsymbol{x}' = -\boldsymbol{n}\boldsymbol{x}\boldsymbol{n}$.

vector $\boldsymbol{a}$ A vector $\boldsymbol{a} = a_1 e_1 + a_2 e_2 + a_3 e_3$ is reflected to

$$\begin{aligned}\pi \boldsymbol{a} \pi^\dagger &= -(\boldsymbol{n} + h e_\infty)\boldsymbol{a}(\boldsymbol{n} + h e_\infty) = -\boldsymbol{n}\boldsymbol{a}\boldsymbol{n} - h\boldsymbol{n}\boldsymbol{a} e_\infty - h e_\infty \boldsymbol{a}\boldsymbol{n} - h^2 e_\infty \boldsymbol{a} e_\infty \\ &= -\boldsymbol{n}\boldsymbol{a}\boldsymbol{n} - h\boldsymbol{n}\boldsymbol{a} e_\infty - h\boldsymbol{a}\boldsymbol{n} e_\infty + h^2 \boldsymbol{a} e_\infty^2 = -\boldsymbol{n}\boldsymbol{a}\boldsymbol{n} - 2h\langle \boldsymbol{n}, \boldsymbol{a} \rangle e_\infty. \qquad (8.86)\end{aligned}$$

If $h = 0$ in particular, this is $-\boldsymbol{n}\boldsymbol{a}\boldsymbol{n}$, which agrees with the reflection at the origin, as shown in Chapter 6.

infinity e_∞ The infinity e_∞ is reflected to

$$\pi e_\infty \pi^\dagger = -(\boldsymbol{n} + h e_\infty) e_\infty (\boldsymbol{n} + h e_\infty) = -\boldsymbol{n} e_\infty \boldsymbol{n} - h\boldsymbol{n} e_\infty^2 - h e_\infty^2 \boldsymbol{n} - h^2 e_\infty^3 = \boldsymbol{n}^2 e_\infty = e_\infty. \qquad (8.87)$$

Thus, the infinity e_∞ is unaltered by reflection.

From these results, reflection of point $p = e_0 + \boldsymbol{x} + \|\boldsymbol{x}\|^2 e_\infty/2$ at $\boldsymbol{x}$ is computed by applying the reflector term by term as follows:

$$\begin{aligned}\pi p \pi^\dagger &= \pi e_0 \pi^\dagger + \pi \boldsymbol{x} \pi^\dagger + \frac{1}{2}\|\boldsymbol{x}\|^2 \pi e_\infty \pi^\dagger \\ &= \left(e_0 + 2h\boldsymbol{n} + \frac{1}{2}\|2h\boldsymbol{n}\|^2 e_\infty\right) - \left(\boldsymbol{n}\boldsymbol{x}\boldsymbol{n} + 2h\langle \boldsymbol{n}, \boldsymbol{x} \rangle e_\infty\right) + \frac{1}{2}\|\boldsymbol{x}\|^2 e_\infty \\ &= e_0 + (2h\boldsymbol{n} + \boldsymbol{x}') + \frac{1}{2}\|(2h\boldsymbol{n} + \boldsymbol{x}')\|^2 e_\infty. \qquad (8.88)\end{aligned}$$

Here, we put $\boldsymbol{x}' = -\boldsymbol{n}\boldsymbol{x}\boldsymbol{n}$ and use the equalities $\|\boldsymbol{x}'\|^2 = \|\boldsymbol{x}\|^2$ and $\langle 2h\boldsymbol{n}, \boldsymbol{x}' \rangle = \langle 2h\boldsymbol{n}, \boldsymbol{x}$ Thus, we see that the point p is reflected to the position $2h\boldsymbol{n} + \boldsymbol{x}'$ (Fig. 8.7).

The importance of the reflector is in the fact that a translation is described by a composition of reflections. In fact, if a reflection with respect to the plane $\pi = \boldsymbol{n} + h e_\infty$ is followed by a reflection with respect to the plane $\pi' = \boldsymbol{n} + h' e_\infty$, the net effect is

$$\begin{aligned}\pi' \pi &= (\boldsymbol{n} + h' e_\infty)(\boldsymbol{n} + h e_\infty) = \boldsymbol{n}^2 + h\boldsymbol{n} e_\infty + h' e_\infty \boldsymbol{n} + hh' e_\infty^2 \\ &= 1 + h\boldsymbol{n} e_\infty - h'\boldsymbol{n} e_\infty = 1 - \frac{1}{2}(2(h' - h)\boldsymbol{n}) e_\infty, \qquad (8.89)\end{aligned}$$

which is a translator by $\boldsymbol{t} = 2(h' - h)\boldsymbol{n}$. Thus, *consecutive reflections with respect to two parallel planes equal a translation by twice the distance between them.* On the other hand, we showed in Sec. 6.7.1 in Chapter 6 that *consecutive reflections with respect to two intersecting planes equal a rotation by twice the angle between them around their intersection line*

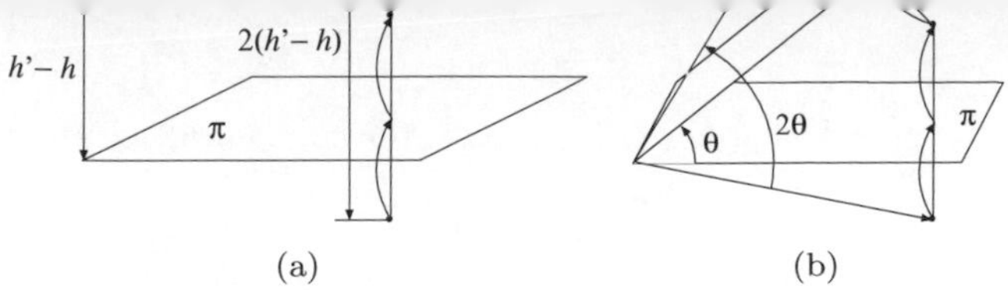

FIGURE 8.8 (a) Composition of reflections with respect to parallel planes in distance $h' - h$ results in a translation by distance $2(h' - h)$. (b) Composition of reflections with respect to intersecting planes with angle θ results in a rotation by angle 2θ.

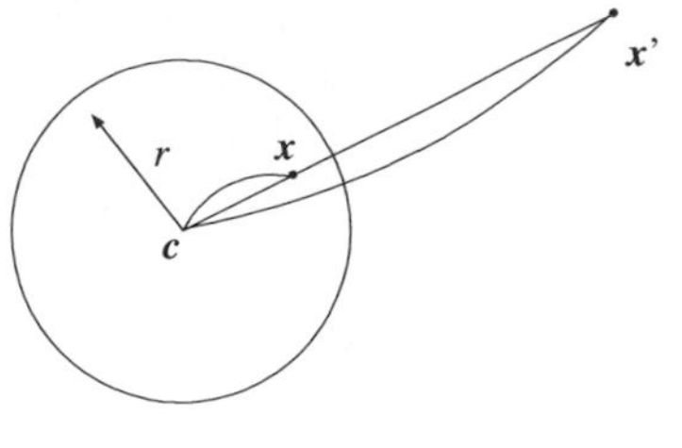

FIGURE 8.9 For a sphere of center $\boldsymbol{c}$ and radius r, a point $\boldsymbol{x}$ is inverted to a point $\boldsymbol{x}'$ on the line passing through the center $\boldsymbol{c}$ and the point $\boldsymbol{x}$ such that $\|\boldsymbol{x}-\boldsymbol{c}\|\|\boldsymbol{x}'-\boldsymbol{c}\|=r^2$.

(Fig. 8.8(b) $\hookrightarrow$ Fig. 6.2 in Chapter 6). Thus, we can interpret a translation to be *a rotation around an axis located infinitely far away*. We also see from this that both the translator and the rotor are versors of grade 2 and hence the motor is a versor of grade 4.

8.6.3 Invertors

One of the most fundamental operations that generate conformal transformations is *inversion* with respect to a sphere. By inversion of a point $\boldsymbol{x}$ with respect to a sphere of center $\boldsymbol{c}$ and radius r, we mean the point $\boldsymbol{x}'$ on the line passing through $\boldsymbol{c}$ and $\boldsymbol{x}$ such that $\|\boldsymbol{x} - \boldsymbol{c}\|\|\boldsymbol{x}' - \boldsymbol{c}\| = r^2$ (Fig. 8.9). We now show that $\sigma = c - r^2 e_\infty/2$, which is the dual representation of that sphere ($\hookrightarrow$ Eq. (8.18), is at the same time the versor for this inversion. It is called the *invertor*, and its inverse is given by

$$\sigma^{-1} = \frac{\sigma}{r^2}. \tag{8.90}$$

In fact,

$$\sigma^2 = (c - \frac{r^2}{2}e_\infty)(c - \frac{r^2}{2}e_\infty) = c^2 - \frac{r^2}{2}ce_\infty - \frac{r^2}{2}e_\infty c + \frac{r^4}{4}e_\infty^2 - \frac{r^2}{2}(ce_\infty + e_\infty c) = r^2, \tag{8.91}$$

where we note that $c^2 = 0$ from Eq. (8.57) and use the following identity:

$$\begin{aligned} ce_\infty + c_\infty c &= (e_0 + \boldsymbol{c} + \frac{\|\boldsymbol{c}\|^2}{2}e_\infty)e_\infty + e_\infty(e_0 + \boldsymbol{c} + \frac{\|\boldsymbol{c}\|^2}{2}e_\infty) \\ &= (e_0 e_\infty + e_\infty e_0) + (\boldsymbol{c}e_\infty + e_\infty \boldsymbol{c}) + \|\boldsymbol{c}\|^2 e_\infty^2 = -2 + (\boldsymbol{c}e_\infty - \boldsymbol{c}e_\infty) = -2. \end{aligned} \tag{8.92}$$

$$\sigma_0 = e_0 - \frac{r^2}{2}e_\infty. \tag{8.93}$$

Indeed, a sphere σ whose center c is at $\boldsymbol{c}$ is expressed using the translator in the form

$$\sigma = c - \frac{r^2}{2}e_\infty = \mathcal{T}_{\boldsymbol{c}}e_0\mathcal{T}_{\boldsymbol{c}}^\dagger - \frac{r^2}{2}\mathcal{T}_{\boldsymbol{c}}e_\infty\mathcal{T}_{\boldsymbol{c}}^\dagger = \mathcal{T}_{\boldsymbol{c}}\Big(e_0 - \frac{r^2}{2}e_\infty\Big)\mathcal{T}_{\boldsymbol{c}}^\dagger = \mathcal{T}_{\boldsymbol{c}}\sigma_0\mathcal{T}_{\boldsymbol{c}}^\dagger. \tag{8.94}$$

Hence, inversion of p with respect to σ is given by

$$\sigma p\sigma^\dagger = (\mathcal{T}_{\boldsymbol{c}}\sigma_0\mathcal{T}_{\boldsymbol{c}}^\dagger)p(\mathcal{T}_{\boldsymbol{c}}\sigma_0\mathcal{T}_{\boldsymbol{c}}^\dagger)^\dagger = \mathcal{T}_{\boldsymbol{c}}\sigma_0\mathcal{T}_{\boldsymbol{c}}^\dagger p\mathcal{T}_{\boldsymbol{c}}\sigma_0^\dagger\mathcal{T}_{\boldsymbol{c}}^\dagger = \mathcal{T}_{\boldsymbol{c}}(\sigma_0(\mathcal{T}_{-\boldsymbol{c}}p\mathcal{T}_{-\boldsymbol{c}}^\dagger)\sigma_0^\dagger)\mathcal{T}_{\boldsymbol{c}}^\dagger. \tag{8.95}$$

In other words, inversion of p with respect to σ is computed by first translating p by $-\boldsymbol{c}$, then inverting it with respect to σ_0, and finally translating the result by $\boldsymbol{c}$.

To see how a given point is inverted, it is sufficient to know how the origin e_0, a vector $\boldsymbol{a}$, and the infinity e_∞ are inverted.

origin e_0 The origin e_0 is inverted with respect to σ_0 to

$$\begin{aligned}\sigma_0 e_0\sigma_0^\dagger &= -\frac{1}{r^2}(e_0 - \frac{r^2}{2}e_\infty)e_0(e_0 - \frac{r^2}{2}e_\infty) = -\frac{r^2}{4}e_\infty e_0 e_\infty \\ &= -\frac{r^2}{4}e_\infty(-2 - e_\infty e_0) = \frac{r^2}{2}e_\infty.\end{aligned} \tag{8.96}$$

Since the conformal space is homogeneous, i.e., multiplication by a scalar has the same meaning, this point also represents infinity. Hence, *the origin e_0 is inverted to the infinity e_∞*.

vector $\boldsymbol{a}$ A vector $\boldsymbol{a} = a_1e_1 + a_2e_2 + a_3e_3$ is inverted to

$$\begin{aligned}\sigma_0\boldsymbol{a}\sigma_0^\dagger &= -\frac{1}{r^2}(e_0 - \frac{r^2}{2}e_\infty)\boldsymbol{a}(e_0 - \frac{r^2}{2}e_\infty) = \frac{1}{r^2}e_0^2\boldsymbol{a} - \frac{1}{2}e_0e_\infty\boldsymbol{a} - \frac{1}{2}e_\infty e_0\boldsymbol{a} - \frac{r^2}{4}e_\infty^2\boldsymbol{a} \\ &= -\frac{1}{2}(e_0e_\infty + e_\infty e_0)\boldsymbol{a} = \boldsymbol{a},\end{aligned} \tag{8.97}$$

i.e., vectors are unaltered by inversion.

infinity e_∞ The infinity e_∞ is inverted to

$$\begin{aligned}\sigma_0 e_\infty\sigma_0^\dagger &= -\frac{1}{r^2}(e_0 - \frac{r^2}{2}e_\infty)e_\infty(e_0 - \frac{r^2}{2}e_\infty) = -\frac{1}{r^2}e_0e_\infty e_0 \\ &= -\frac{1}{r^2}e_0(-2 - e_0e_\infty) = \frac{2}{r^2}e_0.\end{aligned} \tag{8.98}$$

Due to the homogeneity of the space, this represents the origin. Hence, *the infinity e_∞ is inverted to the origin e_0*.

From these results, inversion of point $p = e_0 + \boldsymbol{x} + \|\boldsymbol{x}\|^2e_\infty/2$ is computed by applying the invertor term by term as follows:

$$\begin{aligned}\sigma p\sigma^\dagger &= \sigma_0e_0\sigma_0^\dagger + \sigma_0\boldsymbol{x}\sigma_0^\dagger + \frac{1}{2}\|\boldsymbol{x}\|^2\sigma_0e_\infty\sigma_0^\dagger = \frac{r^2}{2}e_\infty + \boldsymbol{x} + \frac{1}{2}\|\boldsymbol{x}\|^2(\frac{2}{r^2}e_0) \\ &= \frac{\|\boldsymbol{x}\|^2}{r^2}\Big(e_0 + r^2\boldsymbol{x}^{-1} + \frac{1}{2}\|r^2\boldsymbol{x}^{-1}\|^2e_\infty\Big).\end{aligned} \tag{8.99}$$

Here, we put $\boldsymbol{x}^{-1} = \boldsymbol{x}/\|\boldsymbol{x}\|^2$. Since the space is homogeneous, this represents the same point as $e_0 + r^2\boldsymbol{x}^{-1} + \|r^2\boldsymbol{x}^{-1}\|^2e_\infty/2$. Thus, a point at distance $\|\boldsymbol{x}\|$ from the sphere center is inverted to a point at distance $r^2/\|\boldsymbol{x}\|$ in the same direction ($\hookrightarrow$ Exercise 8.5).

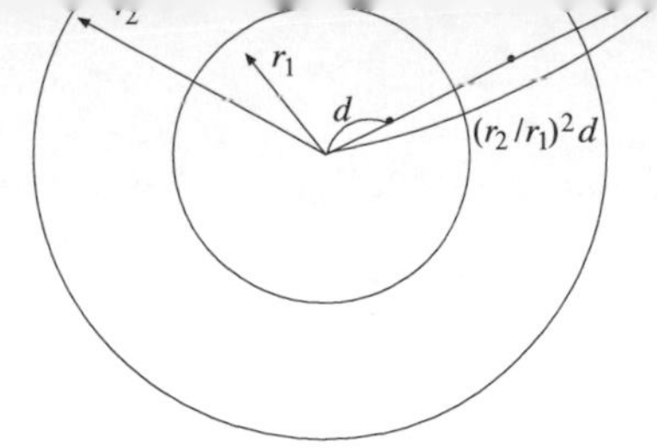

FIGURE 8.10 Consecutive reflections with respect to spheres of radii r_1 and r_2 result in a dilation by $(r_2/r_1)^2$.

8.6.4 Dilator

Just as consecutive reflections with respect to parallel planes give a translation by twice the distance between them and consecutive reflections with respect to intersecting planes give a rotation by twice the angle between them, consecutive inversions give a dilation. In fact, for the sphere $\sigma_1 = e_0 - r_1^2 e_\infty/2$ of radius r_1 centered at the origin e_0, a point at distance d from e_0 is inverted to the point at distance r_1^2/d, and if this point is inverted by the sphere $\sigma_2 = e_0 - r_2^2 e_\infty/2$ of radius r_2 also centered at e_0, it moves to a point at distance $r_2^2/(r_1^2/d)$ $= (r_2/r_1)^2 d$ (Fig. 8.10). This is a dilation by *the square of the ratio of the radii of the two spheres*. The versor

$$\mathcal{D} = \frac{1}{r_1 r_2}\sigma_2\sigma_1 \tag{8.100}$$

that describes this operation is called the *dilator*, where the coefficient $1/(r_1 r_2)$, which does not affect the versor operation, is introduced merely for simplifying the subsequent expressions. However, we need some tricks to obtain a useful form of the dilator. First, we note that

$$\begin{aligned}\mathcal{D} &= \frac{1}{r_1 r_2}\left(e_0 - \frac{r_2^2}{2}e_\infty\right)\left(e_0 - \frac{r_1^2}{2}e_\infty\right) = \frac{1}{r_1 r_2}\left(e_0^2 - \frac{r_1^2}{2}e_0e_\infty - \frac{r_2^2}{2}e_\infty e_0 + \frac{r_1^2 r_2^2}{4}e_\infty^2\right)\\ &= -\frac{r_1/r_2}{2}e_0e_\infty - \frac{r_2/r_1}{2}e_\infty e_0,\end{aligned} \tag{8.101}$$

which is rewritten as follows:

$$\mathcal{D} = -\frac{r_1/r_2}{2}(-2-e_\infty e_0) - \frac{r_2/r_1}{2}(-2-e_0e_\infty) = \frac{r_1}{r_2} + \frac{r_1/r_2}{2}e_\infty e_0 + \frac{r_2}{r_1} + \frac{r_2/r_1}{2}e_0e_\infty. \tag{8.102}$$

Adding Eqs. (8.101) and (8.102) and dividing the sum by 2, we obtain

$$\begin{aligned}\mathcal{D} &= \frac{r_1/r_2 + r_2/r_1}{2} + \frac{r_1/r_2}{2}\frac{e_\infty e_0 - e_0 e_\infty}{2} + \frac{r_2/r_1}{2}\frac{e_0e_\infty - e_\infty e_0}{2}\\ &= \frac{r_1/r_2 + r_2/r_1}{2} + \frac{r_1/r_2}{2}e_\infty \wedge e_0 + \frac{r_2/r_1}{2}e_0 \wedge e_\infty\\ &= \frac{r_2/r_1 + r_1/r_2}{2} + \frac{r_2/r_1 - r_1/r_2}{2}e_0 \wedge e_\infty = \frac{r_2/r_1 + r_1/r_2}{2} + \frac{r_2/r_1 - r_1/r_2}{2}\mathcal{O},\end{aligned} \tag{8.103}$$

where we put

$$\mathcal{O} = e_0 \wedge e_\infty. \tag{8.104}$$

$$\gamma = \log(\frac{r_2}{r_1})^2 \tag{8.105}$$

for the parameter of dilation, we can write

$$\frac{r_2}{r_1} = e^{\gamma/2}, \tag{8.106}$$

so Eq. (8.103) is written in the form

$$\mathcal{D} = \frac{e^{\gamma/2} + e^{-\gamma/2}}{2} + \frac{e^{\gamma/2} - e^{-\gamma/2}}{2}\mathcal{O} = \cosh\frac{\gamma}{2} + \mathcal{O}\sinh\frac{\gamma}{2} = \exp\frac{\gamma}{2}\mathcal{O}, \tag{8.107}$$

where the exponential function "exp" is defined via Taylor expansion of Eq. (8.73). In fac
if we note that

$$\mathcal{O}^2 = 1 \tag{8.108}$$

(↪ Exercise 8.1(6)), we see that

$$\exp\frac{\gamma}{2}\mathcal{O} = 1 + \frac{\gamma}{2}\mathcal{O} + \frac{1}{2!}(\frac{\gamma}{2}\mathcal{O})^2 + \frac{1}{3!}(\frac{\gamma}{2}\mathcal{O})^3 + \cdots$$
$$= \Big(1 + \frac{1}{2!}(\frac{\gamma}{2})^2 + \frac{1}{4!}(\frac{\gamma}{2})^4 + \cdots\Big) + \Big(\frac{\gamma}{2} + \frac{1}{3!}(\frac{\gamma}{2})^3 + \cdots\Big)\mathcal{O} = \cosh\frac{\gamma}{2} + \mathcal{O}\sinh\frac{\gamma}{2}. \tag{8.109}$$

The inverse of Eq. (8.107) is given by

$$\mathcal{D}^{-1} = \cosh\frac{\gamma}{2} - \mathcal{O}\sinh\frac{\gamma}{2} = \exp(-\frac{\gamma}{2}\mathcal{O}). \tag{8.110}$$

This is obvious from Eq. (8.105) if we note that the reciprocal of the magnification rat $(r_2/r_1)^2$ corresponds to the sign reversal of γ, but this can be confirmed as follows:

$$\Big(\cosh\frac{\gamma}{2} - \mathcal{O}\sinh\frac{\gamma}{2}\Big)\Big(\cosh\frac{\gamma}{2} + \mathcal{O}\sinh\frac{\gamma}{2}\Big)$$
$$= \cosh^2\frac{\gamma}{2} + \mathcal{O}\cosh\frac{\gamma}{2}\sinh\frac{\gamma}{2} - \mathcal{O}\cosh\frac{\gamma}{2}\sinh\frac{\gamma}{2} - \mathcal{O}^2\sinh^2\frac{\gamma}{2} = \cosh^2\frac{\gamma}{2} - \sinh^2\frac{\gamma}{2} = 1. \tag{8.111}$$

To see how a given point is dilated, it is sufficient to know how the origin e_0, a vect $\boldsymbol{a}$, and the infinity e_∞ are dilated. From the identities

$$\mathcal{O}e_0 = -e_0 = -e_0\mathcal{O}, \qquad \mathcal{O}e_\infty = e_\infty = -e_\infty\mathcal{O} \tag{8.112}$$

(↪ Exercise 8.6(2)), we see that

$$\mathcal{D}e_0 = \Big(\cosh\frac{\gamma}{2} + \mathcal{O}\sinh\frac{\gamma}{2}\Big)e_0 = e_0\Big(\cosh\frac{\gamma}{2} - \mathcal{O}\sinh\frac{\gamma}{2}\Big) = e_0\mathcal{D}^{-1}, \tag{8.113}$$

$$\mathcal{D}e_\infty = \Big(\cosh\frac{\gamma}{2} + \mathcal{O}\sinh\frac{\gamma}{2}\Big)e_\infty = e_\infty\Big(\cosh\frac{\gamma}{2} - \mathcal{O}\sinh\frac{\gamma}{2}\Big) = e_\infty\mathcal{D}^{-1}. \tag{8.114}$$

Using these and noting that $\mathcal{D}$ has grade 2 so that $\mathcal{D}^\dagger = \mathcal{D}^{-1}$, we obtain the following:

name	grade	expression
reflector	1	$\pi = \boldsymbol{n} + he_\infty$
invertor	1	$\sigma = c - r^2 e_\infty/2$
translator	2	$\mathcal{T}_t = 1 - \boldsymbol{t}e_\infty/2 = \exp(-\boldsymbol{t}e_\infty/2)$
		consecutive reflections for parallel planes
rotor	2	$\mathcal{R} = \cos\Omega/2 - \mathcal{I}\sin\Omega/2 = \exp(-\mathcal{I}\Omega/2)$
		consecutive reflections for intersecting planes
dilator	2	$\mathcal{D} = \cosh\gamma/2 + \mathcal{O}\sin\gamma/2 = \exp\mathcal{O}\gamma/2$
		consecutive inversions for concentric spheres
motor	4	$\mathcal{M} = \mathcal{T}_t\mathcal{R}$
		composition of rotation and translation

origin e_0 The origin e_0 is dilated to

$$\begin{aligned}\mathcal{D}e_0\mathcal{D}^\dagger &= \mathcal{D}^2 e_0 = \left(\exp\frac{\gamma}{2}\mathcal{O}\right)^2 e_0 = \exp\gamma\mathcal{O}e_0 = (\cosh\gamma + \mathcal{O}\sinh\gamma)e_0 \\ &= e_0\cosh\gamma - e_0\sinh\gamma = e^{-\gamma}e_0. \end{aligned} \tag{8.115}$$

Since the space is homogeneous, this still represents the origin.

vector $\boldsymbol{a}$ Since the basis elements e_i, $i = 1, 2, 3$, are anticommutative with e_0 and e_∞, they are commutative with $\mathcal{O} = (e_0e_\infty - e_\infty e_0)/2$. Hence, they are also commutative with $\mathcal{D}$ and $\mathcal{D}^{-1}$, so we see that

$$\mathcal{D}\boldsymbol{a}\mathcal{D}^\dagger = \mathcal{D}\mathcal{D}^{-1}\boldsymbol{a} = \boldsymbol{a}, \tag{8.116}$$

i.e., vectors are invariant to dilation.

infinity e_∞ The infinity e_∞ is dilated to

$$\begin{aligned}\mathcal{D}e_\infty\mathcal{D}^\dagger &= \mathcal{D}^2 e_\infty = \left(\exp\frac{\gamma}{2}\mathcal{O}\right)^2 e_0 = \exp\gamma\mathcal{O}e_\infty = (\cosh\gamma + \mathcal{O}\sinh\gamma)e_\infty \\ &= e_\infty\cosh\gamma + e_\infty\sinh\gamma = e^{\gamma}e_\infty. \end{aligned} \tag{8.117}$$

Due to the homogeneity of the space, this still represents infinity.

From these results, dilation of point $p = e_0 + \boldsymbol{x} + \|\boldsymbol{x}\|^2 e_\infty/2$ is computed by applying the dilator term by term as follows:

$$\begin{aligned}\mathcal{D}p\mathcal{D}^\dagger &= \mathcal{D}e_0\mathcal{D}^\dagger + \mathcal{D}\boldsymbol{x}\mathcal{D}^\dagger + \frac{1}{2}\|\boldsymbol{x}\|^2\mathcal{D}e_\infty\mathcal{D}^\dagger = e^{-\gamma}e_0 + \boldsymbol{x} + \frac{1}{2}\|\boldsymbol{x}\|^2 e^{\gamma}e_\infty \\ &= e^{-\gamma}\left(e_0 + e^{\gamma}\boldsymbol{x} + \frac{1}{2}\|e^{\gamma}\boldsymbol{x}\|^2 e_\infty\right). \end{aligned} \tag{8.118}$$

Since the space is homogeneous, this represents the same position as $e_0 + e^\gamma\boldsymbol{x} + \|e^\gamma\boldsymbol{x}\|^2 e_\infty/2$, i.e., the position $\boldsymbol{x}$ is dilated to the position $e^\gamma\boldsymbol{x}$.

8.6.5 Versors and conformal transformations

The versors introduced so far are summarized in Table 8.2. The important property of versors is that *the geometric product is preserved* up to sign. What we mean by this is that

$$(\mathcal{V}x\mathcal{V}^\dagger)(\mathcal{V}y\mathcal{V}^\dagger) = \mathcal{V}x(\mathcal{V}^\dagger\mathcal{V})y\mathcal{V}^\dagger = (-1)^k\mathcal{V}xy\mathcal{V}^\dagger, \tag{8.119}$$

where we note from the definition of Eq. (8.81) that

$$\mathcal{V}^\dagger\mathcal{V} = \mathcal{V}\mathcal{V}^\dagger = (-1)^k. \tag{8.120}$$

Since the outer product $x \wedge y$ is defined by antisymmetrization $(xy - yx)/2$ of the geometric product, we obtain the following ($\hookrightarrow$ Sec. 6.8 in Chapter 6):

Proposition 8.1 (Versors and outer product) *After a versor of grade k is applied, the outer product of elements x and y is preserved up to sign:*

$$(\mathcal{V}x\mathcal{V}^\dagger) \wedge (\mathcal{V}y\mathcal{V}^\dagger) = (-1)^k\mathcal{V}(x \wedge y)\mathcal{V}^\dagger. \tag{8.121}$$

From this, we observe that

Proposition 8.2 (Versors and spheres) *A versor $\mathcal{V}$ maps a sphere $p_1 \wedge p_2 \wedge p_3 \wedge p_4$ that passes through four points p_i, i = 1, 2, 3, 4, to the sphere $p'_1 \wedge p'_2 \wedge p'_3 \wedge p'_4$ that passes through the transformed points $p'_i = \mathcal{V}p_i\mathcal{V}^\dagger$.*

In fact, if a point p satisfies the equation of the sphere $p_1 \wedge p_2 \wedge p_3 \wedge p_4$, i.e.,

$$p \wedge (p_1 \wedge p_2 \wedge p_3 \wedge p_4) = 0, \tag{8.122}$$

and if p is transformed to $p' = \mathcal{V}p\mathcal{V}^\dagger$, application of the versor $\mathcal{V}$ to the above equation yields

$$\begin{aligned} 0 &= \mathcal{V}(p \wedge p_1 \wedge p_2 \wedge p_3 \wedge p_4)\mathcal{V}^\dagger = (\mathcal{V}p\mathcal{V}^\dagger) \wedge (\mathcal{V}p_1\mathcal{V}^\dagger) \wedge (\mathcal{V}p_2\mathcal{V}^\dagger) \wedge (\mathcal{V}p_3\mathcal{V}^\dagger) \wedge (\mathcal{V}p_4\mathcal{V}^\dagger) \\ &= p' \wedge (p'_1 \wedge p'_2 \wedge p'_3 \wedge p'_4). \end{aligned} \tag{8.123}$$

Hence, p' satisfies the equation of the sphere $p'_1 \wedge p'_2 \wedge p'_3 \wedge p'_4$. Note that here the sign $(-1)^k$ arises four times, so it has no effect (in general, it is irrelevant for expressions that are 0). Thus, we conclude that versors map a sphere to a sphere.

We can easily confirm that angle is preserved by reflection, translation, rotation, and inversion. It is known that conformal mappings that map a sphere to sphere and preserve angle are generated by composing the reflector, the translator, the rotor, and the invertor.

Since the inner product $\langle x, y\rangle$ is defined by symmetrization $(xy + yx)/2$ of the geometric product and since the inner product is a scalar, we see from Eq. (8.120) that

$$\langle \mathcal{V}x\mathcal{V}^\dagger, \mathcal{V}y\mathcal{V}^\dagger\rangle = (-1)^k\mathcal{V}\langle x, y\rangle\mathcal{V}^\dagger = (-1)^k\langle x, y\rangle\mathcal{V}\mathcal{V}^\dagger = \langle x, y\rangle. \tag{8.124}$$

Thus, we observe

Proposition 8.3 (Versors and inner product) *The inner product of elements x and y is preserved by transformation by a versor $\mathcal{V}$:*

$$\langle \mathcal{V}x\mathcal{V}^\dagger, \mathcal{V}y\mathcal{V}^\dagger\rangle = \langle x, y\rangle. \tag{8.125}$$

normalized so that e_0 has coefficient 1 as Eq. (8.8). In the conformal space, the expression

$$\tilde{p} = \alpha e_0 + \alpha \boldsymbol{x} + \frac{\alpha}{2}\|\boldsymbol{x}\|^2 e_\infty \tag{8.126}$$

multiplied by a nonzero scalar α also represents the same position as p. In this case, we see from $\langle e_0, e_\infty \rangle = -1$ that

$$\langle \tilde{p}, e_\infty \rangle = \alpha \langle e_0, e_\infty \rangle = -\alpha. \tag{8.127}$$

Thus, we have $\alpha = -\langle \tilde{p}, e_\infty \rangle$. Hence, for normalizing $\tilde{p}$ so that e_0 has coefficient 1, we need to write $-\tilde{p}/\langle \tilde{p}, e_\infty \rangle$, which corresponds to the form of Eq. (8.8). Hence, extension of Eq. (8.11) to the case of a non-unit coefficient of e_0 is given by

$$\|\boldsymbol{x} - \boldsymbol{y}\|^2 = \frac{-2\langle p, q \rangle}{\langle p, e_\infty \rangle \langle q, e_\infty \rangle}. \tag{8.128}$$

Using this, we can compute the distance between two the positions $\boldsymbol{x}'$ and $\boldsymbol{y}'$ after transformation by a versor $\mathcal{V}$ as follows:

$$\begin{aligned} \|\boldsymbol{x}' - \boldsymbol{y}'\|^2 &= \frac{-2\langle \mathcal{V}p\mathcal{V}^\dagger, \mathcal{V}q\mathcal{V}^\dagger \rangle}{\langle \mathcal{V}p\mathcal{V}^\dagger, e_\infty \rangle \langle \mathcal{V}q\mathcal{V}^\dagger, e_\infty \rangle} = \frac{-2\langle p, q \rangle}{\langle \mathcal{V}p\mathcal{V}^\dagger, \mathcal{V}(\mathcal{V}^\dagger e_\infty \mathcal{V})\mathcal{V}^\dagger \rangle \langle \mathcal{V}q\mathcal{V}^\dagger, \mathcal{V}(\mathcal{V}^\dagger e_\infty \mathcal{V})\mathcal{V}^\dagger \rangle} \\ &= \frac{-2\langle p, q \rangle}{\langle p, \mathcal{V}^\dagger e_\infty \mathcal{V} \rangle \langle q, \mathcal{V}^\dagger e_\infty \mathcal{V} \rangle}. \end{aligned} \tag{8.129}$$

This equals $\|\boldsymbol{x} - \boldsymbol{y}\|^2$ if $\mathcal{V}^\dagger e_\infty \mathcal{V} = e_\infty$, which is rewritten by multiplication of $\mathcal{V}$ from the left and $\mathcal{V}^\dagger$ from the right into the form $e_\infty = \mathcal{V} e_\infty \mathcal{V}^\dagger$. Hence, we observe that

Proposition 8.4 (Versors and isometry) *The conformal transformation defined by versor $\mathcal{V}$ is an isometry if and only if*

$$\mathcal{V} e_\infty \mathcal{V}^\dagger = e_\infty. \tag{8.130}$$

From Eqs. (8.68), (8.77), and (8.87), we see that Eq. (8.130) holds for the translator $\mathcal{T}_{\boldsymbol{t}}$, the rotor $\mathcal{R}$, and the reflector π. As seen from Eqs. (8.98) and (8.117), however, Eq. (8.130) does not hold for the invertor σ and the dilator $\mathcal{D}$. Hence, we conclude that

Proposition 8.5 (Isometric conformal transformations) *A conformal transformation is an isometry only when it is generated by translations, rotations, and reflections.*

8.7 SUPPLEMENTAL NOTE

Conformal mappings are mappings that preserve angles between tangents. In 2D, they are given by an analytical (or regular or holomorphic) function over a domain of the complex plane. Among them, those defined over all the complex plane including the point at infinity that map to circles have the form

$$z' = \frac{\alpha z + \beta}{\gamma z + \delta}, \qquad \alpha\delta - \beta\gamma \neq 0. \tag{8.131}$$

This linear fractional transformation is called the *Möbius transformation* and is generated by compositions of translation $z' = z + \alpha$, rotation/reflection/dilation $z' = \alpha z$, and inversion

Conformal geometry, which is the study of conformal transformations, has a very
history, but it was the American physicist Hestenes who reorganized conformal geomet
from the viewpoint of Clifford algebra (↪ Supplemental note to Chapter 6). He patent
his Clifford geometrical formulation of conformal geometry [11], but its academic use
not prevented. Clifford himself called his geometry "geometric algebra," while later mat
ematicians call it "Clifford algebra" in his honor. However, Hestenes preferred to call
"geometric algebra" to emphasize its application aspects rather than its pure mathematic
structure.

As stated in Sec. 8.6.1, the composition of two versors is given by their geometric produc
resulting in another versor. Hence, the set of versors forms a group under geometric product
called the *Clifford group*. A versor $\mathcal{V}$ in the form of Eq. (8.80) is said to be *unitary* if i
inverse is given by $\mathcal{V}^{-1} = v_1 v_2 \cdots v_k$ and a *spinor* if the grade is even. The set of spinors
a group under composition, called the *spinor group*. As stated in Sec. 6.8 in Chapter 6,
spinors in 3D are rotors constructed from unit vectors. In the 5D conformal space, whic
is a non-Euclidean space that includes directions of negative norms, spinors have rath
complicated forms.

Using conformal geometry, we can prove many theorems involving spheres, circle
planes, and lines. In this chapter, however, we discussed only those algebraic aspects relat
to Grassmann and Clifford algebras. Application of the conformal geometry to engineeri
problems mostly remains to be seen, but some efforts have been made. For example, Dorst
al. [5] used it for ray tracing computation in computer graphics, while Perwass [16] appli
it to the pose estimation problem in computer vision. They both used the software too
that they themselves created [6, 17] for the computation. Bayro-Corrochano [3] discuss
the possibilities of applying geometric algebra to robot arm control, image processing, a
3D shape modeling for computer vision.

8.8 EXERCISES

8.1. The 5D conformal space is the set of all the elements in the form of Eq. (8.1). If w
represent a 3D point $\boldsymbol{x}$ in the form of Eq. (8.10), what subset does the set of suc
elements form in 5D with $\boldsymbol{x}$ ranging over all 3D space? In other words, what *embeddin*
does Eq. (8.9) define from $\mathbb{R}^3$ to $\mathbb{R}^5$?

8.2. Consider a 5D space spanned by basis $\{e_1, e_2, e_3, e_4, e_5\}$, and introduce the followi
Minkowski metric to define a 5D Minkowski space $\mathbb{R}^{4,1}$:

$$\langle e_1, e_1\rangle = \langle e_2, e_2\rangle = \langle e_3, e_3\rangle = \langle e_4, e_4\rangle = 1, \qquad \langle e_5, e_5\rangle = -1,$$

$$\langle e_i, e_j\rangle = 0, \qquad i \neq j.$$

(1) If we write the elements of this space in the form

$$x = x_1 e_1 + x_2 e_2 + x_3 e_3 + x_4 e_4 + x_5 e_5,$$

what subset do elements such that $\|x\|^2 = 0$ make? (This set is called the *nu*
cone.)

(2) Define e_0 and e_∞ by

$$e_0 = \frac{1}{2}(e_4 + e_5), \qquad e_\infty = e_5 - e_4,$$

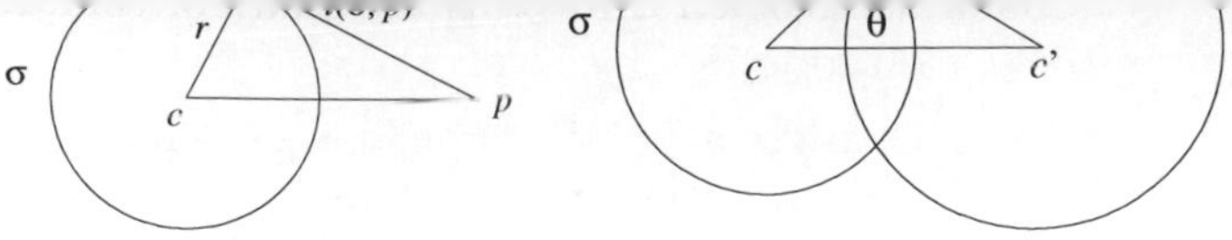

(a) (b)

FIGURE 8.11 (a) Tangential distance. (b) Intersection of spheres.

and show that this 5D space $\mathbb{R}^{4,1}$ is the same as the 5D conformal space in Exercise 8.1.

(3) From this, explain how the conformal space embeds the 3D Euclidean space $\mathbb{R}^3$ in the 5D Minkowski space $\mathbb{R}^{4,1}$.

(4) Define a Clifford algebra in the 5D Minkowski space $\mathbb{R}^{4,1}$ by introducing the following geometric products among the basis elements:

$$e_1^2 = e_2^2 = e_3^2 = e_4^2 = 1, \qquad e_5^2 = -1, \qquad e_i e_j = -e_j e_i, \qquad i \neq j.$$

Show that this coincides with the Clifford algebra of the conformal space.

8.3. As shown in Eq. (8.18), a sphere of radius r centered at $\boldsymbol{c}$ has the dual representation

$$\sigma = c - \frac{r^2}{2} e_\infty,$$

where we let $c = e_0 + \boldsymbol{c} + \|\boldsymbol{c}\|^2 e_\infty / 2$.

(1) Consider a point p outside the sphere σ. Let $t(\sigma, p)$ be the *tangential distance* between the sphere σ and the point p, i.e., the length of the line segment starting from p and tangent to σ at the endpoint (Fig. 8.11(a)). Show that the inner product of σ and p has the form

$$\langle \sigma, p \rangle = -\frac{1}{2} t(p, \sigma)^2,$$

and hence point p is on sphere σ if and only if

$$\langle \sigma, p \rangle = 0.$$

(2) Suppose another sphere

$$\sigma' = c' - \frac{r'^2}{2} e_\infty$$

intersects the sphere σ. Let θ be the angle made by their tangent planes at the intersection (= the angle between the segments starting from the intersection and pointing to the sphere centers) (Fig. 8.11(b)). Show that the inner product of σ and σ' has the form

$$\langle \sigma, \sigma' \rangle = r r' \cos\theta,$$

and hence spheres σ and σ' are orthogonal, i.e., their tangent planes are orthogonal at the intersection, if and only if

$$\langle \sigma, \sigma' \rangle = 0.$$

by applying the inverse rotation $\mathcal{R}^{-1}$ and then translation by t followed by th
rotation $\mathcal{R}$.

(2) Consider a rotation around a point not at the origin that has the same axis ar angle as rotor $\mathcal{R}$. Show that if a point p is rotated around the position $\boldsymbol{t}$ in th way, it is equivalent to applying the motor $\mathcal{R}\mathcal{T}_t\mathcal{R}^{-1}$ to point p.

(3) Show that the above rotor is also written as $\mathcal{T}_{t-\mathcal{R}t\mathcal{R}^{-1}}\mathcal{R}$.

8.5. By inversion with respect to a sphere, the sphere center is mapped to infinity, ar infinity is mapped to the sphere center. Using this fact, show that the center c of th sphere that has dual representation σ is given by

$$c = -\frac{1}{2}\sigma e_\infty \sigma.$$

8.6. For the flat point $\mathcal{O}$ defined by Eq. (8.104), show the following:

(1) Equation (8.108) holds.

(2) Equation (8.112) holds.

Camera Imaging and Conformal Transformations

In Chapters 2–7, we focused on the geometry of lines and planes, to which we added spheres and circles in Chapter 8 and considered their conformal transformations. Conformal transformations include familiar mappings such as translation, rotation, and scale change that frequently appear in many engineering applications, but inversion is a unique mapping involving spheres and circles. In this chapter, we consider camera imaging geometry as a typical geometric problem that involves inversion. First, we describe conventional perspective projection cameras. Then, we turn to fisheye lens cameras. We further analyze the imaging geometry of omnidirectional cameras that use parabolic mirrors. We show that inversion with respect to a sphere plays an essential role in all such cameras. We further describe how we can obtain 3D interpretation of a scene from omnidirectional camera images, and its imaging geometry is compared with those of cameras that use hyperbolic and elliptic mirrors.

9.1 PERSPECTIVE CAMERAS

Figure 9.1(a) simplifies the imaging geometry of conventional perspective cameras: an incoming ray of light through the center of the lens is focused on the receptive surface, producing an upside-down and right-and-left reversed image. The symmetry axis of the lens is called the *optical axis*. Figure 9.1(b) shows its abstraction, using the coordinate system with the origin O at the lens center and the z-axis along the optical axis. The receptive surface is also called the *image plane*. Its intersection with the optical axis is called the *principal point*, and the distance from the lens center O is generally known as the *focal length*. It is not necessarily the same as the optical focal length of the lens itself, but both coincide for scenes infinitely far apart (for near scenes, the exact focus point is determined by solving what is called the *lens equation*). Let θ be the *incidence angle*, i.e., the angle between the optical axis and the ray of light passing through the lens center O. As shown in Fig. 9.1(b), the ray focuses on the image plane at distance d from the principal point given by

$$d = f \tan\theta, \tag{9.1}$$

where f is the focal length. The mapping from the outside scene to the image plane defined in this way is called *perspective projection*. In the perspective projection model of Fig. 7.2 in Chapter 7, the image plane is placed before the lens center, but the geometric relationships are the same.

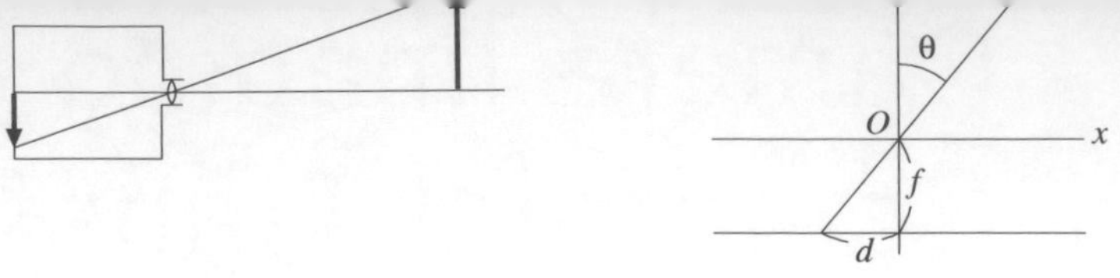

(a) (b)

FIGURE 9.1 (a) Camera imaging geometry. (b) Perspective projection.

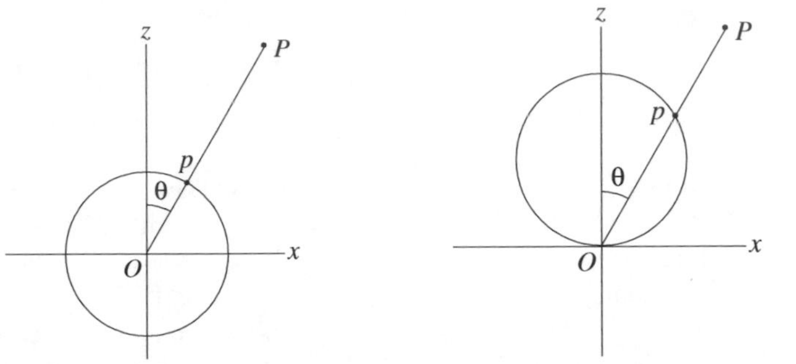

(a) (b)

FIGURE 9.2 Spherical camera models. (a) Central image sphere for omnidirectional cameras. (b Perspective image sphere for perspective cameras.

Since all points on an incoming ray are projected to the same point, a camera is essen tially a device to record incoming rays. Hence, wherever the image plane is placed, or eve if it is a nonplanar shape, the recorded information is the same. From this point of viev the simplest mathematical model is to consider a sphere surrounding the lens center O an regard the (color or intensity) value of the ray as recorded at its intersection with the spher (Fig. 9.2(a)). We call such a sphere the *image sphere*.

We should note that for the usual camera only those rays incoming from the fror are recorded, while for this spherical camera model incoming rays from all directions ar recorded. This means that Fig. 9.2(a) is a mathematical idealization of *omnidirectional* (c *catadioptric*) cameras. For recording only the front rays, we place the sphere so that passes through the lens center O (Fig. 9.2(b)). Hereafter, we call the sphere of Fig. 9.2(a the *central image sphere* and the sphere of Fig. 9.2(b) the *perspective image sphere*.

Let f be the radius of the image sphere in Fig. 9.2(b). If we consider an image plan that passes through the center of the sphere and is orthogonal to the optical axis, th correspondence between the image sphere and the image plane is 1 to 1 and given b *stereographic projection*, as shown in Fig. 9.3(a); a point p on the image sphere is mappe to the intersection p' of the image plane $z = f$ with the line passing through p and th origin O (↪ Fig. 4.2 in Chapter 4 and Fig 8.2 in Chapter 8). This stereographic projectio is actually an inversion with respect to a sphere surrounding the origin O with radius $\sqrt{2}$ (the dotted circle in Fig. 9.3), which we call the *inversion sphere*. If a point p on the imag sphere is inverted with respect to this sphere, it is mapped, by definition, to a point p' o

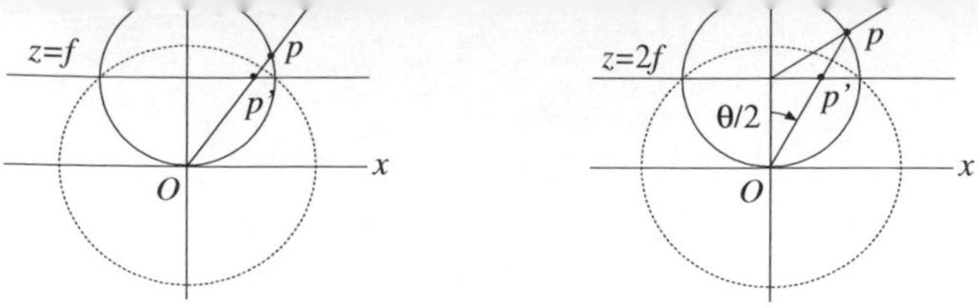

(a) (b)

FIGURE 9.3 Stereographic projection of a sphere onto a plane. (a) Perspective camera model. (b) Fisheye lens camera model.

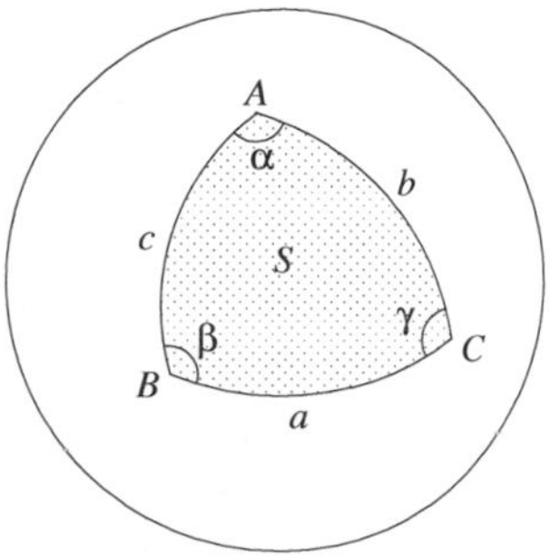

FIGURE 9.4 Spherical triangle ABC on a unit sphere.

the line Op such that $|Op'| = 2f^2/|Op|$. We can see that this point is on the image plane by the following reasoning.

The inversion sphere, the image sphere, and the image plane share a circle C of radius f as their intersection. The circle C is on the inversion sphere, so it is unchanged by the inversion. Since a sphere is inverted to a sphere and since the origin O is inverted to infinity, a sphere that passes through O is inverted to a sphere of an infinite radius, i.e., a plane, that contains the circle C. Hence, the inversion of the image sphere coincides with the image plane. Thus, the stereographic projection results in an inversion ($\hookrightarrow$ Exercise 9.1).

Traditional World 9.1 (Spherical trigonometry) As is well known, the shortest path that connects two points on a sphere is the great circle (= circle that has the same radius as the sphere) passing through them. A *spherical triangle* is obtained by connecting three points on a sphere by great circles, and the study of spherical triangles is known as *spherical trigonometry*. Let α, β, and γ be the interior angles (= the angles made by the tangents to the great circles) at vertex A, B, and C, respectively, of a spherical triangle on a unit sphere. Let a, b, and c be the lengths (= the angles made by the vectors from the sphere center) of the sides opposite to A, B, and C, respectively (Fig. 9.4). The following relationships are well known:

$$\frac{\sin\alpha}{\sin a} = \frac{\sin\beta}{\sin b} = \frac{\sin\gamma}{\sin c}, \tag{9.2}$$

$$\cos a = \cos b\cos c + \sin b\sin c\cos\alpha, \qquad \cos\alpha = -\cos\beta\cos\gamma + \sin\beta\sin\gamma\cos a. \tag{9.3}$$

$$\frac{}{a} = \frac{}{b} = \frac{}{c}, \qquad (9.$$

and the *law of cosines*,

$$a^2 = b^2 + c^2 - 2bc\cos\alpha, \qquad (9.$$

of a planar triangle, where α, β, and γ are the interior angles of the vertices and a, b, ar c are the lengths of their opposite sides, respectively. Hence, Eqs. (9.2) and (9.3) are calle the law of sines and the law of cosines, respectively, of a spherical triangle. As is intuitive evident, the sum of the interior angles of a spherical triangle is larger than π. It is know that the area S of this spherical triangle is given by

$$S = \alpha + \beta + \gamma - \pi. \qquad (9.$$

If the sphere has radius r, the area S is magnified r^2 times.

9.2 FISHEYE LENS CAMERAS

The angle of camera view is around 100° for ordinary lenses, and wider fileds are viewe using *wide-angle lenses*. Today, we can view around 180° or more using a *fisheye len* We consider the central image sphere of Fig. 9.2(a) for modeling such fisheye lenses. Th resulting spherical image can be mapped to a plane by stereographic projection, as describe earlier. For this mapping, we let the radius of the image sphere be $2f$ and regard the plan $z = 2f$ that passes through the sphere center as the image plane (Fig. 9.3(b)). From th definition of the central image sphere, the incoming ray with incidence angle θ is recorded a point p on the sphere, as shown in the Fig. 9.3(b). If this point is stereographically projecte from the south pole O onto the image plane $z = 2f$, it is projected to point p' that make angle $\theta/2$ from the optical axis, due to the relationship between the central and inscribe angles. Hence, the distance d of p' from the principal point is

$$d = 2f\tan\frac{\theta}{2} \qquad (9.$$

(Fig. 9.5(a)). This stereographic projection is also an inversion. The radius of the inversic sphere is $2f$ (the dotted circle in Fig. 9.3(b)). The inversion sphere shares a circle C o radius $2f$ with the image sphere and the image plane. By the reasoning stated earlier, th image sphere that passes through O is inverted to a sphere of infinite radius, i.e., a plan that contains the circle C and infinity.

Many commercially available fisheye lenses satisfy Eq. (9.7) with high accuracy. Th constant f is often called the "focal length" of the fisheye lens for convenience. The reasc that we let the image sphere radius be $2f$ and write the coefficient in Eq. (9.7) as $2f$ that $2f\tan(\theta/2) \approx f\tan\theta$ for $\theta \approx 0$, hence f has the same meaning as the focal length f o the perspective projection of Eq. (9.1) in the neighborhood of the principal point. As see from Eq. (9.7), the scene in front of the camera with a 180° angle of view is imaged withi a circle of radius $2f$ around the principal point, and the outside is the image of the scen behind the camera (Fig. 9.5(b)). Many of today's fisheye lenses available on the market ca view around a 200° angle (↪ Exercise 9.2).

Figure 9.6 shows a fisheye lens image of an outdoor scene. A fisheye lens image lik this can be rectified into a perspective image. We first map the planar image to the imag sphere, as shown in Fig. 9.3(b), and rotate the sphere so that the part of interest comes nea

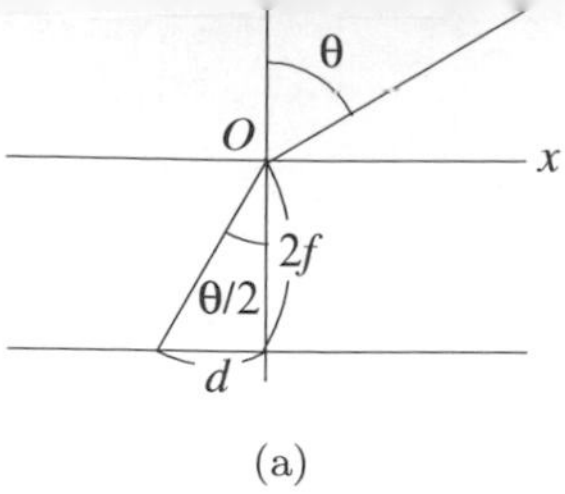

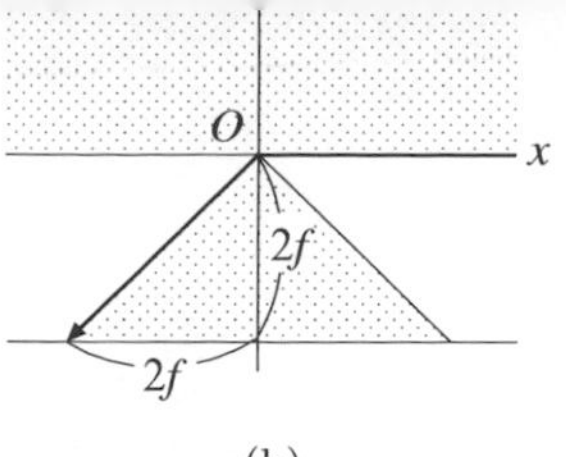

FIGURE 9.5 Imaging geometry by a fisheye lens. (a) The ray with incidence angle θ is imaged at a point of distance $d = 2f \tan \theta/2$ from the principal point. (b) The scene in front with a 180° angle of view is imaged within a circle of radius $2f$ around the principal point; the outside is the image of the scene behind.

FIGURE 9.6 A fisheye lens image of an outdoor scene.

the north pole. Then, the image on the sphere is expanded so that the northern hemisphere part covers the entire sphere by doubling the latitude of each point from the north pole. By stereographically projecting the resulting spherical image from the south pole, we obtain an image as if taken by a perspective camera of focal length f. However, this is only a hypothetical process; we need not actually produce spherical images. We can directly map the planar fisheye lens image to a planar perspective image by a computation that simulates this process (↪ Exercise 9.4). Figure 9.7 shows perspective images transformed from the fisheye lens image of Fig. 9.6 in this way: the front image and the images as if obtained by rotating the camera by 90° to left, right, up, and down. This process can be used to assist vehicle driving by mounting a fisheye lens camera in front, warning the driver of approaching cars from right or left. If multiple fisheye lens cameras are used, we can generate an image of the ground surface around the vehicle as if seen from high above. Such techniques are currently used in various ways for vehicle-mounted camera applications.

The fisheye lens for which Eq. (9.7) holds is called the *stereographic lens* and is widely

FIGURE 9.7 Perspective images transformed from the fisheye lens image in Fig. 9.6. The fro image and the images obtained by virtually rotating the camera by 90° to left, right, up, ar down.

used for various applications, but there exist other types of projection models, as listed i Table 9.1. They have different functional relationships between the incidence angle θ ar the distance d of the imaged point from the principal point; Fig. 9.8 shows their plots.

9.3 OMNIDIRECTIONAL CAMERAS

A typical camera that can cover nearly all the scene around over 360° uses a parabol mirror. Its principle is depicted in Fig. 9.9(a). Let

$$z = -\frac{1}{4f}(x^2 + y^2) \quad (9.8$$

be the equation of the mirror. The point F at $(0, 0, -f)$ is called its *focus*, and f the *foc length*. An incoming ray toward F intersects the mirror at p and reflects there upward. we take an image of the mirror from above, almost all incoming rays toward the focus are captured.

Incidentally, this geometry is interpreted alternatively: a ray coming upward from belo intersects with the mirror at point p and reflects there toward the focus F. This is th principle of the microwave dish antenna and the parabolic sound reflector.

Since the parabolic surface is symmetric around the optical axis (= z-axis), it suffices t consider the relationship in the zx plane. Suppose a ray with incidence angle θ reflects a p on the parabolic mirror. Let d be the distance of the reflected ray from the optical axi Then, the reflection point p is at $(d, 0, -d^2/4f)$ (Fig. 9.9(b)). We see that

$$\tan\theta = \frac{d}{f - d^2/4f}. \quad (9.9$$

called the "focal length."

	name	projection equation
1.	perspective projection	$d = f \tan\theta$
2.	stereographic projection	$d = 2f \tan\theta/2$
3.	orthogonal projection	$d = f \sin\theta$
4.	equisolid angle projection	$d = 2f \sin\theta/2$
5.	equidistance projection	$d = f\theta$

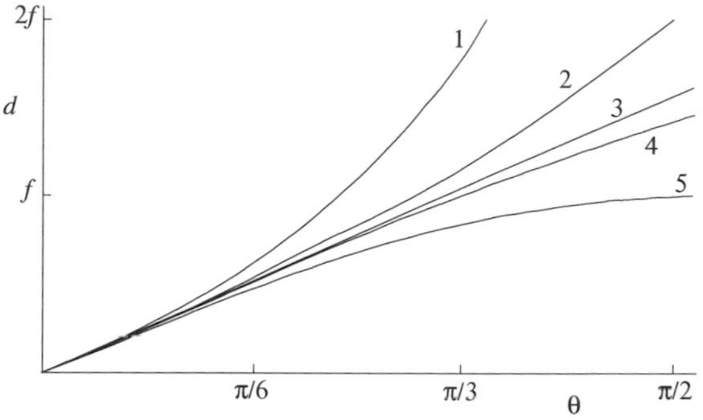

FIGURE 9.8 The incidence angle θ vs. the distance d of the imaged point from the principal point for different projection models. 1. Perspective projection. 2. Stereographic projection. 3. Orthogonal projection. 4. Equisolid angle projection. 5. Equidistance projection.

Comparing this with the double-angle formula of the tangent,

$$\tan\theta = \frac{2\tan\theta/2}{1-\tan^2\theta/2}, \tag{9.10}$$

we find

$$d = 2f\tan\frac{\theta}{2}. \tag{9.11}$$

This coincides with Eq. (9.7). Hence, *the same relationship holds* for the parabolic mirror and the fisheye lens, and the focal length of the parabolic mirror and the focal length of the fisheye lens have the same meaning ($\hookrightarrow$ Exercise 9.3).

This means that the omnidirectional camera image can be viewed as a stereographic projection of the image sphere surrounding the focus of the parabolic mirror. This is illustrated in Fig. 9.10. Consider an image sphere of radius $2f$ surrounding the focus F. Suppose the image plane is placed so that it passes through F and is orthogonal to the optical axis. A ray with incidence angle θ reflects at point p on the mirror. This point corresponds to point q on the image sphere. It is stereographically projected from the south pole onto the image plane at p'. We can see that p' is right below the point p on the mirror. This is because the line starting from the south pole and passing through p' makes angle $\theta/2$ with the optical axis due to the relationship between the central and inscribed angles, and hence p' has distance $2f\tan\theta/2$ from F on the image plane. This stereographic projection is also an inversion; the inversion sphere is around the south pole with radius $2\sqrt{2}f$ (the dotted circle in Fig. 9.10). This inversion sphere shares a circle of radius $2f$ with the image sphere, the image plane, and the parabolic mirror as their intersection. The scene within angle 180°

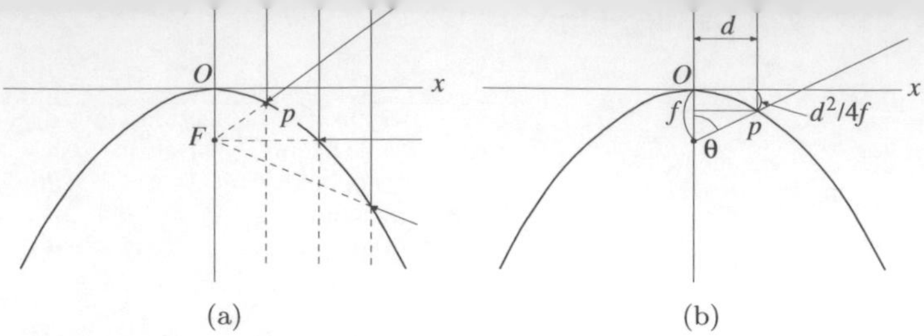

FIGURE 9.9 (a) Incoming rays toward the focus F of a parabolic mirror are reflected as parall rays upward. (b) The relationship between the incident angle θ and the distance d of the reflect ray from the optical axis.

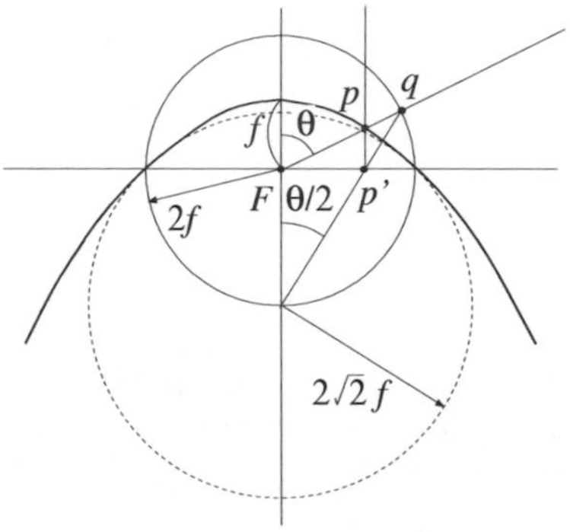

FIGURE 9.10 If the omnidirectional camera image is identified as a plane that passes through t focus F of the parabolic mirror and is orthogonal to the optical axis, it can be thought of as stereographic projection from the south pole of an image sphere around the focus with radius 2

in front of the camera (in the direction of the optical axis) is imaged inside this circle; th outside is the image of the scene behind the camera.

9.4 3D ANALYSIS OF OMNIDIRECTIONAL IMAGES

Let us call the image taken by a fisheye lens camera or an omnidirectional camera with parabolic mirror simply an *omnidirectional image*. We have seen that it is regarded as stereographic projection of the central image sphere shown in Fig. 9.2(a). One of the mos important consequences of this is that *lines in the 3D scene are projected to circles*. Th reasoning is as follows.

Given a line L in the scene, consider the plane passing through L and the center of the image sphere. The intersection of this plane with the image sphere is a great circl (Fig. 9.11(a)). The stereographic projection of the image sphere to the image plane is a inversion, which is a conformal mapping. Since a circle is mapped to a circle by a conform mapping, all lines in the scene are imaged as circles on omnidirectional images.

A 3D outdoor scene usually contains many parallel lines such as horizontal and vertic boundaries. Such parallel lines are projected to the omnidirectional image as circles inte secting at a common point. For example, parallel lines on an infinitely large horizontal plan are imaged as in Fig. 9.11(b). The intersection of circles resulting from parallel line image

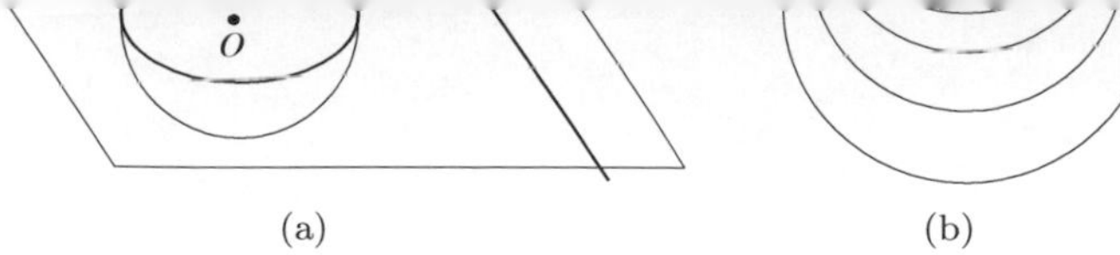

FIGURE 9.11 (a) Lines in the scene are mapped to great circles on the image sphere. (b) Parallel lines in the scene are imaged as circles intersecting at common vanishing points on the omnidirectional image.

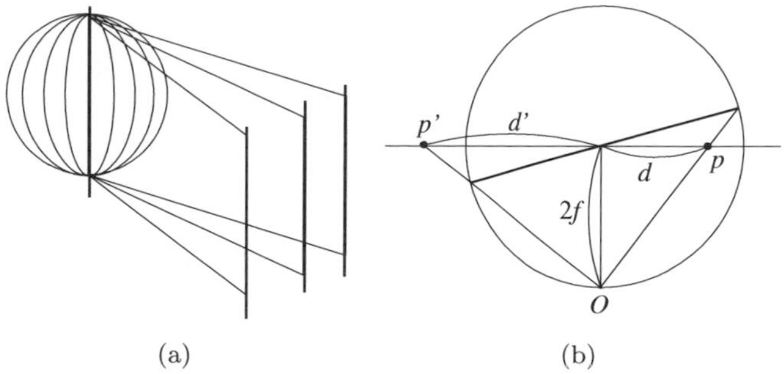

FIGURE 9.12 (a) The vanishing points on the omnidirectional image indicate the 3D direction of the parallel lines. (b) The focal length of the omnidirectional image can be computed from the locations p and p' of the vanishing points.

is called their *vanishing point*. The position on the image sphere that corresponds to the vanishing point indicates the 3D direction of the parallel lines, because the planes defined by the center of the image sphere and the individual parallel lines have a common intersection line passing through the positions that correspond to the vanishing points (Fig. 9.12(a)).

This fact enables us to estimate the vanishing points on the omnidirectional image by fitting circles to images of parallel lines in the scene and detecting their intersection, thereby computing the 3D direction of the parallel lines in the scene. For this computation, we need to know the focal length f, which can be estimated from the locations of the vanishing points. Let pp' be the vanishing point pair. It can be regarded as a stereographic projection of the diametric point pair on the image sphere of radius $2f$ that corresponds to the 3D direction of the parallel lines (Fig. 9.12). Let d and d' be the distances of p and p', respectively, from the principal point, and O the south pole of the image sphere. Then, $\triangle Opp'$ is a right triangle, for which $|Op| = \sqrt{d^2 + 4f^2}$, $|Op'| = \sqrt{d'^2 + 4f^2}$, and $|pp'| = d + d'$ hold. From $|Op|^2 + |Op'|^2 = |pp'|^2$, we obtain

$$f = \frac{\sqrt{dd'}}{2}. \tag{9.12}$$

Thus, *the geometric mean* of the distances of the vanishing points from the principal point equals $2f$. This computation requires knowledge of the principal point position. If it is unknown, we can estimate it as *the intersection of two segments of vanishing point pairs* if we can observe images of two sets of parallel lines with different orientations.

The 3D interpretation of the scene from an omnidirectional image can be done by first transforming it to a perspective image, as described in Sec. 9.2, and then applying the well-established computer vision techniques. However, by exploiting the knowledge that

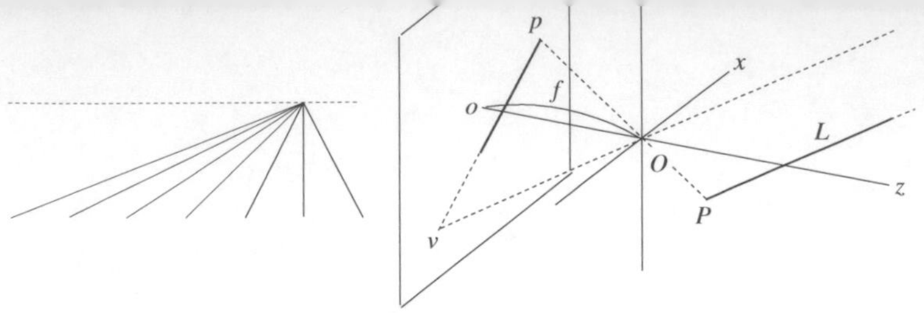

(a) (b)

FIGURE 9.13 (a) In a perspective image, projections of parallel lines in the scene converge at vanishing point. (b) The 3D orientation of a line L can be computed from the vanishing point p its projection.

an omnidirectional image is a stereographic projection, which is a conformal mappin of the central image sphere, we can do the same analysis directly without transformi omnidirectional images to perspective images (↪ Exercise 9.4).

Traditional World 9.2 (Perspective image analysis) 3D analysis of perspective car era images has a long history and is now a well-established domain of computer visic research. The best known is the fact that projections of parallel lines in the 3D scene inte sect on the image plane at a common *vanishing point*.

If we take an image of parallel lines on an infinitely large plane, we obtain an ima like Fig. 9.13(a). The vanishing point indicates the 3D orientation of the lines in the scen As illustrated in Fig. 9.12(b), a half line L in the scene starting from point P is project onto the image plane as a line segment connecting the projection p of P and the vanishi point v, which is the projection of a point infinitely far from P on the half line L. The li connecting the vanishing point v and the lens center O indicates the direction of L, which given by $\overrightarrow{vO} = (a, b, f)^\top$, where f is the focal length and (a, b) the location of the vanishi point v.

This computation requires knowledge of the focal length f. If it is unknown, we ca compute it if we detect two vanishing points corresponding to two sets of parallel lines th are orthogonal in the scene. Let (a, b) and (a', b') be the locations of the vanishing point The corresponding 3D directions are $(a, b, f)^\top$ and $(a', b', f)^\top$, respectively. Since they a orthogonal, they satisfy $aa' + bb' + f^2 = 0$. Hence,

$$f = \sqrt{-aa' - bb'}. \quad (9.1$$

This computation is done with respect to an image coordinate system with origin o the principal point. If the principal point o is unknown, we can estimate it if we obser vanishing points of three sets of mutually orthogonal parallel lines, e.g., east-west, nort south, and up-down. Let v_1, v_2, and v_3 be the three vanishing points (Fig. 9.14). Th principal point o is at the *orthocenter* of $\Delta v_1 v_2 v_3$. This is a consequence of the fact that t three directions $\overrightarrow{Ov_1}$, $\overrightarrow{Ov_2}$, and $\overrightarrow{Ov_3}$ from the lens center O are mutually orthogonal and th $\overrightarrow{Oo}$ is perpendicular to the image plane. To do this type of analysis in practice, however, w need to detect points and lines in the image precisely, which is very difficult, since all imag processing algorithms entail errors to some extent. In principle, omnidirectional images wit a large field of view are more suited to 3D interpretation computation.

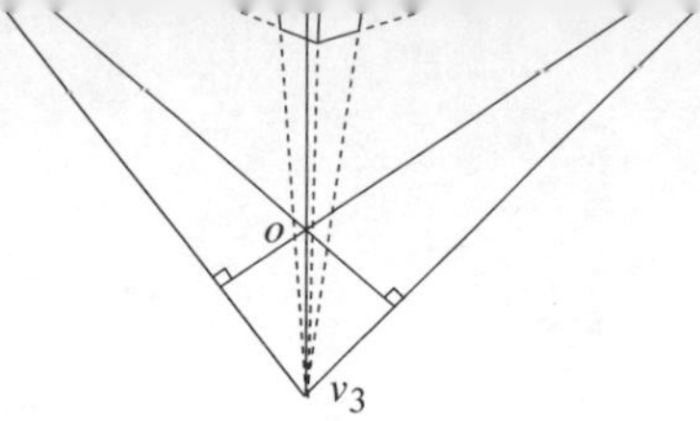

FIGURE 9.14 If the vanishing points v_1, v_2, and v_3 of three sets of mutually orthogonal lines are detected, the principal point o is at the orthocenter of the triangle they make.

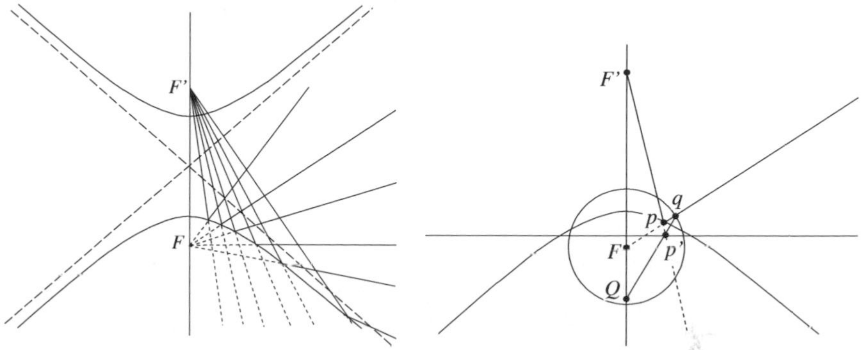

(a) (b)

FIGURE 9.15 (a) Incoming rays toward one focus F are reflected so that they converge to the other focus F'. (b) The observed image is obtained by projecting the image sphere surrounding the focus from a certain point Q onto the image plane placed in a certain position.

9.5 OMNIDIRECTIONAL CAMERAS WITH HYPERBOLIC AND ELLIPTIC MIRRORS

Omnidirectional cameras are also obtained by using hyperbolic mirrors. A hyperbolic mirror has two focuses, and rays incoming toward one focus F are reflected in such a way that they converge to the other focus F' (Fig. 9.15). It follows that if we place a perspective camera such that its lens center is at F', all incoming rays can be captured. Figure 9.16(a) is an indoor scene image taken by such a camera, and Fig. 9.16(b) shows its parts transformed to perspective images. Note that we can view only the outside of the cone defined by the asymptotes of the hyperbolic surface (dashed lines in Fig. 9.15(a)). The omnidirectional camera with a parabolic mirror can be thought of as the limit of moving the focus to which reflected rays converge infinitely far away.

We can also consider a hypothetical image sphere surrounding the focus F toward which rays enter. We can obtain an omnidirectional image by projecting the sphere onto a plane. As shown in Fig. 9.15(b), the ray reflected at point p on the hyperbolic mirror intersects with the image plane at point p'. This ray is recorded on the image sphere at point q. It is known that there exists a point Q on the optical axis inside the image sphere such that the point q is projected to the point p' on the image plane from Q. The position of such Q depends on the shape of the hyperbolic mirror. If the mirror shape is close to a parabola, the point Q is close to the south pole of the image sphere, and the image plane is close to

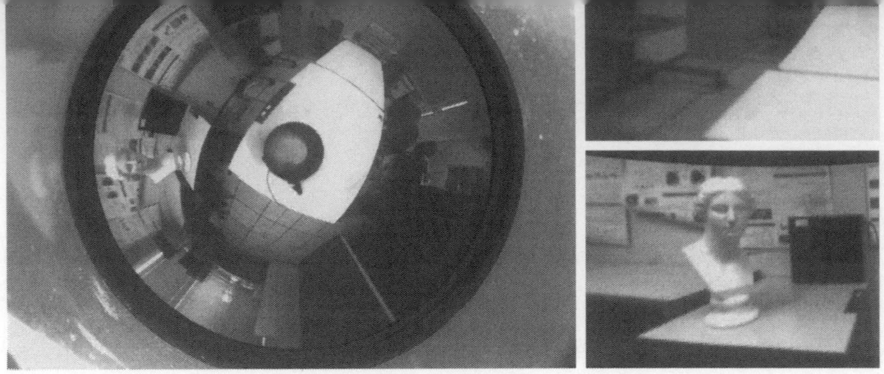

(a) (b)

FIGURE 9.16 (a) An indoor scene image taken by an omnidirectional camera with a hyperbol mirror. (b) Perspectively transformed partial images.

the focus F. Hence, the projection is approximately stereographic. For general hyperbol mirrors, however, the projection is not stereographic, so it is not an inversion with respe to a sphere, hence not a conformal mapping. As a result, lines in the scene are general imaged as ellipses and hyperbolas. Still, parallel lines in the scene are imaged as curv meeting at a common vanishing point, and the line segment connecting it with the focus indicates the 3D orientation of the parallel lines.

The above observations also apply to elliptic mirrors. An elliptic surface also has tw focuses, and rays incoming toward one focus F are reflected as if diverging from the oth focus F' (Fig. 9.17(a)). By capturing the diverging rays through a lens system, we can obtai an image of a large field of view. In this case, too, we can consider a hypothetical imag sphere surrounding the focus F toward which rays enter. We can obtain an omnidirection image by projecting the sphere onto a plane. As shown in Fig. 9.17(b), the ray reflecte at point p on the elliptic mirror looks as if it is emanating from the focus F' through p on the image plane. This ray is recorded on the image sphere at point q. It is known th there exists a point Q on the optical axis inside the image sphere such that the point q projected to the point p' on the image plane from Q. The position of such Q depends on th shape of the elliptic mirror. This projection is not stereographic, so it is not an inversio with respect to a sphere, hence not a conformal mapping. As a result, lines in the scene ar generally imaged as ellipses and hyperbolas. Still, parallel lines in the scene are imaged a curves meeting at a common vanishing point, and the line segment connecting it with th focus F indicates the 3D orientation of the parallel lines. It is also known that for a give omnidirectional camera with an elliptic mirror, we can define an omnidirectional camer with a hyperbolic mirror such that the resulting images are the same.

Traditional World 9.3 (Ellipse, hyperbola, and parabola) An ellipse centered at th origin O with major and minor axes aligned to the x- and y-axes is given by

$$\frac{x^2}{a^2} + \frac{y^2}{b^2} = 1, \qquad (9.14$$

where a and b, the lengths of the major and minor semi-axes, are often referred to simply a

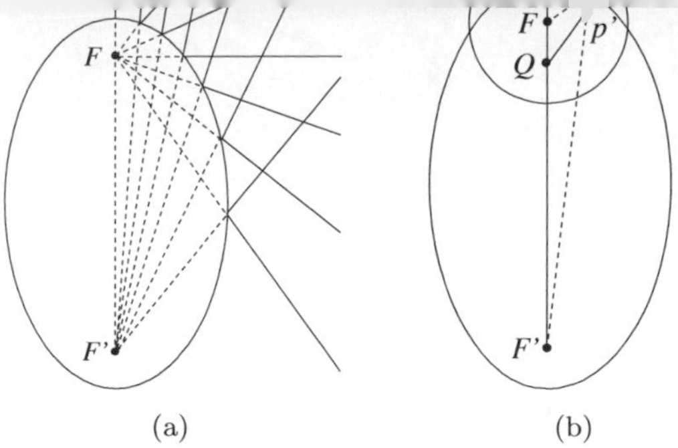

FIGURE 9.17 (a) Incoming rays toward one focus F are reflected as if diverging from the other focus F'. (b) The observed image is obtained by projecting the image sphere surrounding the focus from a certain point Q onto the image plane placed in a certain position.

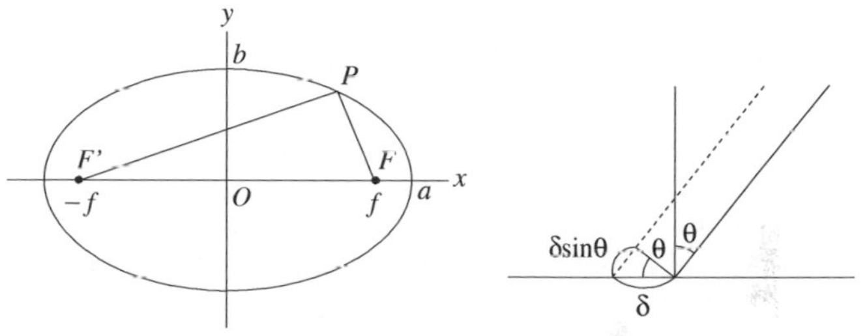

FIGURE 9.18 (a) An ellipse is defined as a locus of points for which the sum of the distances from the two focuses F and F' is constant. (b) If the incidence point of a ray with incidence angle θ is infinitesimally displaced by δ, the optical path length increases by $\delta \sin\theta$.

the major and minor axes. Its *focuses* (or *foci*) F and F' are on the major axis at distance f from the origin O, where

$$f = \sqrt{a^2 - b^2}. \tag{9.15}$$

For any point P on this ellipse, the equality

$$|FP| + |PF'| = 2a \tag{9.16}$$

holds (Fig. 9.18(a)). Namely, an ellipse is a locus of points for which *the sum of the distances from the two focuses is constant*. Equation (9.14) is obtained from this (↪ Exercise 9.5(1)). As a consequence, light rays emanating from one focus reflect on the ellipse to converge to the other. This can be confirmed by computing the tangent direction at P by differentiating Eq. (9.14) to see that the angles of incidence and reflection for FP and PF are equal. However, this can be more easily understood from *Fermat's principle*, well known in physics, that *light propagates along the shortest optical path*. Mathematically, this is formulated as a *variational principle*: if the optical path is infinitesimally displaced, or *perturbed*, the path length is stationary, by which we mean "constant up to higher order terms of the perturbation." Equation (9.16) implies that the path length is constant if the reflection point is perturbed. The fact that FPF' is indeed the optical path is reasoned as follows. If

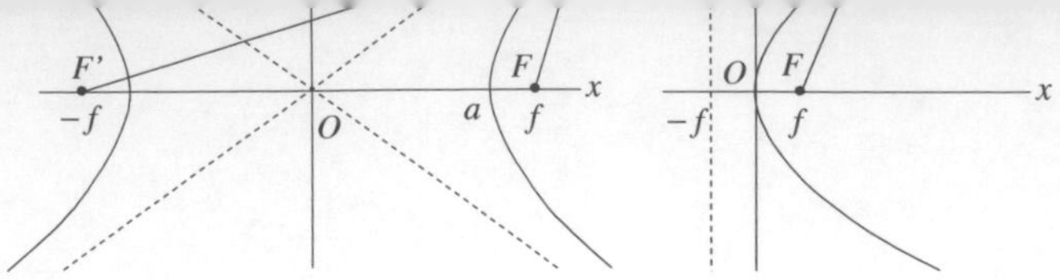

(a) (b)

FIGURE 9.19 (a) A hyperbola is defined as a locus of points for which the difference of the distanc from the two focuses F and F' is constant. (b) A parabola is defined as a locus of points for whi the distances from the focus F and from the directrix are equal.

the reflection point of a ray with incidence angle θ is displaced by an infinitesimal distan δ, the incident ray increases its path length by $\delta \sin\theta$ (Fig. 9.18(b)). In order that the tot optical path length is the same, the reflected ray must decrease its path length by the san amount $\delta \sin\theta$ with the same reflection angle θ. In other words, the incidence and reflectio angles must be equal, defining a physical optical path. The *eccentricity* of the ellipse Eq. (9.14) is defined to be

$$e = \frac{f}{a} \quad (< 1). \tag{9.17}$$

By definition, $e = 0$ for circles; the ellipse becomes flatter as e approaches 1.

A hyperbola centered at the origin O and symmetric with respect to the x- and y-ax is given by

$$\frac{x^2}{a^2} - \frac{y^2}{b^2} = 1, \tag{9.18}$$

or interchanging the terms of x and y. The *focuses* F and F' of the hyperbola of Eq. (9.18) are on the x-axis at distance f from the origin O, where

$$f = \sqrt{a^2 + b^2}. \tag{9.19}$$

For any point P on this hyperbola, the equality

$$|FP| - |PF'| = \pm 2a \tag{9.20}$$

holds (Fig. 9.19(a)). Namely, a hyperbola is a locus of points for which *the difference the distances from the two focuses is constant*. Equation (9.18) is obtained from this (Exercise 9.5(2)). As a consequence, light rays emanating from one focus reflect on th hyperbola as if diverging from the other. This can be confirmed by computing the tanger direction by differentiation, but this is also seen from Fermat's principle as in the ellips case: Eq. (9.20) implies that the path length is constant, and if the reflection point infinitesimally displaced, the incidence and reflection angles must be equal so that incider and reflected rays should cancel the increase and the decrease of the optical paths. Th eccentricity of the hyperbola of Eq. (9.18) is defined to be

$$e = \frac{f}{a} \quad (> 1). \tag{9.21}$$

As e approaches 1, the two curves of the hyperbola become flatter along the x-axis. As becomes larger, they become more linear in the direction of y. In the part where $|x|$ an

$$y = \pm\frac{\ }{a}x, \qquad (9.22)$$

called the *asymptotes* of the hyperbola of Eq. (9.18).

A parabola passing through the origin O and symmetric with respect to the x-axis is given by

$$y^2 = 4fx. \qquad (9.23)$$

The point F at $(f, 0)$ on the x-axis is called its *focus*; the line $x = -f$ is called its *directrix*. If we let H be the foot of the perpendicular line to the directrix from a point P on the parabola, the equality

$$|HP| = |PF| \qquad (9.24)$$

holds (Fig. 9.19(b)). Namely, a parabola is a locus of points for which *the distances from the directrix and from the focus are equal*. Equation (9.23) is obtained from this (↪ Exercise 9.5(3)). From Eq. (9.24), we can show that light rays coming from the left parallel to the x-axis are reflected as if diverging from the focus F; rays coming in from the right are reflected so as to converge to F. This can be explained by computing the tangent to the parabola and also by Fermat's principle. This is intuitively obvious if we regard a parabola as the limit of moving one of the focuses of an ellipse infinitely far away. The eccentricity of all parabolas is defined to be

$$e = 1, \qquad (9.25)$$

since a parabola is regarded as the limit of an ellipse whose eccentricity approaches 1 and at the same time as the limit of a hyperbola whose eccentricity approaches 1.

9.6 SUPPLEMENTAL NOTE

Geometric analysis of perspective camera imaging and 3D analysis based on it are a central theme of computer vision study [7, 10, 13, 14]. For fisheye lens camera calibration based on stereographic projection and its applications, see Kanatani [15]. The principle and applications of omnidirectional cameras are described in detail in Benosman and Kang [1]. Geyer and Daniilidis [9] give a detailed analysis of the hypothetical projection point and the image plane location of omnidirectional cameras using hyperbolic and elliptic mirrors for modeling them as a projection of a spherical image. Perwass [16] and Bayro-Corrochano [3] describe the imaging geometry of omnidirectional cameras in terms of geometric algebra equations. Ellipses, hyperbolas, and parabolas are collectively called *conics* or *conic loci* and are studied in the framework of projective geometry. Semple and Roth [19] is a classical textbook on this. The omnidirectional camera images in Fig. 9.16 are provided by the courtesy of Yasushi Kanazawa of the Toyohashi University of Technology, Japan.

9.7 EXERCISES

9.1. Let (x, y) be the stereographic projection of point (X, Y, Z) on a unit sphere surrounding the origin O from the south pole $(0, 0, -1)$ onto the xy plane (Fig. 9.20).

(1) Show that point (X, Y, Z) and point (x, y) are related by the following relations:

$$x = \frac{X}{1+Z}, \qquad y = \frac{Y}{1+Z},$$

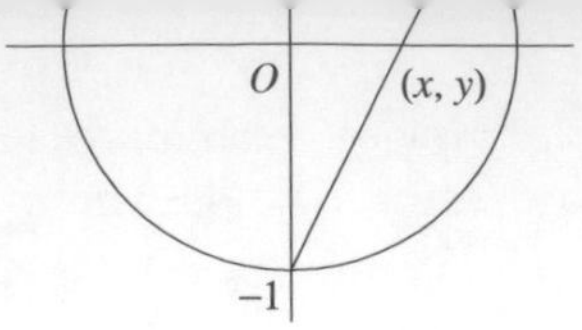

FIGURE 9.20 Stereographic projection of a unit sphere from the south pole.

$$X = \frac{2x}{1+x^2+y^2}, \qquad Y = \frac{2y}{1+x^2+y^2}, \qquad Z = \frac{1-x^2-y^2}{1+x^2+y^2}.$$

(2) Show that this correspondence is an inversion with respect to a sphere of radi $\sqrt{2}$ surrounding the south pole $(0, 0, -1)$.

9.2. Consider an xyz coordinate system associated with the fisheye lens camera that sati fies Eq. (9.7) with the origin O at its lens center and the z-axis along its optical axi Show that a point (X, Y, Z) on a unit sphere surrounding the origin is imaged at point (x, y) given by

$$x = \frac{2fX}{1+Z}, \qquad y = \frac{2fY}{1+Z},$$

where we assume that the image origin is at the principal point and the x- and y-ax are aligned to the X- and Y-axes in the same directions (Fig. 9.21).

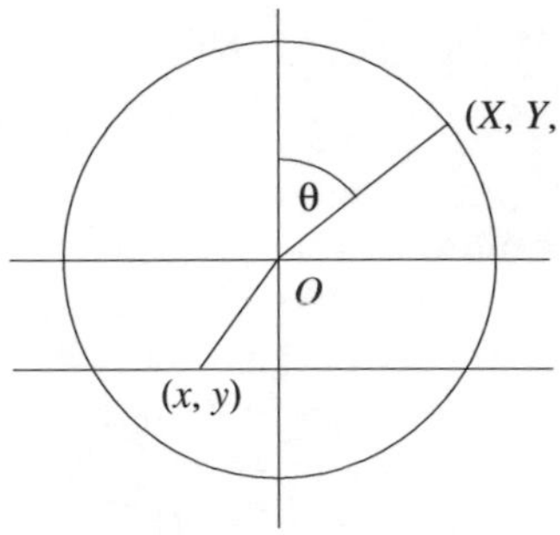

FIGURE 9.21 Fisheye lens camera imaging geometry.

9.3. Consider an xyz coordinate system associated with the omnidirectional camera wit a parabolic mirror that satisfies Eq. (9.7) with the origin O at its lens center and th z-axis along its optical axis. Show that a point (X, Y, Z) on a unit sphere surroundin the origin is imaged at a point (x, y) given by

$$x = \frac{2fX}{1+Z}, \qquad y = \frac{2fY}{1+Z},$$

where we assume that the image origin is at the principal point and the x- and y-axe are aligned to the X- and Y-axes in the same directions.

9.4. Camara imaging can be regarded as recording incoming rays of light regardless of th

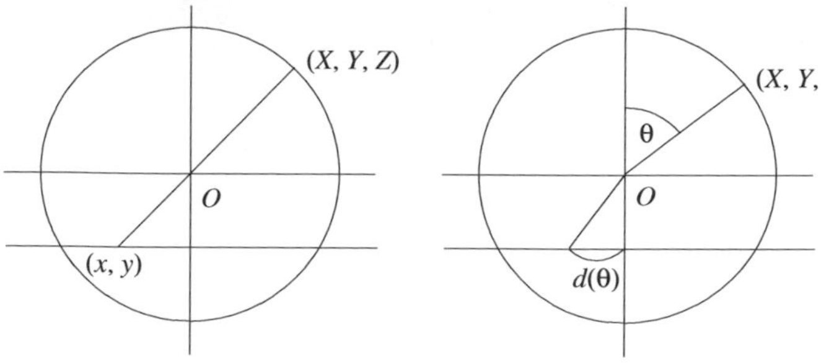

(a) (b)

FIGURE 9.22 (a) Perspective camera imaging. (b) General camera imaging.

(1) Compute the position (X, Y, Z) on the celestial sphere from its perspective projection image (x, y) on the image plane (Fig. 9.22(a)), where the x- and y-axes of the image coordinate system are assumed to be aligned to the X- and Y-axes in the same directions.

(2) Suppose the image we are observing is not taken by a perspective camera but by a general camera for which the relation $d = d(\theta)$ holds for the incidence angle θ of the incoming ray and the distance d of its image from the principal point (Fig. 9.22(b)). Describe the computational procedure for transforming the image to a perspective view as if taken by a perspective camera of focal length f.

(3) Using the same principle, describe the procedure for transforming the image to a perspective view as if taken by a perspective camera of focal length f whose optical axis is oriented in a specified direction.

9.5. (1) Derive Eq. (9.14) from Eq. (9.16).

(2) Derive Eq. (9.18) from Eq. (9.20).

(3) Derive Eq. (9.23) from Eq. (9.24).

Answers

Chapter 2

2.1. Consider vectors $\boldsymbol{a}$ and $\boldsymbol{b}$ starting from the origin O. Let A and B be their respective endpoints (Fig. A2.1). For $\triangle ABC$, the law of cosines

$$AB^2 = OA^2 + OB^2 - 2OA \cdot OB \cos\theta$$

holds. This is obtained by letting H be the foot of the perpendicular line from B to OA and applying the Pythagorean theorem to the right triangles $\triangle OHB$ and $\triangle HAB$. From the above law of cosines follows

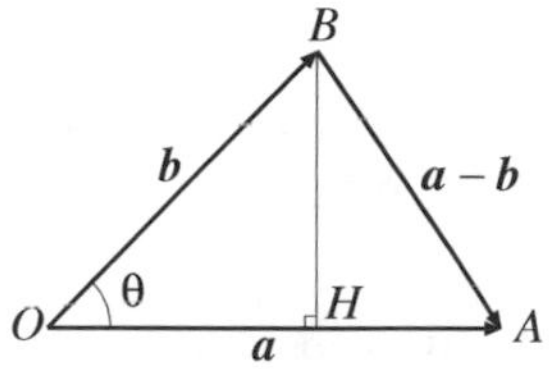

FIGURE A2.1

$$\|\boldsymbol{a} - \boldsymbol{b}\|^2 = \|\boldsymbol{a}\|^2 + \|\boldsymbol{b}\|^2 - 2\|\boldsymbol{a}\|\|\boldsymbol{b}\|\cos\theta.$$

Since the left side is $\langle \boldsymbol{a} - \boldsymbol{b}, \boldsymbol{a} - \boldsymbol{b}\rangle = \langle \boldsymbol{a}, \boldsymbol{a}\rangle - 2\langle \boldsymbol{a}, \boldsymbol{b}\rangle + \langle \boldsymbol{b}, \boldsymbol{b}\rangle = \|\boldsymbol{a}\|^2 - 2\langle \boldsymbol{a}, \boldsymbol{b}\rangle + \|\boldsymbol{b}\|^2$, we obtain Eq. (2.10).

2.2. Consider a function $f(t) = \|\boldsymbol{a} - t\boldsymbol{b}\|^2$ of t. By definition, $f(t) \geq 0$ for all t. Expanding the right side, we have

$$f(t) = \langle \boldsymbol{a} - t\boldsymbol{b}, \boldsymbol{a} - t\boldsymbol{b}\rangle = \langle \boldsymbol{a}, \boldsymbol{a}\rangle - 2t\langle \boldsymbol{a}, \boldsymbol{b}\rangle + t^2\langle \boldsymbol{b}, \boldsymbol{b}\rangle = \|\boldsymbol{b}\|^2 t^2 - 2\langle \boldsymbol{a}, \boldsymbol{b}\rangle t + \|\boldsymbol{a}\|^2.$$

For $\boldsymbol{b} \neq 0$, this is a quadratic polynomial in t, and the condition that $f(t) \geq 0$ for all t is that the quadratic equation $f(t) = 0$ either has no real roots or has one multiple root (Fig. A2.2). This is the case if and only if the discriminant D is 0 or negative:

$$D = \langle \boldsymbol{a}, \boldsymbol{b}\rangle^2 - \|\boldsymbol{a}\|^2\|\boldsymbol{b}\|^2 \leq 0.$$

The Schwartz inequality of Eq. (2.12) is obtained from this. If $\boldsymbol{b} = 0$, the equality of Eq. (2.12) holds. The equality also holds for $\boldsymbol{a} = 0$. Otherwise, the equality holds only when $f(t) = 0$ for some t, i.e., $\boldsymbol{a} = t\boldsymbol{b}$ for some t.

2.3. From the Schwarz inequality, we obtain

$$\begin{aligned}\|\boldsymbol{a} + \boldsymbol{b}\|^2 &= \langle \boldsymbol{a} + \boldsymbol{b}, \boldsymbol{a} + \boldsymbol{b}\rangle = \langle \boldsymbol{a}, \boldsymbol{a}\rangle + 2\langle \boldsymbol{a}, \boldsymbol{b}\rangle + \langle \boldsymbol{b}, \boldsymbol{b}\rangle \\ &\leq \|\boldsymbol{a}\|^2 + 2\|\boldsymbol{a}\|\|\boldsymbol{b}\| + \|\boldsymbol{b}\|^2 = (\|\boldsymbol{a}\| + \|\boldsymbol{b}\|)^2,\end{aligned}$$

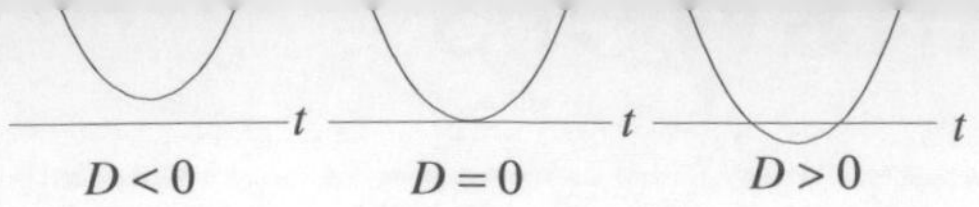

FIGURE A2.2

from which Eq. (2.13) is obtained. The equality holds in the same case that th equality holds for the Schwarz inequality. If we consider a triangle made by $\boldsymbol{a}$ and by making the endpoint of $\boldsymbol{a}$ and the starting point of $\boldsymbol{b}$ coincide, Eq. (2.13) stat that "one side of a triangle is shorter than the sum of the lengths of the other side (Fig. A2.3). This is the origin of the name "triangle inequality."

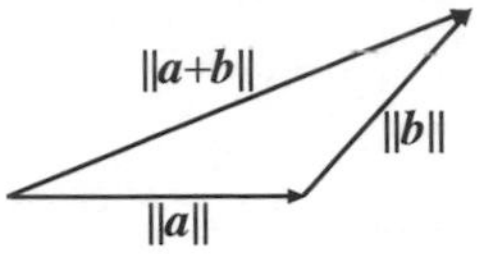

FIGURE A2.3

2.4. Since $\boldsymbol{a} \times \alpha\boldsymbol{b} = \alpha\boldsymbol{a} \times \boldsymbol{b}$ and $\boldsymbol{a} \times \beta\boldsymbol{c} = \beta\boldsymbol{a} \times \boldsymbol{c}$ from the definition of the vector produc the linearity $\boldsymbol{a} \times (\alpha\boldsymbol{b} + \beta\boldsymbol{c}) = \alpha\boldsymbol{a} \times \boldsymbol{b} + \beta\boldsymbol{a} \times \boldsymbol{c}$ is seen if the distributivity $\boldsymbol{a} \times (\boldsymbol{b} +$ $= \boldsymbol{a} \times \boldsymbol{b} + \boldsymbol{a} \times \boldsymbol{c}$ is shown.

Consider, first, the case where $\boldsymbol{b}$ and $\boldsymbol{c}$ are orthogonal to $\boldsymbol{a}$. Since $\boldsymbol{b}$ and $\boldsymbol{c}$ are include in the plane Π orthogonal to $\boldsymbol{a}$, the sum $\boldsymbol{b}+\boldsymbol{c}$ is also in Π. Since vector products of a vectors with $\boldsymbol{a}$ are orthogonal to $\boldsymbol{a}$, vectors $\boldsymbol{a} \times \boldsymbol{b}$, $\boldsymbol{a} \times \boldsymbol{c}$, and $\boldsymbol{a} \times (\boldsymbol{b} + \boldsymbol{c})$ are all in I From the definition of the vector product, they are also orthogonal to $\boldsymbol{b}$, $\boldsymbol{c}$, and $\boldsymbol{b}+$ within Π, and their lengths are $\|\boldsymbol{a}\|\|\boldsymbol{b}\|$, $\|\boldsymbol{a}\|\|\boldsymbol{c}\|$, and $\|\boldsymbol{a}\|\|\boldsymbol{b}+\boldsymbol{c}\|$, respectively. Henc $\boldsymbol{a} \times \boldsymbol{b}$, $\boldsymbol{a} \times \boldsymbol{c}$, and $\boldsymbol{a} \times (\boldsymbol{b} + \boldsymbol{c})$ are obtained by magnifying $\boldsymbol{b}$, $\boldsymbol{c}$, and $\boldsymbol{b}+\boldsymbol{c}$, respectivel by $\|\boldsymbol{a}\|$ and rotating them by 90° within Π (Fig. A2.4(a) shows the plane Π viewe from above). Hence, $\boldsymbol{a} \times (\boldsymbol{b} + \boldsymbol{c}) = \boldsymbol{a} \times \boldsymbol{b} + \boldsymbol{a} \times \boldsymbol{c}$ holds.

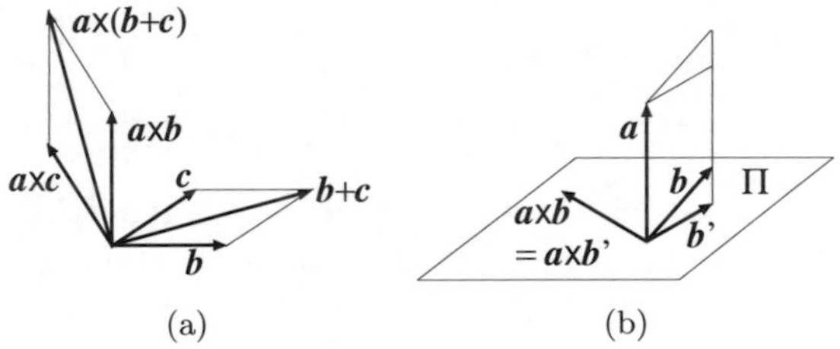

FIGURE A2.4

Next, consider the case where $\boldsymbol{b}$ and $\boldsymbol{c}$ are not necessarily orthogonal to $\boldsymbol{a}$. Let Π b the plane orthogonal to $\boldsymbol{a}$, and let $\boldsymbol{b}'$ be the projection of $\boldsymbol{b}$ onto Π. Then, $\boldsymbol{a}$ and $\boldsymbol{b}$ spa the same plane as $\boldsymbol{a}$ and $\boldsymbol{b}'$ (Fig. A2.4(b)). From the definition of the vector produc vectors $\boldsymbol{a} \times \boldsymbol{b}$ and $\boldsymbol{a} \times \boldsymbol{b}'$ are both orthogonal to that plane. Also, the parallelogra made by $\boldsymbol{a}$ and $\boldsymbol{b}$ has the same area as the parallelogram made by $\boldsymbol{a}$ and $\boldsymbol{b}'$. Henc $\boldsymbol{a} \times \boldsymbol{b} = \boldsymbol{a} \times \boldsymbol{b}'$. Similarly, $\boldsymbol{a} \times \boldsymbol{c} = \boldsymbol{a} \times \boldsymbol{c}'$. Since $\boldsymbol{b}' + \boldsymbol{c}'$ is the projection of $\boldsymbol{b} + \boldsymbol{c}$ ont Π, the equality $\boldsymbol{a} \times (\boldsymbol{b} + \boldsymbol{c}) = \boldsymbol{a} \times (\boldsymbol{b}' + \boldsymbol{c}')$ holds. Since $\boldsymbol{a} \times (\boldsymbol{b}' + \boldsymbol{c}') = \boldsymbol{a} \times \boldsymbol{b}' + \boldsymbol{a} \times \boldsymbol{c}$ we obtain $\boldsymbol{a} \times (\boldsymbol{b} + \boldsymbol{c}) = \boldsymbol{a} \times \boldsymbol{b} + \boldsymbol{a} \times \boldsymbol{c}$.

Hence, the area is a_2b_1. If b is along the y-axis, we can write $b = b_2e_2$, and the height of $\boldsymbol{a}$ measured from the y-axis is a_1 (Fig. A2.5(b)). Hence, the area is a_1b_2. Thus, the formula is correct in these cases.

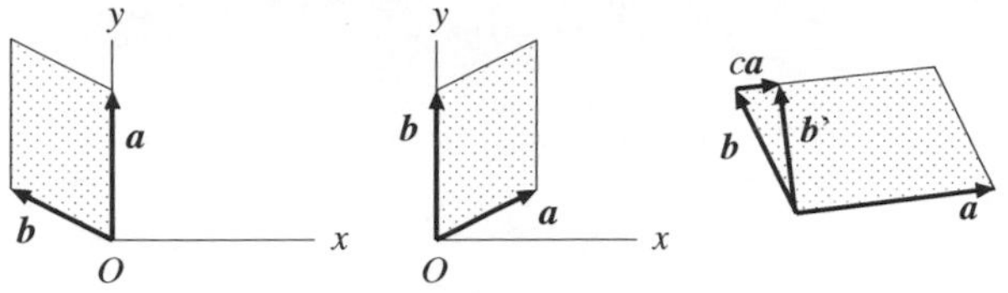

(a) (b) (c)

FIGURE A2.5

Suppose neither $\boldsymbol{a}$ nor $\boldsymbol{b}$ is along the y-axis. Regard $\boldsymbol{a}$ as the base of the parallelogram. Its area is unchanged if $\boldsymbol{b}$ is moved with its height fixed. Hence, the parallelogram made by $\boldsymbol{a}$ and $\boldsymbol{b}$ and the parallelogram made by $\boldsymbol{a}$ and $\boldsymbol{b}' = \boldsymbol{b} + c\boldsymbol{a}$ have the same area for any c (Fig. A2.5(c)). Since $a_1 \neq 0$ from the assumption that $\boldsymbol{a}$ is not along the y-axis, we can let $c = -b_1/a_1$. Then,

$$\boldsymbol{b}' = b_1e_1 + b_2e_2 - \frac{b_1}{a_1}(a_1e_1 + a_2e_2) = \left(b_2 - \frac{a_2b_1}{a_1}\right)e_2.$$

Thus, this vector $\boldsymbol{b}'$ is along the y-axis, and the area defined by $\boldsymbol{a}$ and $\boldsymbol{b}'$ is

$$S = a_1\left(b_2 - \frac{a_2b_1}{a_1}\right) = a_1b_2 - a_2b_1.$$

2.6. If vectors $\boldsymbol{a}$ and $\boldsymbol{b}$ make angle θ, the area of the parallelogram they define is

$$S = \|\boldsymbol{a}\|\|\boldsymbol{b}\|\sin\theta = \|\boldsymbol{a}\|\|\boldsymbol{b}\|\sqrt{1-\cos^2\theta} = \|\boldsymbol{a}\|\|\boldsymbol{b}\|\sqrt{1 - \frac{\langle \boldsymbol{a}, \boldsymbol{b}\rangle^2}{\|\boldsymbol{a}\|^2\|\boldsymbol{b}\|^2}}$$
$$= \sqrt{\|\boldsymbol{a}\|^2\|\boldsymbol{b}\|^2 - \langle \boldsymbol{a}, \boldsymbol{b}\rangle^2}.$$

We can confirm that the given formula is equal to this from the following identity:

$$(a_2b_3 - a_3b_2)^2 + (a_3b_1 - a_1b_3)^2 + (a_1b_2 - a_2b_1)^2$$
$$= (a_1^2 + a_2^2 + a_3^2)(b_1^2 + b_2^2 + b_3^2) - (a_1b_1 + a_2b_2 + a_3b_3)^2.$$

2.7. The projections of vectors $\boldsymbol{a} = a_1e_1 + a_2e_2 + a_3e_3$ and $\boldsymbol{b} = b_1e_1 + b_2e_2 + b_3e_3$ onto the yz plane are $a_2e_2 + a_3e_3$ and $b_2e_2 + b_3e_3$, respectively. From Exercise 2.5, the area of the parallelogram made by them is $S_{yz} = |a_2b_3 - a_3b_2|$. Similarly, the projections onto the zx and xy planes have areas $S_{zx} = |a_3b_1 - a_1b_3|$ and $S_{xy} = |a_1b_2 - a_2b_13|$, respectively. Hence, the area S of the parallelogram made by $\boldsymbol{a}$ and $\boldsymbol{b}$ is

$$S = \|\boldsymbol{a} \times \boldsymbol{b}\| = \sqrt{(a_2b_3 - a_3b_2)^2 + (a_3b_1 - a_1b_3)^2 + (a_1b_2 - a_2b_1)^2}$$
$$= \sqrt{S_{yz}^2 + S_{zx}^2 + S_{xy}^2}.$$

The inner product of this $\boldsymbol{c}$ with $\boldsymbol{a}$ is

$$\langle \boldsymbol{c}, \boldsymbol{a} \rangle = (a_2b_3 - a_3b_2)a_1 + (a_3b_1 - a_1b_3)a_2 + (a_1b_2 - a_2b_1)a_3 = 0.$$

The inner product of $\boldsymbol{c}$ with $\boldsymbol{b}$ is

$$\langle \boldsymbol{c}, \boldsymbol{a} \rangle = (a_2b_3 - a_3b_2)b_1 + (a_3b_1 - a_1b_3)b_2 + (a_1b_2 - a_2b_1)b_3 = 0.$$

Hence, $\boldsymbol{c}$ is orthogonal to both $\boldsymbol{a}$ and $\boldsymbol{b}$.

2.9. Regard the parallelogram made by vectors $\boldsymbol{a}$ and $\boldsymbol{b}$ as the base of the parallelepipe The base area is $S = \|\boldsymbol{a} \times \boldsymbol{b}\|$. Since $\boldsymbol{a} \times \boldsymbol{b}$ is orthogonal to the base, its unit surfa normal is

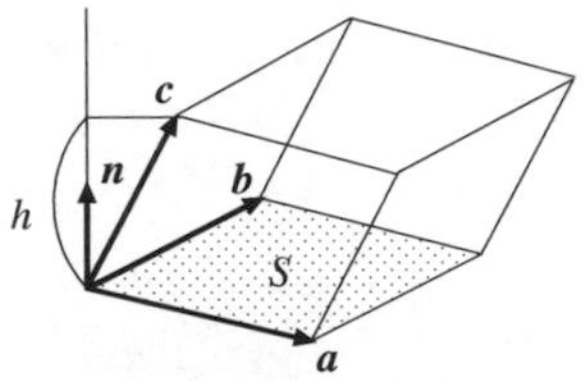

FIGURE A2.6

$$\boldsymbol{n} = \frac{\boldsymbol{a} \times \boldsymbol{b}}{\|\boldsymbol{a} \times \boldsymbol{b}\|}.$$

The height of the parallelepiped made by $\boldsymbol{a}$, $\boldsymbol{b}$, and $\boldsymbol{c}$ equals the projection of $\boldsymbol{c}$ ont the direction of $\boldsymbol{n}$ (Fig. A2.6). Hence,

$$h = \langle \boldsymbol{n}, \boldsymbol{c} \rangle = \langle \frac{\boldsymbol{a} \times \boldsymbol{b}}{\|\boldsymbol{a} \times \boldsymbol{b}\|}, \boldsymbol{c} \rangle = \frac{\langle \boldsymbol{a} \times \boldsymbol{b}, \boldsymbol{c} \rangle}{\|\boldsymbol{a} \times \boldsymbol{b}\|}.$$

This is positive when $\boldsymbol{c}$ is on the side of $\boldsymbol{n}$ and negative if it is on the other sid Hence, the signed volume is given by

$$V = hS = \langle \boldsymbol{a} \times \boldsymbol{b}, \boldsymbol{c} \rangle.$$

2.10. From Eq. (2.20), the following identities hold:

$$\begin{aligned} (\boldsymbol{a} \times \boldsymbol{b}) \times \boldsymbol{c} &= \langle \boldsymbol{a}, \boldsymbol{c} \rangle \boldsymbol{b} - \langle \boldsymbol{b}, \boldsymbol{c} \rangle \boldsymbol{a}, & \boldsymbol{a} \times (\boldsymbol{b} \times \boldsymbol{c}) &= \langle \boldsymbol{a}, \boldsymbol{c} \rangle \boldsymbol{b} - \langle \boldsymbol{a}, \boldsymbol{b} \rangle \boldsymbol{c}, \\ (\boldsymbol{b} \times \boldsymbol{c}) \times \boldsymbol{a} &= \langle \boldsymbol{b}, \boldsymbol{a} \rangle \boldsymbol{c} - \langle \boldsymbol{c}, \boldsymbol{a} \rangle \boldsymbol{b}, & \boldsymbol{b} \times (\boldsymbol{c} \times \boldsymbol{a}) &= \langle \boldsymbol{b}, \boldsymbol{a} \rangle \boldsymbol{c} - \langle \boldsymbol{b}, \boldsymbol{c} \rangle \boldsymbol{a}, \\ (\boldsymbol{c} \times \boldsymbol{a}) \times \boldsymbol{b} &= \langle \boldsymbol{c}, \boldsymbol{b} \rangle \boldsymbol{a} - \langle \boldsymbol{a}, \boldsymbol{b} \rangle \boldsymbol{c}, & \boldsymbol{c} \times (\boldsymbol{a} \times \boldsymbol{b}) &= \langle \boldsymbol{c}, \boldsymbol{b} \rangle \boldsymbol{a} - \langle \boldsymbol{c}, \boldsymbol{a} \rangle \boldsymbol{b}. \end{aligned}$$

Evidently, the sum of the three expressions on the left and the sum of the thre expressions on the right both vanish.

2.11. The equality is obtained as follows:

$$\begin{aligned} \langle \boldsymbol{x} \times \boldsymbol{y}, \boldsymbol{a} \times \boldsymbol{b} \rangle &= |\boldsymbol{x}, \boldsymbol{y}, \boldsymbol{a} \times \boldsymbol{b}| = \langle \boldsymbol{x}, \boldsymbol{y} \times (\boldsymbol{a} \times \boldsymbol{b}) \rangle = \langle \boldsymbol{x}, \langle \boldsymbol{y}, \boldsymbol{b} \rangle \boldsymbol{a} - \langle \boldsymbol{y}, \boldsymbol{a} \rangle \boldsymbol{b} \rangle \\ &= \langle \boldsymbol{x}, \boldsymbol{a} \rangle \langle \boldsymbol{y}, \boldsymbol{b} \rangle - \langle \boldsymbol{x}, \boldsymbol{b} \rangle \langle \boldsymbol{y}, \boldsymbol{a} \rangle. \end{aligned}$$

$$= \Big(\cos\Omega + l_1^2(1-\cos\Omega)\Big)a_1 + \Big(l_1l_2(1-\cos\Omega) - l_3\sin\Omega\Big)a_2$$
$$+\Big(l_1l_3(1-\cos\Omega) + l_2\sin\Omega\Big)a_3,$$
$$a_2' = a_2\cos\Omega + (l_3a_1 - l_1a_3)\sin\Omega + (a_1l_1 + a_2l_2 + a_3l_3)l_2(1-\cos\Omega)$$
$$= \Big(l_2l_1(1-\cos\Omega) + l_3\sin\Omega\Big)a_1 + \Big(\cos\Omega + l_2^2(1-\cos\Omega)\Big)a_2$$
$$+\Big(l_2l_3(1-\cos\Omega) - l_1\sin\Omega\Big)a_3,$$
$$a_3' = a_3\cos\Omega + (l_1a_2 - l_2a_1)\sin\Omega + (a_1l_1 + a_2l_2 + a_3l_3)l_3(1-\cos\Omega)$$
$$= \Big(l_3l_1(1-\cos\Omega) - l_2\sin\Omega\Big)a_1 + \Big(l_3l_2(1-\cos\Omega) + l_1\sin\Omega\Big)a_2$$
$$+\Big(\cos\Omega + l_3^2(1-\cos\Omega)\Big)a_3.$$

2.13. Suppose two lines l and l' are parallel. They are coplanar. Let H and H' be their supporting points. Since their position vectors $\overrightarrow{OH}$ and $\overrightarrow{OH'}$ starting from the origin O are orthogonal to l and l', respectively, the vector $\overrightarrow{HH'}$ is orthogonal to both l and l'. Hence, its length is the distance between l and l', and

$$d = \|\overrightarrow{HH'}\| = \left\|\frac{\boldsymbol{m}\times\boldsymbol{n}}{\|\boldsymbol{m}\|^2} - \frac{\boldsymbol{m}'\times\boldsymbol{n}'}{\|\boldsymbol{m}'\|^2}\right\|.$$

Since the two lines are parallel, there is a constant α ($\neq 0$) such that $\boldsymbol{m}' = \alpha\boldsymbol{m}$. Hence, d is rewritten as

$$d = \left\|\frac{\boldsymbol{m}\times\boldsymbol{n}}{\|\boldsymbol{m}\|^2} - \frac{\boldsymbol{m}\times\boldsymbol{n}'}{\alpha\|\boldsymbol{m}\|^2}\right\| = \frac{\|\boldsymbol{m}\times(\boldsymbol{n}-\boldsymbol{n}'/\alpha)\|}{\|\boldsymbol{m}\|^2}.$$

Since $\boldsymbol{n}$ and $\boldsymbol{n}'$ are both orthogonal to $\boldsymbol{m}$ ($= \boldsymbol{m}'/\alpha$), the vector $\boldsymbol{n} - \boldsymbol{n}'/\alpha$ is also orthogonal to $\boldsymbol{m}$. Hence, the above expression is further rewritten as

$$d = \frac{\|\boldsymbol{m}\|\|\boldsymbol{n}-\boldsymbol{n}'/\alpha\|}{\|\boldsymbol{m}\|^2} = \frac{\|\boldsymbol{n}-\boldsymbol{n}'/\alpha\|}{\|\boldsymbol{m}\|} = \left\|\frac{\boldsymbol{n}}{\|\boldsymbol{m}\|} - \frac{\boldsymbol{n}'}{\|\alpha\boldsymbol{m}\|}\right\| = \left\|\frac{\boldsymbol{n}}{\|\boldsymbol{m}\|} - \frac{\boldsymbol{n}'}{\|\boldsymbol{m}'\|}\right\|.$$

2.14. The unit surface normal $\boldsymbol{n}$ to the plane Π is orthogonal to both the direction $\boldsymbol{m}$ of the line l and the unit vector $\boldsymbol{u}$. Hence, it is written as

$$\boldsymbol{n} = \frac{\boldsymbol{m}\times\boldsymbol{u}}{\|\boldsymbol{m}\times\boldsymbol{u}\|}.$$

The distance h of the plane Π from the origin O equals the projected length of the position vector $\boldsymbol{x}_H = \boldsymbol{m}\times\boldsymbol{n}_l/\|\boldsymbol{m}\|^2$ (see Eq. (2.63)) of the supporting point of l onto the direction along the unit surface normal $\boldsymbol{n}$ of Π. Hence,

$$h = \langle\boldsymbol{n}, \boldsymbol{x}_H\rangle = \langle\frac{\boldsymbol{m}\times\boldsymbol{u}}{\|\boldsymbol{m}\times\boldsymbol{u}\|}, \frac{\boldsymbol{m}\times\boldsymbol{n}_l}{\|\boldsymbol{m}\|^2}\rangle = \frac{\langle\boldsymbol{m}\times\boldsymbol{u}, \boldsymbol{m}\times\boldsymbol{n}_l\rangle}{\|\boldsymbol{m}\times\boldsymbol{u}\|\|\boldsymbol{n}\|\|\boldsymbol{m}\|^2} = \frac{|\boldsymbol{m}, \boldsymbol{u}, \boldsymbol{m}\times\boldsymbol{n}_l|}{\|\boldsymbol{m}\times\boldsymbol{u}\|\|\boldsymbol{m}\|^2}$$
$$= \frac{\langle\boldsymbol{m}, \boldsymbol{u}\times(\boldsymbol{m}\times\boldsymbol{n}_l)\rangle}{\|\boldsymbol{m}\times\boldsymbol{u}\|\|\boldsymbol{m}\|^2} = \frac{\langle\boldsymbol{m}, \langle\boldsymbol{u}, \boldsymbol{n}_l\rangle\boldsymbol{m}\rangle}{\|\boldsymbol{m}\times\boldsymbol{u}\|\|\boldsymbol{m}\|^2} = \frac{\langle\boldsymbol{n}_l, \boldsymbol{u}\rangle\langle\boldsymbol{m}, \boldsymbol{m}\rangle}{\|\boldsymbol{m}\times\boldsymbol{u}\|\|\boldsymbol{m}\|^2} = \frac{\langle\boldsymbol{n}_l, \boldsymbol{u}\rangle}{\|\boldsymbol{m}\times\boldsymbol{u}\|},$$

where Eq. (2.20) for the vector triple product is used.

$$n = \frac{}{\|u \times v\|}.$$

The distance h of the plane Π from the origin equals the projected length of the vect p onto the direction along the unit surface normal n. Hence,

$$h = \langle p, n \rangle = \langle p, \frac{u \times v}{\|u \times v\|} \rangle = \frac{\langle p, u \times v \rangle}{\|u \times v\|} = \frac{|p, u, v|}{\|u \times v\|}.$$

Chapter 3

3.1. If vectors a, b, and c are regarded as a basis, its reciprocal basis is given by

$$a' = \frac{b \times c}{|a, b, c|} \qquad b' = \frac{c \times a}{|a, b, c|} \qquad c' = \frac{a \times b}{|a, b, c|}.$$

Hence, if vector x is expressed in the form $x = aa + bb + cc$, the coefficients a, b, ar c are given from Eq. (3.7) by

$$a = \langle a', x \rangle = \frac{\langle b \times c, x \rangle}{|a, b, c|} = \frac{|x, b, c|}{|a, b, c|},$$
$$b = \langle b', x \rangle = \frac{\langle c \times a, x \rangle}{|a, b, c|} = \frac{|a, x, c|}{|a, b, c|},$$
$$c = \langle c', x \rangle = \frac{\langle a \times b, x \rangle}{|a, b, c|} = \frac{|a, b, x|}{|a, b, c|}.$$

3.2. Since the Kronecker delta δ_i^j takes 1 for $i = j$ and 0 otherwise, the following identiti hold:

$$\delta_i^j a^i = \delta_1^j a^1 + \delta_2^j a^2 + \delta_3^j a^3 = a^j, \qquad \delta_i^j a_j = \delta_i^1 a_1 + \delta_i^2 a_2 + \delta_i^3 a_3 = a_i.$$

3.3. (1) Rewriting the dummy index i as k in Eq. (3.29) and multiplying both sides b g^{ij}, we obtain

$$g^{ij} g_{kj} a^k = g^{ij} a_j.$$

(Recall that the summation is implied.) From Eq. (3.32), the left side reduces t

$$g_{kj} g^{ji} a^k = \delta_k^i a^k = a^i,$$

resulting in Eq. (3.30). Rewriting the dummy index j as k in Eq. (3.30) an multiplying both sides by g_{ij}, we obtain

$$g_{ij} a^i = g_{ij} g^{ik} a_k.$$

From Eq. (3.32), the right side reduces to

$$g_{ji} g^{ik} a_k = \delta_j^k a_k = a_j,$$

resulting in Eq. (3.29).

From Eq. (3.32), the left side reduces to

$$g_{ki}g^{ij}e^k = \delta^j_k e^k = e^j,$$

resulting in Eq. (3.36) after interchanging the indices i and j. Rewriting the dummy index j as k in Eq. (3.36) and multiplying both sides by g_{ij}, we obtain

$$g_{ij}e^i = g_{ij}g^{ik}e_k.$$

From Eq. (3.32), the right side reduces to

$$g_{ji}g^{ik}e_k = \delta^k_j e_k = e_j,$$

resulting in Eq. (3.35) after interchanging the indices i and j.

3.4. (1) By differentiation, we obtain

$$\begin{aligned} e_r &= e_1 \sin\theta\cos\phi + e_2\sin\theta\sin\phi + e_3\cos\theta, \\ e_\theta &= e_1 r\cos\theta\cos\phi + e_2 r\cos\theta\sin\phi - e_3 r\sin\theta, \\ e_\phi &= -e_1 r\sin\theta\sin\phi + e_2 r\sin\theta\cos\phi. \end{aligned}$$

Hence, the metric tensor has the form

$$\begin{aligned} g_{rr} &= \langle e_r, e_r\rangle = \sin^2\theta\cos^2\phi + \sin^2\theta\sin^2\phi + \cos^2\theta \\ &= \sin^2\theta(\cos^2\phi + \sin^2\phi) + \cos^2\theta = 1, \\ g_{\theta\theta} &= \langle e_\theta, e_\theta\rangle = r^2\cos^2\theta\cos^2\phi + r^2\cos^2\theta\sin^2\phi + r^2\sin^2\theta \\ &= r^2\cos^2\theta(\cos^2\phi + \sin^2\phi) + r^2\sin^2\theta = r^2, \\ g_{\phi\phi} &= \langle e_\phi, e_\phi\rangle = r^2\sin^2\theta\sin^2\phi + r^2\sin^2\theta\cos^2\phi \\ &= r^2\sin^2\theta(\sin^2\phi + \cos^2\phi) = r^2\sin^2\phi, \\ g_{r\theta} &= \langle e_r, e_\theta\rangle = r\sin\theta\cos\theta\cos^2\theta + r\sin\theta\cos\theta\sin^2\theta - r\cos\theta\sin\theta \\ &= r\cos\theta\sin\theta(\cos^2\theta + \sin^2\theta - 1) = 0, \\ g_{r\phi} &= \langle e_r, e_\phi\rangle = -r\sin^2\theta\cos\theta\sin\theta + r\sin^2\theta\sin\theta\cos\theta = 0, \\ g_{\theta\phi} &= -r^2\cos\theta\sin\theta\cos\phi\sin\phi + r^2\cos\theta\sin\theta\sin\phi\cos\phi = 0. \end{aligned}$$

(2) If g_{ij}, i, $j = r$, θ, ϕ, is regarded as a matrix, its determinant is

$$g = 1\cdot r^2\cdot r^2\sin^2\theta = r^4\sin^2\theta.$$

From the definition of the angles θ and ϕ, the vectors $\{e_r,\ e_\theta,\ e_\phi\}$ are a right-handed system. Hence, from Eq. (3.45), the volume element is given by

$$I_{r\theta\phi} = \sqrt{g} = r^2\sin\theta.$$

Using this, the volume V of a sphere of radius R is computed as follows:

$$\begin{aligned} V &= \int_0^{2\pi}\int_0^{\pi}\int_0^{R} I_{r\theta\phi}\,dr d\theta d\phi = \int_0^{2\pi} d\phi \int_0^{\pi}\sin\theta d\theta \int_0^{R} r^2 dr \\ &= 2\pi\Big[-\cos\theta\Big]_0^{\pi}\left[\frac{r^3}{3}\right]_0^{r} = 2\pi(1+1)\frac{R^3}{3} = \frac{4}{3}\pi R^3. \end{aligned}$$

Hence, the metric tensor has the form

$$\begin{aligned} g_{rr} &= \langle e_r, e_r \rangle = \cos^2\theta + \sin^2\theta = 1, \\ g_{\theta\theta} &= \langle e_\theta, e_\theta \rangle = r^2 \sin^2\theta + r^2 \cos^2\theta = r^2, \\ g_{r\theta} &= \langle e_r, e_\theta \rangle = -r\cos\theta\sin\theta + r\sin\theta\cos\theta = 0, \\ g_{zz} &= \langle e_z, e_z \rangle = 1, \quad g_{rz} = \langle e_r, e_z \rangle = 0, \quad g_{\theta z} = \langle e_\theta, e_z \rangle = 0. \end{aligned}$$

(2) If g_{ij}, i, $j = r$, θ, ϕ, is regarded as a matrix, its determinant is

$$g = 1 \cdot r^2 \cdot 1 = r^2.$$

From the definition of the angle θ, the vectors $\{e_r,\ e_\theta,\ e_z\}$ are a right-hand system. Hence, from (3.45), the volume element is given by

$$I_{r\theta z} = \sqrt{g} = r.$$

Using this, the volume V of a cylinder of height h and radius R is computed as follows:

$$\begin{aligned} V &= \int_0^h \int_0^{2\pi} \int_0^R I_{r\theta z} dr d\theta dz = \int_0^h dz \int_0^{2\pi} d\theta \int_0^R r dr \\ &= h \cdot 2\pi \left[\frac{r^2}{2} \right]_0^r = 2h\pi \frac{R^2}{2} = \pi R^2 h. \end{aligned}$$

3.6. (1) Rewriting the dummy index i as k in the first equation of Eq. (3.51) and multiplying both sides by $A_i^{i'}$, we obtain

$$A_i^{i'} e_{i'} = A_i^{i'} A_{i'}^k e_k.$$

(Recall that the summation is implied.) From Eq. (3.50), the right side reduces to

$$A_{i'}^k A_i^{i'} e_k = \delta_i^k e_k = e_i,$$

resulting in the second equation. Rewriting the dummy index i' as k' in the second equation and multiplying both sides by $A_{i'}^i$, we obtain

$$A_{i'}^i e_i = A_{i'}^i A_i^{k'} e_{k'}.$$

From Eq. (3.50), the right side reduces to

$$A_i^{k'} A_{i'}^i e_{k'} = \delta_{i'}^{k'} e_{k'} = e_{i'},$$

resulting in the first equation.

(2) Rewriting the dummy index i as k in the first equation of Eq. (3.52) and multiplying both sides by $A_{i'}^i$, we obtain

$$A_{i'}^i a^{i'} = A_{i'}^i A_k^{i'} a^k.$$

resulting in the second equation. Rewriting the dummy index i' as k' in the second equation and multiplying both sides by $A_i^{i'}$, we obtain

$$A_i^{i'} a^i = A_i^{i'} A_{k'}^i a^{k'}.$$

From Eq. (3.50), the right side reduces to

$$\delta_{k'}^{i'} a^{k'} = a^{i'},$$

resulting in the first equation.

(3) Rewriting the dummy index i as k in the first equation of Eq. (3.55) and multiplying both sides by $A_{i'}^i$, we obtain

$$A_{i'}^i e^{i'} = A_{i'}^i A_k^{i'} e^k.$$

From Eq. (3.50), the right side reduces to

$$\delta_k^i e^k = e^i,$$

resulting in the second equation. Rewriting the dummy index i' as k' in the second equation and multiplying both sides by $A_i^{i'}$, we obtain

$$A_i^{i'} e^i = A_i^{i'} A_{k'}^i e^{k'}.$$

From Eq. (3.50), the right side reduces to

$$\delta_{k'}^{i'} e^{k'} = e^{i'},$$

resulting in the first equation.

(4) Rewriting the dummy index i as k in the first equation of Eq. (3.57) and multiplying both sides by $A_i^{i'}$, we obtain

$$A_i^{i'} a_{i'} = A_i^{i'} A_{i'}^k a_k.$$

From Eq. (3.50), the right side reduces to

$$A_{i'}^k A_i^{i'} a_k = \delta_i^k a_k = a_i,$$

resulting in the second equation. Rewriting the dummy index i' as k' in the second equation and multiplying both sides by $A_{i'}^i$, we obtain

$$A_{i'}^i a_i = A_{i'}^i A_i^{k'} a_{k'}.$$

From Eq. (3.50), the right side reduces to

$$A_i^{k'} A_{i'}^i a_{k'} = \delta_{i'}^{k'} a_{k'} = a_{i'},$$

resulting in the first equation.

$$A^{i'}_{i} A^{j'}_{j} g_{i'j'} = A^{i'}_{i} A^{j'}_{j} A^{k}_{i'} A^{l}_{j'} g_{kl}.$$

From Eq. (3.50), the right side reduces to

$$A^{k}_{i'} A^{i'}_{i} A^{l}_{j'} A^{j'}_{j} g_{kl} = \delta^{k}_{i} \delta^{l}_{j} g_{kl} = g_{ij},$$

resulting in the second equation. Rewriting the dummy indices i' and j' as and l', respectively, in the second equation and multiplying both sides by $A^{i}_{i'} A^{j}_{j'}$ we obtain

$$A^{i}_{i'} A^{j}_{j'} g_{ij} = A^{i}_{i'} A^{j}_{j'} A^{k'}_{i} A^{l'}_{j} g_{k'l'}.$$

From Eq. (3.50), the right side reduces to

$$A^{k'}_{i} A^{i}_{i'} A^{l'}_{j} A^{j}_{j'} g_{k'l'} = \delta^{k'}_{i'} \delta^{l'}_{j'} g_{k'l'} = g_{i'j'},$$

resulting in the first equation.

(6) Rewriting the dummy indices i and j as k and l, respectively, in the first equation of Eq. (3.62) and multiplying both sides by $A^{i}_{i'} A^{j}_{j'}$, we obtain

$$A^{i}_{i'} A^{j}_{j'} g^{i'j'} = A^{i}_{i'} A^{j}_{j'} A^{i'}_{k} A^{j'}_{l} g^{kl}.$$

From Eq. (3.50), the right side reduces to

$$A^{i}_{i'} A^{i'}_{k} A^{j}_{j'} A^{j'}_{l} g^{kl} = \delta^{i}_{k} \delta^{j}_{l} g^{kl} = g^{ij},$$

resulting in the second equation. Rewriting the dummy indices i' and j' as and l', respectively, in the second equation and multiplying both sides by $A^{i'}_{i} A^{j'}_{j}$ we obtain

$$A^{i'}_{i} A^{j'}_{j} g^{ij} = A^{i'}_{i} A^{j'}_{j} A^{i}_{k'} A^{j}_{l'} g^{k'l'}.$$

From Eq. (3.50), the right side reduces to

$$A^{i'}_{i} A^{i}_{k'} A^{j'}_{j} A^{j}_{l'} g^{k'l'} = \delta^{i'}_{k'} \delta^{j'}_{l'} g^{k'l'} = g^{i'j'},$$

resulting in the first equation.

Chapter 4

4.1. If we write $\boldsymbol{q} = q_1 i + q_2 j + q_3 k$ for the vector part of the quaternion q, Eq. (4.21) written as

$$\boldsymbol{a}' = (q_0 + \boldsymbol{q})\boldsymbol{a}(q_0 - \boldsymbol{q}) = q_0^2 \boldsymbol{a} + q_0(\boldsymbol{qa} - \boldsymbol{aq}) - \boldsymbol{qaq}.$$

From Eq. (4.10), the expression $\boldsymbol{qa} - \boldsymbol{aq}$ reduces to

$$\boldsymbol{qa} - \boldsymbol{aq} = \boldsymbol{q} \times \boldsymbol{a} - \boldsymbol{a} \times \boldsymbol{q} = 2\boldsymbol{q} \times \boldsymbol{a} = 2(q_2 a_3 - q_3 a_2) i + 2(q_3 a_1 - q_1 a_3) j + 2(q_1 a_2 - q_2 a_1) k$$

By a manipulation similar to Eq. (4.26), the expression $\boldsymbol{qaq}$ reduces to

$$\boldsymbol{qaq} = \|\boldsymbol{q}\|^2 \boldsymbol{a} - 2\langle \boldsymbol{q}, \boldsymbol{a} \rangle \boldsymbol{q}.$$

$$
\begin{aligned}
&= q_0^2 a_1 + 2q_0q_2a_3 - 2q_0q_3a_2 - \|\boldsymbol{q}\|^2 a_1 + 2(q_1a_1 + q_2a_2 + q_3a_3)q_1 \\
&= (q_0^2 - \|\boldsymbol{q}\|^2 + 2q_1^2)a_1 + (-2q_0q_3 + 2q_2q_1)a_2 + (2q_0q_2 + 2q_3q_1)a_3 \\
&= (q_0^2 + q_1^2 - q_2^2 - q_3^2)a_1 + 2(q_1q_2 - q_0q_3)a_2 + 2(q_1q_3 + q_0q_2)a_3.
\end{aligned}
$$

Similarly, we obtain a_2' and a_3' in the form

$$
\begin{aligned}
a_2' &= 2(q_2q_1 + q_0q_3)a_1 + (q_0^2 - q_1^2 + q_2^2 - q_3^2)a_2 + 2(q_2q_3 - q_0q_1)a_3, \\
a_3' &= 2(q_3q_1 - q_0q_2)a_1 + 2(q_3q_2 + q_0q_1)a_2 + (q_0^2 - q_1^2 - q_2^2 + q_3^2)a_3.
\end{aligned}
$$

4.2. The inverse cosine function $\cos^{-1} x$ has singularities at $x = \pm 1$. Since computer calculation is based on finite length data, the value of $\cos^{-1} x$ cannot be precisely computed for $x \approx \pm 1$ on many machines. Similarly, the inverse sine function $\sin^{-1} x$ has a singularity at $x = \pm 1$, so $\sin^{-1} x$ cannot be precisely computed for $x \approx \pm 1$. In view of this, we should divide the computation into the following two cases so that the argument is close to 0 (either applies to $q_0 = 0.5$).

$$
\Omega = \begin{cases} 2\cos^{-1} q_0 & |q_0| \leq 0.5 \\ 2\sin^{-1} \sqrt{q_1^2 + q_2^2 + q_3^2} & |q_0| \geq 0.5. \end{cases}
$$

4.3. If we determine the angle Ω by the above procedure and compute the axis from Eq. (4.24) in the form

$$
\boldsymbol{l} = \frac{q_1 i + q_2 j + q_3 k}{\sqrt{q_1^2 + q_2^2 + q_3^2}},
$$

the case of $\Omega > \pi$ occurs for $q_0 < 0$. Hence, if $q_0 < 0$, we replace the values by

$$
\Omega \leftarrow 2\pi - \Omega, \qquad \boldsymbol{l} \leftarrow -\boldsymbol{l}.
$$

4.4. Since $i^\dagger = -i$, we have $iii^\dagger = -i^3 = i$, $iji^\dagger = -iji = -ki = -j$, and $iki^\dagger = -iki = ji = -k$. Hence,

$$
i\boldsymbol{a}i^\dagger = i(a_1 i + a_2 j + a_3 k)i^\dagger = a_1 i - a_2 j - a_3 k.
$$

This represents reflection of $\boldsymbol{a}$ with respect to the x-axis, i.e., 180° rotation around it. Similarly, j and k, respectively, act as reflection with respect to the y-axis and the z-axis. Since a unit vector $\boldsymbol{l}$ can be regarded as a quaternion with $q_0 = 0$, the expression $\boldsymbol{l}q\boldsymbol{l}^\dagger$ represents reflection of $\boldsymbol{a}$ with respect to $\boldsymbol{l}$, i.e., 180° rotation around it.

4.5. For a scalar α, we have $\boldsymbol{a}' = \alpha\boldsymbol{a}\alpha^* = \alpha^2\boldsymbol{a}$, which represents α^2 times magnification of the vector $\boldsymbol{a}$ (or reduction for $\alpha^2 < 1$). If we let $\tilde{q} = q/\|q\|$ for $q \neq 0$, this is a unit quaternion. Hence, the action of q on vector $\boldsymbol{a}$ is

$$
\boldsymbol{a}' = q\boldsymbol{a}q^\dagger = \|q\|\tilde{q}\boldsymbol{a}\tilde{q}^\dagger\|q\| = \|q\|^2\tilde{q}\boldsymbol{a}\tilde{q}^\dagger.
$$

This represents rotation by the unit quaternion $\tilde{q}$ followed by magnification by $\|q\|^2$, also referred to as "scaled rotation."

$$z = \frac{\quad}{\alpha' + \beta' z'},$$

we obtain

$$z'' = \frac{\gamma' + \delta'(\gamma+\delta z)/(\alpha+\beta z)}{\alpha' + \beta'(\gamma+\delta z)/(\alpha+\beta z)} = \frac{\gamma'(\alpha+\beta z) + \delta'(\gamma+\delta z)}{\alpha'(\alpha+\beta z) + \beta'(\gamma+\delta z)} = \frac{\gamma'\alpha + \delta'\gamma + (\gamma'\beta + \delta'\delta)z}{\alpha'\alpha + \beta'\gamma + (\alpha'\beta + \beta'\delta)z}.$$

If the composite linear fractional transformation is written in the form

$$z'' = \frac{\gamma'' + \delta'' z}{\alpha'' + \beta'' z},$$

the parameters α'', β'', γ'', and δ'' are given by

$$\alpha'' = \alpha'\alpha + \beta'\gamma, \qquad \beta'' = \alpha'\beta + \beta'\delta, \qquad \gamma'' = \gamma'\alpha + \delta'\gamma, \qquad \delta'' = \gamma'\beta + \delta'\delta.$$

If the parameters before composition are expressed in the matrix form of Eq. (4.46 the matrix product is

$$\begin{pmatrix} \alpha' & \beta' \\ \gamma' & \delta' \end{pmatrix} \begin{pmatrix} \alpha & \beta \\ \gamma & \delta \end{pmatrix} = \begin{pmatrix} \alpha'\alpha + \beta'\gamma & \alpha'\beta + \beta'\delta \\ \gamma'\alpha + \delta'\gamma & \gamma'\beta + \delta'\delta \end{pmatrix},$$

which equals the matrix representation

$$\begin{pmatrix} \alpha'' & \beta'' \\ \gamma'' & \delta'' \end{pmatrix}$$

of the parameters α'', β'', γ'', and δ'' after the composition.

Chapter 5

5.1. In terms of the basis $\{e_1, e_2, e_3\}$, the sum of any number of bivectors reduces to th form

$$x e_2 \wedge e_3 + y e_3 \wedge e_1 + z e_1 \wedge e_2.$$

In order to express this in the form

$$\begin{aligned} \boldsymbol{a} \wedge \boldsymbol{b} &= (a_1 e_1 + a_2 e_2 + a_3 e_3) \wedge (b_1 e_1 + b_2 e_2 + b_3 e_3) \\ &= (a_2 b_3 - a_3 b_2) e_2 \wedge e_3 + (a_3 b_1 - a_1 b_3) e_3 \wedge e_1 + (a_1 b_2 - a_2 b_1) e_1 \wedge e_2 \end{aligned}$$

for some vectors $\boldsymbol{a} = a_1 e_1 + a_2 e_2 + a_3 e_3$ and $\boldsymbol{b} = b_1 e_1 + b_2 e_2 + b_3 e_3$, all we need is t find a_1, a_2, a_3, b_1, b_2, and b_3 that satisfy

$$x = a_2 b_3 - a_3 b_2, \qquad y = a_3 b_1 - a_1 b_3, \qquad z = a_1 b_2 - a_2 b_1,$$

for given x, y, and z. This is equivalent to finding vectors $\boldsymbol{a}$ and $\boldsymbol{b}$ such that (i) the are orthogonal to $\boldsymbol{x} = x e_1 + y e_2 + z e_3$, (ii) the three vectors $\{\boldsymbol{a}, \boldsymbol{b}, \boldsymbol{x}\}$ form a righ handed system, and (iii) the parallelogram made by $\boldsymbol{a}$ and $\boldsymbol{b}$ has area $\|\boldsymbol{x}\|$. There exis infinitely many such $\boldsymbol{a}$ and $\boldsymbol{b}$.

Hence, the given equality expresses the following vector triple product identity:

$$\boldsymbol{x} \times (\boldsymbol{a} \times \boldsymbol{b}) = \langle \boldsymbol{x}, \boldsymbol{b} \rangle \boldsymbol{a} - \langle \boldsymbol{x}, \boldsymbol{a} \rangle \boldsymbol{b}.$$

(2) From $\boldsymbol{a} \wedge \boldsymbol{b} = -(\boldsymbol{a} \times \boldsymbol{b})^*$ and $(\boldsymbol{a} \wedge \boldsymbol{b} \wedge \boldsymbol{c})^* = |\boldsymbol{a}, \boldsymbol{b}, \boldsymbol{c}|$, we can write

$$\begin{aligned} -\boldsymbol{x} \wedge \boldsymbol{y} \cdot (\boldsymbol{a} \times \boldsymbol{b})^* &= -((\boldsymbol{x} \wedge \boldsymbol{y}) \wedge (\boldsymbol{a} \times \boldsymbol{b}))^* \\ &= -(\boldsymbol{x} \wedge \boldsymbol{y} \wedge (\boldsymbol{a} \times \boldsymbol{b}))^* = -|\boldsymbol{x}, \boldsymbol{y}, \boldsymbol{a} \times \boldsymbol{b}| = -\langle \boldsymbol{x} \times \boldsymbol{y}, \boldsymbol{a} \times \boldsymbol{b} \rangle. \end{aligned}$$

Hence, the given equality express the following identity:

$$\langle \boldsymbol{x} \times \boldsymbol{y}, \boldsymbol{a} \times \boldsymbol{b} \rangle = \langle \boldsymbol{x}, \boldsymbol{a} \rangle \langle \boldsymbol{y}, \boldsymbol{b} \rangle - \langle \boldsymbol{x}, \boldsymbol{b} \rangle \langle \boldsymbol{y}, \boldsymbol{a} \rangle.$$

5.3. (1) The (signed) magnitude of the bivector $\overrightarrow{OA} \wedge \overrightarrow{OC}$ equals the (signed) area of the parallelogram made by vectors $\overrightarrow{OA}$ and $\overrightarrow{OC}$, which is twice the area of the (signed) area of the triangle $\triangle OAC$. The same holds for other bivectors. If the line l is at distance h from the origin O, the area of the triangle $\triangle OAC$ is $h \cdot AC/2$. Hence, the cross ratio is written as the ratio $(AC/BC)/(AD/BD)$ of the lengths along the line l.

(2) Let $\boldsymbol{a}$, $\boldsymbol{b}$, $\boldsymbol{c}$, and $\boldsymbol{d}$ be the unit vectors obtained by dividing $\overrightarrow{OA}$, $\overrightarrow{OB}$, and $\overrightarrow{OC}$, $\overrightarrow{OD}$ by their respective lengths $|OA|$, $|OB|$, $|OC|$, and $|OD|$:

$$\boldsymbol{a} = \frac{\overrightarrow{OA}}{|OA|}, \qquad \boldsymbol{b} = \frac{\overrightarrow{OB}}{|OB|}, \qquad \boldsymbol{c} = \frac{\overrightarrow{OC}}{|OC|}, \qquad \boldsymbol{d} = \frac{\overrightarrow{OD}}{|OD|}.$$

Then, the cross ratio $[A, B; C, D]$ is written as

$$[A, B; C, D] = \frac{|OA|\boldsymbol{a} \wedge |OC|\boldsymbol{c}}{|OB|\boldsymbol{b} \wedge |OC|\boldsymbol{c}} \Big/ \frac{|OA|\boldsymbol{a} \wedge |OD|\boldsymbol{d}}{|OB|\boldsymbol{b} \wedge |OD|\boldsymbol{d}} = \frac{\boldsymbol{a} \wedge \boldsymbol{c}}{\boldsymbol{b} \wedge \boldsymbol{c}} \Big/ \frac{\boldsymbol{a} \wedge \boldsymbol{d}}{\boldsymbol{b} \wedge \boldsymbol{d}}.$$

In other words, the cross ratio of four points depends only on their directions from the origin O. It follows that the cross ratio has the same value if computed on any line that intersects the four rays OA, OB, OC, and OD.

Chapter 6

6.1. By interchanging neighboring symbols after sign change, we can rewrite i^2 as follows:

$$i^2 = (e_3 \underbrace{e_2)(e_3}_{} e_2) = -\underbrace{e_3 e_3}_{1} \underbrace{e_2 e_2}_{1} = -1.$$

Similarly, we can see that $j^2 = -1$ and $k^2 = -1$. For ij and ji, we interchange neighboring symbols after sign change to obtain

$$ij = (e_3 e_2) \underbrace{(e_1 e_3)}_{} = -e_3 \underbrace{e_2 e_3}_{} e_1 = \underbrace{e_3 e_3}_{1} \underbrace{e_2 e_1}_{k} = k,$$

$$ji = (e_1 \underbrace{e_3)(e_3}_{1} e_2) = \underbrace{e_1 e_2}_{} = -\underbrace{e_2 e_1}_{k} = -k.$$

Similarly, we can see that $jk = i = -kj$ and $ki = j = -jk$.

$$\boldsymbol{b}\wedge\boldsymbol{a} = \frac{\boldsymbol{ba}-\boldsymbol{ab}}{2} = \boldsymbol{ba} - \frac{\boldsymbol{ab}+\boldsymbol{ba}}{2} = \boldsymbol{ba} - \langle\boldsymbol{a},\boldsymbol{b}\rangle.$$

From these, we obtain

$$\begin{aligned}(\boldsymbol{a}\wedge\boldsymbol{b})(\boldsymbol{b}\wedge\boldsymbol{a}) &= (\boldsymbol{ab}-\langle\boldsymbol{a},\boldsymbol{b}\rangle)(\boldsymbol{ba}-\langle\boldsymbol{a},\boldsymbol{b}\rangle) = \boldsymbol{ab}^2\boldsymbol{a} - \langle\boldsymbol{a},\boldsymbol{b}\rangle\boldsymbol{ab} - \langle\boldsymbol{a},\boldsymbol{b}\rangle\boldsymbol{ba} + \langle\boldsymbol{a},\boldsymbol{b}\rangle\\ &= \|\boldsymbol{b}\|^2\boldsymbol{a}^2 - \langle\boldsymbol{a},\boldsymbol{b}\rangle(\boldsymbol{ab}+\boldsymbol{ba}) + \langle\boldsymbol{a},\boldsymbol{b}\rangle^2 = \|\boldsymbol{a}\|^2\|\boldsymbol{b}\|^2 - 2\langle\boldsymbol{a},\boldsymbol{b}\rangle^2 + \langle\boldsymbol{a},\boldsymbol{b}\rangle^2\\ &= \|\boldsymbol{a}\|^2\|\boldsymbol{b}\|^2 - \langle\boldsymbol{a},\boldsymbol{b}\rangle^2 = \|\boldsymbol{a}\wedge\boldsymbol{b}\|^2.\end{aligned}$$

6.3. Using Eq. (6.30), we can express $\boldsymbol{x}$ as

$$\begin{aligned}\boldsymbol{x} &= \boldsymbol{x}(\boldsymbol{a}\wedge\boldsymbol{b})(\boldsymbol{a}\wedge\boldsymbol{b})^{-1} = (\boldsymbol{x}\cdot\boldsymbol{a}\wedge\boldsymbol{b}+\boldsymbol{x}\wedge\boldsymbol{a}\wedge\boldsymbol{b})(\boldsymbol{a}\wedge\boldsymbol{b})^{-1}\\ &= (\boldsymbol{x}\cdot\boldsymbol{a}\wedge\boldsymbol{b})(\boldsymbol{a}\wedge\boldsymbol{b})^{-1} + \boldsymbol{x}\wedge\boldsymbol{a}\wedge\boldsymbol{b}(\boldsymbol{a}\wedge\boldsymbol{b})^{-1}.\end{aligned}$$

The first term is parallel to the plane $\boldsymbol{a}\wedge\boldsymbol{b}$, and the second term is orthogonal to i This is shown as follows. If $\boldsymbol{x}$ is orthogonal to the plane $\boldsymbol{a}\wedge\boldsymbol{b}$, we see from Eq. (5.3 in Chapter 5 that

$$\boldsymbol{x}\cdot\boldsymbol{a}\wedge\boldsymbol{b} = \langle\boldsymbol{x},\boldsymbol{a}\rangle\boldsymbol{b} - \langle\boldsymbol{x},\boldsymbol{b}\rangle\boldsymbol{a} = 0.$$

Hence, the first term vanishes and the second term is $\boldsymbol{x}$ itself. If, on the other han $\boldsymbol{x}$ is included in the plane $\boldsymbol{a}\wedge\boldsymbol{b}$, we have $\boldsymbol{x}\wedge\boldsymbol{a}\wedge\boldsymbol{b} = 0$ by the property of the out product. Hence, the first term is $\boldsymbol{x}$. Thus, we conclude that

$$\boldsymbol{x}_{\|} = (\boldsymbol{x}\cdot\boldsymbol{a}\wedge\boldsymbol{b})(\boldsymbol{a}\wedge\boldsymbol{b})^{-1}, \qquad \boldsymbol{x}_{\perp} = \boldsymbol{x}\wedge\boldsymbol{a}\wedge\boldsymbol{b}(\boldsymbol{a}\wedge\boldsymbol{b})^{-1}.$$

The reflection $\boldsymbol{x}_{\top}$ of $\boldsymbol{x}$ with respect to the plane $\boldsymbol{a}\wedge\boldsymbol{b}$ is obtained by subtracting fro $\boldsymbol{x}$ twice the rejection $\boldsymbol{x}\wedge\boldsymbol{a}\wedge\boldsymbol{b}(\boldsymbol{a}\wedge\boldsymbol{b})^{-1}$. From

$$(\boldsymbol{b}\wedge\boldsymbol{c})\boldsymbol{a} = -\boldsymbol{a}\cdot\boldsymbol{b}\wedge\boldsymbol{c} + \boldsymbol{a}\wedge\boldsymbol{b}\wedge\boldsymbol{c},$$

which is obtained in the same way as Eq. (6.30), we can write $\boldsymbol{x}_{\top}$ in the form

$$\begin{aligned}\boldsymbol{x}_{\top} &= \boldsymbol{x} - 2\boldsymbol{x}\wedge\boldsymbol{a}\wedge\boldsymbol{b}(\boldsymbol{a}\wedge\boldsymbol{b})^{-1} = (\boldsymbol{x}\cdot\boldsymbol{a}\wedge\boldsymbol{b})(\boldsymbol{a}\wedge\boldsymbol{b})^{-1} - \boldsymbol{x}\wedge\boldsymbol{a}\wedge\boldsymbol{b}(\boldsymbol{a}\wedge\boldsymbol{b})^{-1}\\ &= (\boldsymbol{x}\cdot\boldsymbol{a}\wedge\boldsymbol{b} - \boldsymbol{x}\wedge\boldsymbol{a}\wedge\boldsymbol{b})(\boldsymbol{a}\wedge\boldsymbol{b})^{-1} = -(-\boldsymbol{x}\cdot\boldsymbol{a}\wedge\boldsymbol{b}+\boldsymbol{x}\wedge\boldsymbol{a}\wedge\boldsymbol{b})(\boldsymbol{a}\wedge\boldsymbol{b})^{-1}\\ &= -(\boldsymbol{a}\wedge\boldsymbol{b})\boldsymbol{x}(\boldsymbol{a}\wedge\boldsymbol{b})^{-1}.\end{aligned}$$

6.4. The surface element of the plane spanned by vectors $\boldsymbol{a}$ and $\boldsymbol{b}$ is given by Eq. (6.68 Hence, we obtain the following identities:

$$\mathcal{I}\boldsymbol{a} = \frac{\boldsymbol{aba}-\boldsymbol{baa}}{2\|\boldsymbol{a}\|\|\boldsymbol{b}\|\sin\theta} = \frac{\boldsymbol{aba}-\|\boldsymbol{a}\|^2\boldsymbol{b}}{2\|\boldsymbol{a}\|\|\boldsymbol{b}\|\sin\theta}, \qquad \boldsymbol{a}\mathcal{I} = \frac{\boldsymbol{aab}-\boldsymbol{aba}}{2\|\boldsymbol{a}\|\|\boldsymbol{b}\|\sin\theta} = \frac{\|\boldsymbol{a}\|^2\boldsymbol{b}-\boldsymbol{aba}}{2\|\boldsymbol{a}\|\|\boldsymbol{b}\|\sin\theta},$$

$$\mathcal{I}\boldsymbol{b} = \frac{\boldsymbol{abb}-\boldsymbol{bab}}{2\|\boldsymbol{a}\|\|\boldsymbol{b}\|\sin\theta} = \frac{\|\boldsymbol{b}\|^2\boldsymbol{a}-\boldsymbol{bab}}{2\|\boldsymbol{a}\|\|\boldsymbol{b}\|\sin\theta}, \qquad \boldsymbol{b}\mathcal{I} = \frac{\boldsymbol{bab}-\boldsymbol{bba}}{2\|\boldsymbol{a}\|\|\boldsymbol{b}\|\sin\theta} = \frac{\boldsymbol{bab}-\|\boldsymbol{b}\|^2\boldsymbol{a}}{2\|\boldsymbol{a}\|\|\boldsymbol{b}\|\sin\theta}.$$

Hence,

$$\mathcal{I}\boldsymbol{a} = -\boldsymbol{a}\mathcal{I}, \qquad \mathcal{I}\boldsymbol{b} = -\boldsymbol{b}\mathcal{I}.$$

In other words, $\boldsymbol{a}$ and $\boldsymbol{b}$ are anticommutative with $\mathcal{I}$. Since any vector $\boldsymbol{u}$ on this plar is written in the form of $\alpha\boldsymbol{a}+\beta\boldsymbol{b}$, it satisfies $\mathcal{I}\boldsymbol{u} = -\boldsymbol{u}\mathcal{I}$. Hence, any vector on th plane specified by the surface element $\mathcal{I}$ is anticommutative with it.

$$
\begin{aligned}
u' = RuR^\dagger &= \Big(\cos\frac{\Omega}{2} - \mathcal{I}\sin\frac{\Omega}{2}\Big)u\Big(\cos\frac{\Omega}{2} + \mathcal{I}\sin\frac{\Omega}{2}\Big) \\
&= u\cos^2\frac{\Omega}{2} + (u\mathcal{I} - \mathcal{I}u)\cos\frac{\Omega}{2}\sin\frac{\Omega}{2} - \mathcal{I}u\mathcal{I}\sin^2\frac{\Omega}{2} \\
&= u\cos^2\frac{\Omega}{2} + 2u\mathcal{I}\cos\frac{\Omega}{2}\sin\frac{\Omega}{2} - u\sin^2\frac{\Omega}{2} \\
&= u\Big(\cos^2\frac{\Omega}{2} - \sin^2\frac{\Omega}{2}\Big) + 2u\mathcal{I}\cos\frac{\Omega}{2}\sin\frac{\Omega}{2} = u\cos\Omega + u\mathcal{I}\sin\Omega,
\end{aligned}
$$

where we have noted that u is on this plane and hence is anticommutative with $\mathcal{I}$ and that $\mathcal{I}^2 = -1$. Similarly, we obtain

$$
v' = v\cos\Omega + v\mathcal{I}\sin\Omega.
$$

Since u and v are mutually orthogonal unit vectors, we can write the surface element $\mathcal{I}$ as

$$
\mathcal{I} = u \wedge v
$$

from the definition of the relative orientation of u and v. From the orthogonality of u and v, we can write from (6.29)

$$
uv = u \wedge v.
$$

Hence,

$$
u\mathcal{I} = u(u\wedge v) = uuv = \|u\|^2 v = v.
$$

Due to the orthogonality of u and v, they are anticommutative with each other, so

$$
v\mathcal{I} = v(u\wedge v) = vuv = -vvu = -\|v\|^2 u = -u.
$$

Hence, we obtain

$$
u' = u\cos\Omega + v\sin\Omega, \qquad v' = -u\sin\Omega + v\cos\Omega.
$$

6.6. From Eq. (6.22), we can see that

$$
\begin{aligned}
&-(nan^{-1})\wedge(nbn^{-1})\wedge(ncn^{-1}) \\
&= -\frac{1}{6}\Big((nan^{-1})(nbn^{-1})(ncn^{-1}) + (nbn^{-1})(ncn^{-1})(nan^{-1}) \\
&\quad +(ncn^{-1})(nan^{-1})(nbn^{-1}) - (ncn^{-1})(nbn^{-1})(nan^{-1}) \\
&\quad -(nbn^{-1})(nan^{-1})(ncn^{-1}) - (nan^{-1})(ncn^{-1})(nbn^{-1})\Big) \\
&= -\frac{1}{6}n(abc + bca + cab - cba - bac - acb)n^{-1} = -n(a\wedge b\wedge c)n^{-1}.
\end{aligned}
$$

Chapter 7

7.1. The line passing through two points x_2 and x_3 is given by $p_2 \wedge p_3$, and the equation of this line is $p \wedge (p_2 \wedge p_3) = 0$. Hence, point x_1 is on this line if and only if $p_1 \wedge (p_2 \wedge p_3) = 0$.

and only if $p_1 \wedge (p_2 \wedge p_3 \wedge p_4) = 0$.

7.3. If we let $x = x_0e_0 + x_1e_1 + x_2e_2 + x_3e_3$ and $y = y_0e_0 + y_1e_1 + y_2e_2 + y_3e_3$, their out product is

$$x \wedge y = (x_0y_1 - x_1y_0)e_0 \wedge e_1 + (x_0y_2 - x_2y_0)e_0 \wedge e_2 + (x_0y_3 - x_3y_0)e_0 \wedge e_3$$
$$+(x_2y_3 - x_3y_2)e_2 \wedge e_3 + (x_3y_1 - x_1y_3)e_3 \wedge e_1 + (x_2y_3 - x_3y_2)e_2 \wedge e_3.$$

Hence, factorizing L is equivalent to finding x_0, x_1, x_2, x_3, y_0, y_1, y_2, and y_3 such th

$$m_1 = x_0y_1 - x_1y_0, \qquad m_2 = x_0y_2 - x_2y_0, \qquad m_3 = x_0y_3 - x_3y_0,$$

$$n_1 = x_2y_3 - x_3y_2, \qquad n_2 = x_3y_1 - x_1y_3, \qquad n_3 = x_2y_3 - x_3y_2,$$

for given m_1, m_2, m_3, n_1, n_2, and n_3. If we let

$$\boldsymbol{m} = m_1e_1 + m_2e_2 + m_3e_3, \qquad \boldsymbol{n} = n_1e_1 + n_2e_2 + n_3e_3,$$

$$\boldsymbol{x} = x_1e_1 + x_2e_2 + x_3e_3, \qquad \boldsymbol{y} = y_1e_1 + y_2e_2 + y_3e_3,$$

this is equivalent to finding x_0, y_0, $\boldsymbol{x}$, and $\boldsymbol{y}$ such that

$$\boldsymbol{m} = x_0\boldsymbol{y} - y_0\boldsymbol{x}, \qquad \boldsymbol{n} = \boldsymbol{x} \times \boldsymbol{y},$$

for given $\boldsymbol{m}$ and $\boldsymbol{n}$. Evidently, $\langle \boldsymbol{m}, \boldsymbol{n} \rangle = 0$ holds if such x_0, y_0, $\boldsymbol{x}$, and $\boldsymbol{y}$ exist. Co versely, suppose $\langle \boldsymbol{m}, \boldsymbol{n} \rangle = 0$. We can choose two vectors $\boldsymbol{x}$ and $\boldsymbol{y}$ that are orthogon to $\boldsymbol{n}$ such that $\boldsymbol{n} = \boldsymbol{x} \times \boldsymbol{y}$. Since $\boldsymbol{m}$ is orthogonal to $\boldsymbol{n}$, we can express $\boldsymbol{m}$ as a linea combination of such $\boldsymbol{x}$ and $\boldsymbol{y}$ in the form $\boldsymbol{m} = \alpha\boldsymbol{x} + \beta\boldsymbol{y}$. Then, we can choose x_0 an y_0 to be $x_0 = -\alpha$ and $y_0 = \beta$. Hence, L is factorized if and only if $\langle \boldsymbol{m}, \boldsymbol{n} \rangle = 0$.

7.4. Let $\boldsymbol{n} = n_1e_1 + n_2e_2 + n_3e_3$. Choose two vectors $\boldsymbol{a}$ and $\boldsymbol{b}$ that are orthogonal to such that

$$\boldsymbol{n} = \boldsymbol{a} \times \boldsymbol{b}.$$

Define vectors $\boldsymbol{x} = x_1e_1+x_2e_2+x_3e_3$, $\boldsymbol{y} = y_1e_1+y_2e_2+y_3e_3$, and $\boldsymbol{z} = z_1e_1+z_2e_2+z_3e$ as follows:

$$\boldsymbol{x} = \frac{h}{\|\boldsymbol{n}\|}\boldsymbol{n}, \qquad \boldsymbol{y} = \boldsymbol{x} + \boldsymbol{a}, \qquad \boldsymbol{z} = \boldsymbol{x} + \boldsymbol{b}.$$

Then, we see that

$$\boldsymbol{z} \wedge \boldsymbol{x} = \boldsymbol{b} \wedge \boldsymbol{x}, \qquad \boldsymbol{x} \wedge \boldsymbol{y} = \boldsymbol{x} \wedge \boldsymbol{a},$$

$$\boldsymbol{y} \wedge \boldsymbol{z} = \boldsymbol{x} \wedge \boldsymbol{b} + \boldsymbol{a} \wedge \boldsymbol{x} + \boldsymbol{a} \wedge \boldsymbol{b} = \boldsymbol{x} \wedge (\boldsymbol{b} - \boldsymbol{a}) + \boldsymbol{a} \wedge \boldsymbol{b}.$$

Hence, the following holds:

$$\boldsymbol{y} \wedge \boldsymbol{z} + \boldsymbol{z} \wedge \boldsymbol{x} + \boldsymbol{y} \wedge \boldsymbol{z} = \boldsymbol{a} \wedge \boldsymbol{b} = n_1e_2 \wedge e_3 + n_2e_3 \wedge e_1 + n_3e_1 \wedge e_2.$$

We also obtain the following:

$$\boldsymbol{x} \wedge \boldsymbol{y} \wedge \boldsymbol{z} = \boldsymbol{x} \wedge (\boldsymbol{x} + \boldsymbol{a}) \wedge (\boldsymbol{x} + \boldsymbol{b}) = \boldsymbol{x} \wedge \boldsymbol{a} \wedge \boldsymbol{b} = |\boldsymbol{x}, \boldsymbol{a}, \boldsymbol{b}| e_1 \wedge e_2 \wedge e_3$$
$$= \langle \frac{h}{\|\boldsymbol{n}\|}\boldsymbol{n}, \boldsymbol{a} \times \boldsymbol{b} \rangle e_1 \wedge e_2 \wedge e_3 = \langle \frac{h}{\|\boldsymbol{n}\|}\boldsymbol{n}, \boldsymbol{n} \rangle e_1 \wedge e_2 \wedge e_3 = he_1 \wedge e_2 \wedge e_3.$$

$$
\begin{aligned}
x \wedge y \wedge z &= (e_0 + \boldsymbol{x}) \wedge (e_0 + \boldsymbol{y}) \wedge (e_0 + \boldsymbol{z}) \\
&= e_0 \wedge (\boldsymbol{y} \wedge \boldsymbol{z} + \boldsymbol{z} \wedge \boldsymbol{x} + \boldsymbol{x} \wedge \boldsymbol{y}) + \boldsymbol{x} \wedge \boldsymbol{y} \wedge \boldsymbol{z} \\
&= n_1 e_0 \wedge e_2 \wedge e_3 + n_2 e_0 \wedge e_3 \wedge e_1 + n_3 e_0 \wedge e_1 \wedge e_2 + h e_1 \wedge e_2 \wedge e_3 = \Pi.
\end{aligned}
$$

Thus, an arbitrary trivector Π is always factorized.

7.5. (1) The following holds:

$$
\begin{aligned}
& L \wedge L' \\
&= (m_1 e_0 \wedge e_1 + m_2 e_0 \wedge e_2 + m_3 e_0 \wedge e_3 + n_1 e_2 \wedge e_3 + n_2 e_3 \wedge e_1 + n_3 e_1 \wedge e_2) \\
&\quad \wedge (m_1' e_0 \wedge e_1 + m_2' e_0 \wedge e_2 + m_3' e_0 \wedge e_3 + n_1' e_2 \wedge e_3 + n_2' e_3 \wedge e_1 + n_3' e_1 \wedge e_2) \\
&= m_1 n_1' e_0 \wedge e_1 \wedge e_2 \wedge e_3 + m_2 n_2' e_0 \wedge e_2 \wedge e_3 \wedge e_1 + m_3 n_3' e_0 \wedge e_3 \wedge e_1 \wedge e_2 \\
&\quad + n_1 m_1' e_2 \wedge e_3 \wedge e_0 \wedge e_1 + n_2 m_2' e_3 \wedge e_1 \wedge e_0 \wedge e_2 + n_3 m_3' e_1 \wedge e_2 \wedge e_0 \wedge e_3 \\
&= (m_1 n_1' + m_2 n_2' + m_3 n_3' + n_1 m_1' + n_2 m_2' + n_3 m_3') e_0 \wedge e_1 \wedge e_2 \wedge e_3.
\end{aligned}
$$

(2) We can write L as $L = p_1 \wedge p_2$ if the line L is defined by two points p_1 and p_2. Similarly, we can write L' as $L' = p_3 \wedge p_4$ if the line L' is defined by two points p_3 and p_4. Then, lines L and L' are coplanar if and only if such four points p_1, p_2, p_3, and p_4 are coplanar. This condition is written as $p_1 \wedge p_2 \wedge p_3 \wedge p_4 = 0$, or $L \wedge L' = 0$. From the above (1), this is equivalent to $\langle \boldsymbol{m}, \boldsymbol{n}' \rangle + \langle \boldsymbol{n}, \boldsymbol{m}' \rangle = 0$.

7.6. (1) If we let $L = p_2 \cup p_3$, Eq. (7.79) is written as $\Pi = p_1 \cup L$. From Proposition 7.4, its dual is $\Pi^* = p_1^* \cap L^*$. Since $L^* = p_2^* \cap p_3^*$ by Proposition 7.5, we can write it as $\Pi^* = p_1^* \cap p_2^* \cap p_3^*$. Hence, Eq. (7.81) holds.

(2) If we let $L = \Pi_1 \cap \Pi_2$, Eq. (7.80) is written as $p = L \cap \Pi_3$. From Proposition 7.4, its dual is $p^* = L^* \cup \Pi_3^*$. Since $L^* = \Pi_1^* \cup \Pi_2^*$ by Proposition 7.5, we can write it as $p^* = \Pi_1^* \cup \Pi_2^* \cup \Pi_3^*$. Hence, Eq. (7.82) holds.

Chapter 8

8.1. As shown by Eq. (8.9), all points p satisfy

$$
\|p\|^2 = x_1^2 + x_2^2 + x_3^2 - 2x_\infty = 0,
$$

and $x_0 = 1$. This means that all points p are embedded in a hypersurface

$$
x_\infty = \frac{1}{2}(x_1^2 + x_2^2 + x_3^2),
$$

and in a hyperplane $x_0 = 1$ in 5D, i.e., in a 3D parabolic surface defined as their intersection.

8.2. (1) An element x has the norm

$$
\begin{aligned}
\|x\|^2 &= \langle x_1 e_1 + x_2 e_2 + x_3 e_3 + x_4 e_4 + x_5 e_5, x_1 e_1 \mid x_2 e_2 + x_3 e_3 + x_4 e_4 + x_5 e_5 \rangle \\
&= x_1^2 + x_2^2 + x_3^2 + x_4^2 - x_5^2.
\end{aligned}
$$

Hence, all elements x such that $\|x\|^2 = 0$ form a hypercone around axis e_5 given by

$$
x_5 = \pm\sqrt{x_1^2 + x_2^2 + x_3^2 + x_4^2}.
$$

Hence, all elements x in $\mathbb{R}^{4,1}$ have the following form:

$$\begin{aligned} x &= x_1e_1 + x_2e_2 + x_3e_3 + x_4(e_0 - \frac{1}{2}e_\infty) + x_5(e_0 + \frac{1}{2}e_\infty) \\ &= (x_4 + x_5)e_0 + x_1e_1 + x_2e_2 + x_3e_3 + \frac{1}{2}(x_5 - x_4)e_0. \end{aligned}$$

If we let

$$x_0 = x_4 + x_5, \qquad x_\infty = \frac{1}{2}(x_5 - x_4),$$

all elements x are written in the form of Eq. (8.1). Since x_4 and x_5 are express
in terms of x_0 and x_∞ as

$$x_4 = \frac{1}{2}x_0 - x_\infty, \qquad x_5 = \frac{1}{2}x_0 + x_\infty,$$

the norm $\|x\|^2$ can be written as follows:

$$\begin{aligned} \|x\|^2 &= x_1^2 + x_2^2 + x_3^2 + x_4^2 - x_5^2 = x_1^2 + x_2^2 + x_3^2 + \Big(\frac{1}{2}x_0 - x_\infty\Big)^2 - \Big(\frac{1}{2}x_0 + x_\infty \\ &= x_1^2 + x_2^2 + x_3^2 - 2x_0x_\infty. \end{aligned}$$

This coincides with the definition of the 5D conformal space.

(3) Since $x_0 = 1$ is equivalently written as $x_4 + x_5 = 1$, the conformal space can
regarded as embedding $\mathbb{R}^3$ in the 3D parabolic surface defined as the intersecti
of the hypercone

$$x_5 = \pm\sqrt{x_1^2 + x_2^2 + x_3^2 + x_4^2},$$

and the hyperplane $x_4 + x_5 = 1$ in $\mathbb{R}^{4,1}$.

(4) By definition, the geometric products among e_i, $i = 1, 2, 3$, are given by

$$e_i^2 = 1, \qquad e_ie_j + e_je_i = 0.$$

Next, the geometric products of e_i, $i = 1, 2, 3$, with e_0 and e_∞ are given by

$$\begin{aligned} e_ie_0 + e_0e_i &= e_i\Big(\frac{e_4 + e_5}{2}\Big) + \Big(\frac{e_4 + e_5}{2}\Big)e_i = \frac{1}{2}(e_ie_4 + e_ie_5 + e_4e_i + e_5e_i) = 0, \\ e_ie_\infty + e_\infty e_i &= e_i(e_5 - e_4) + (e_5 - e_4)e_i = e_ie_5 - e_ie_4 + e_5e_i - e_4e_i = 0. \end{aligned}$$

Finally, the geometric products involving e_0 and e_∞ are given by

$$\begin{aligned} e_0^2 &= \Big(\frac{e_4 + e_5}{2}\Big)^2 = \frac{e_4^2 + e_4e_5 + e_5e_4 + e_5^2}{4} = \frac{1 - 1}{4} = 0, \\ e_\infty^2 &= (e_5 - e_4)^2 = e_5^2 - e_5e_4 - e_4e_5 + e_4^2 = -1 + 1 = 0, \\ e_0e_\infty + e_\infty e_0 &= \Big(\frac{e_4 + e_5}{2}\Big)(e_5 - e_4) + (e_5 - e_4)\Big(\frac{e_4 + e_5}{2}\Big) \\ &= \frac{1}{2}(e_4e_5 - e_4^2 + e_5^2 - e_5e_4) + \frac{1}{2}(e_5e_4 + e_5^2 - e_4^2 - e_4e_5) \\ &= \frac{1}{2}(-1 - 1) + \frac{1}{2}(-1 - 1) = -2. \end{aligned}$$

These agree with the rule of Eqs. (8.50) and (8.51).

of p and σ is given by

$$\langle \sigma, p \rangle = \langle c - \frac{r^2}{2}e_\infty, p \rangle = \langle c, p \rangle - \frac{r^2}{2}\langle e_\infty, p \rangle$$
$$= -\frac{1}{2}\|\boldsymbol{c} - \boldsymbol{x}\|^2 - \frac{r^2}{2}\langle e_\infty, e_0 + \boldsymbol{x} + \frac{\|\boldsymbol{x}\|^2}{2}e_\infty \rangle = -\frac{1}{2}\|\boldsymbol{c} - \boldsymbol{x}\|^2 + \frac{r^2}{2}.$$

Since the tangential distance is

$$t(p, \sigma) = \sqrt{\|\boldsymbol{c} - \boldsymbol{x}\|^2 - r^2},$$

we see that $\langle \sigma, p \rangle = -t(p, \sigma)^2/2$. Hence, point p is on sphere σ if and only if $\langle \sigma, p \rangle = 0$.

(2) The inner product of σ and σ' is given as follows:

$$\langle \sigma, \sigma' \rangle = \langle c - \frac{r^2}{2}e_\infty, c' - \frac{r'^2}{2}e_\infty \rangle = \langle c, c' \rangle - \frac{r^2}{2}\langle e_\infty, c' \rangle - \frac{r'^2}{2}\langle c, e_\infty \rangle$$
$$= -\frac{1}{2}\|\boldsymbol{c} - \boldsymbol{c}'\|^2 + \frac{r^2}{2} + \frac{r'^2}{2}.$$

As we see from Fig. 8.11(b), the angle θ satisfies the law of cosines

$$\|\boldsymbol{c} - \boldsymbol{c}'\|^2 = r^2 + r'^2 - 2rr' \cos\theta.$$

Hence, the inner product $\langle \sigma, \sigma' \rangle$ is written as

$$\langle \sigma, \sigma' \rangle = -\frac{1}{2}(r^2 + r'^2 - 2rr' \cos\theta) + \frac{r^2 + r'^2}{2} - rr' \cos\theta.$$

It follows that σ and σ' are orthogonal if and only if $\langle \sigma, \sigma' \rangle = 0$.

8.4. (1) From Eq. (8.58), we obtain

$$\mathcal{R}\mathcal{T}_t\mathcal{R}^{-1} = \mathcal{R}(1 - \frac{1}{2}\boldsymbol{t}e_\infty)\mathcal{R}^{-1} = \mathcal{R}\mathcal{R}^{-1} - \frac{1}{2}\mathcal{R}\boldsymbol{t}e_\infty\mathcal{R}^{-1} = 1 - \frac{1}{2}\mathcal{R}\boldsymbol{t}\mathcal{R}^{-1}\mathcal{R}e_\infty\mathcal{R}^{-1}$$
$$= 1 - \frac{1}{2}\mathcal{R}\boldsymbol{t}\mathcal{R}^{-1}e_\infty = \mathcal{T}_{\mathcal{R}\boldsymbol{t}\mathcal{R}^{-1}},$$

where we have noted that the infinity e_∞ is invariant to rotation around the origin and that $\mathcal{R}e_\infty\mathcal{R}^{-1} = e_\infty$ holds.

(2) If point p is rotated around the origin, it moves to $\mathcal{R}p\mathcal{R}^{-1}$. To compute the same rotation around the position $\boldsymbol{t}$, we first translate p by $-\boldsymbol{t}$, then rotate it around the origin, and finally traslate it by $\boldsymbol{t}$. This composition is given by

$$\mathcal{T}_t(\mathcal{R}(\mathcal{T}_{-t}p\mathcal{T}_{-t}^{-1})\mathcal{R}^{-1})\mathcal{T}_t^{-1} = (\mathcal{T}_t\mathcal{R}\mathcal{T}_t^{-1})p(\mathcal{T}_t\mathcal{R}\mathcal{T}_t^{-1})^{-1}.$$

(3) We see that the following holds:

$$\mathcal{T}_t\mathcal{R}\mathcal{T}_t^{-1} = \mathcal{T}_t\mathcal{R}\mathcal{T}_{-t}\mathcal{R}^{-1}\mathcal{R} = \mathcal{T}_t(\mathcal{R}\mathcal{T}_{-t}\mathcal{R}^{-1})\mathcal{R} = \mathcal{T}_t\mathcal{T}_{-\mathcal{R}t\mathcal{R}^{-1}}\mathcal{R} = \mathcal{T}_{t-\mathcal{R}t\mathcal{R}^{-1}}\mathcal{R}.$$

inversion of e_∞ is given by

$$\sigma e_\infty \sigma^\dagger = -\sigma e_\infty \sigma^{-1} = -\frac{\sigma e_\infty \sigma}{r^2}.$$

From Eq. (8.98), we can infer that this is equal to $2/r^2$ times the center c of the sphe σ. Hence, the center c is written as

$$c = -\frac{1}{2}\sigma e_\infty \sigma.$$

8.6. Equation (8.104) is rewritten as

$$\mathcal{O} = \frac{1}{2}(e_0 e_\infty - e_\infty e_0).$$

Noting that $e_0^2 = e_\infty^2 = 0$ and $e_0 e_\infty + e_\infty e_0 = -2$, we obtain the following:

(1) We see that

$$\begin{aligned}\mathcal{O}^2 &= \Big(\frac{e_0 e_\infty - e_\infty e_0}{2}\Big)^2 = \frac{1}{4}\Big((e_0 e_\infty + e_\infty e_0)^2 - 4e_0 e_\infty e_\infty e_0 - 4e_\infty e_0 e_0 e_\infty\Big) \\ &= \frac{1}{4}\Big((-2)^2 - 4e_0 e_\infty^2 e_0 - 4e_\infty e_0^2 e_\infty\Big) = 1.\end{aligned}$$

(2) The following hold:

$$\mathcal{O} e_0 = \frac{1}{2}(e_0 e_\infty - e_\infty e_0)e_0 = \frac{1}{2}(-2 - e_\infty e_0 - e_\infty e_0)e_0 = \frac{1}{2}(-2e_0) = -e_0,$$

$$e_0\mathcal{O} = \frac{1}{2}e_0(e_0 e_\infty - e_\infty e_0) = \frac{1}{2}e_0(e_0 e_\infty + 2 + e_0 e_\infty) = \frac{1}{2}(2e_0) = e_0,$$

$$\mathcal{O} e_\infty = \frac{1}{2}(e_0 e_\infty - e_\infty e_0)e_\infty = \frac{1}{2}(e_0 e_\infty + 2 + e_0 e_\infty)e_\infty = \frac{1}{2}(2e_\infty) = e_\infty,$$

$$e_\infty\mathcal{O} = \frac{1}{2}e_\infty(e_0 e_\infty - e_\infty e_0) = \frac{1}{2}e_\infty(-2 - e_\infty e_0 - e_\infty e_0) = \frac{1}{2}e_\infty(-2e_\infty) = -e_\infty$$

From these, we obtain Eq. (8.112).

Chapter 9

9.1. (1) Point (X, Y, Z) is on the plane $z = Z$, which is at distane $1 + Z$ from the sout pole $(0, 0, -1)$. This plane is magnified by $1/(1 + Z)$ if projected onto the x plane at distance 1 from the south pole. Hence,

$$x = \frac{X}{1+Z}, \qquad y = \frac{Y}{1+Z}.$$

If follows that

$$x^2 + y^2 = \frac{X^2 + Y^2}{(1+Z)^2} = \frac{1 - Z^2}{(1+Z)^2} = \frac{1-Z}{1+Z},$$

which can be solved in terms of Z in the form

$$Z = \frac{1 - x^2 - y^2}{1 + x^2 + y^2} = \frac{2}{1 + x^2 + y^2} - 1.$$

$$X = (1+Z)x = \frac{\quad}{1+x^2+y^2}, \qquad Y = (1+Z)y = \frac{\quad}{1+x^2+y^2}.$$

(2) The distance of point (X, Y, Z) from the south pole $(0, 0, -1)$ is

$$\begin{aligned}\sqrt{X^2+Y^2+(1+Z)^2} &= \sqrt{\frac{4x^2}{(1+x^2+y^2)^2} + \frac{4y^2}{(1+x^2+y^2)^2} + \frac{4}{(1+x^2+y^2)^2}} \\ &= \frac{2}{\sqrt{1+x^2+y^2}}.\end{aligned}$$

The distance of point (x, y) on the xy plane from $(0, 0, -1)$ is $\sqrt{x^2+y^2+1}$. Since the product of these two distances is 2, the mapping between (X, Y, Z) and (x, y) is an inversion with respect to a sphere of radius $\sqrt{2}$.

9.2. If the incoming ray through (X, Y, Z) toward the lens center at the origin has incidence angle θ, we have

$$Z = \cos\theta.$$

Noting that the half angle formula for tangent is

$$\tan\frac{\theta}{2} = \sqrt{\frac{1-\cos\theta}{1+\cos\theta}},$$

we can write the distance d of the imaged point (x, y) from the origin as follows:

$$d = 2f\tan\frac{\theta}{2} = 2f\sqrt{\frac{1-\cos\theta}{1+\cos\theta}} = 2f\sqrt{\frac{1-Z}{1+Z}} = 2f\sqrt{\frac{1-Z^2}{(1+Z)^2}} = 2f\frac{\sqrt{1-Z^2}}{1+Z}.$$

From this, we see that the point (x, y) is at $(d\cos\phi, d\sin\phi)$, where ϕ is the direction angle from the x-axis. Since this is equal to the direction angle of (X, Y), we obtain

$$\cos\phi = \frac{X}{\sqrt{X^2+Y^2}} = \frac{X}{\sqrt{1-Z^2}}, \qquad \sin\phi = \frac{Y}{\sqrt{X^2+Y^2}} = \frac{Y}{\sqrt{1-Z^2}}.$$

Hence, x and y are expressed as follows:

$$\begin{aligned} x &= d\cos\phi = 2f\frac{\sqrt{1-Z^2}}{1+Z}\frac{X}{\sqrt{1-Z^2}} = \frac{2fX}{1+Z}, \\ y &= d\sin\phi = 2f\frac{\sqrt{1-Z^2}}{1+Z}\frac{Y}{\sqrt{1-Z^2}} = \frac{2fY}{1+Z}. \end{aligned}$$

9.3. If the incoming ray through (X, Y, Z) toward the focus located at the origin has incidence angle θ, we have $Z = \cos\theta$. Since Eq. (9.11) has the same form as Eq. (9.7), we obtain the same relationship as in the preceding problem.

9.4. (1) If a 3D point (X, Y, Z) is imaged by a perspective camera of focal length f, its image (x, y) is given by

$$x = \frac{fX}{Z}, \qquad y = \frac{fY}{Z}.$$

$$X = \frac{x}{x^2+y^2+f^2}, \qquad Y = \frac{y}{x^2+y^2+f^2}, \qquad Z = \frac{f}{x^2+y^2+f^2}.$$

Hence, a point (x, y) in the image can be regarded as a perspective image of th point (X, Y, Z) defined by this relationship.

(2) Define a buffer in which the perspective image to be created is to be store Let (x, y) be the pixel position into which the value is to be written. The poi (X, Y, Z) on the celestial sphere which is to be imaged at (x, y) by a hypothetic perspective camera of focal length f is given by the above (1). The inciden angle θ of the ray through that point is

$$\theta = \tan^{-1}\frac{\sqrt{X^2+Y^2}}{Z}.$$

This ray should be imaged by the real non-perspective camera at $(d\cos\phi, d\sin\phi$ at distance $d = d(\theta)$ from the principal point, where ϕ is the direction angle fro the x-axis. Since (X, Y) has the same direction angle, we see that

$$\cos\phi = \frac{X}{\sqrt{X^2+Y^2}}, \qquad \sin\phi = \frac{Y}{\sqrt{X^2+Y^2}}.$$

Hence, all we need is to copy the image value of the point with coordinat $(dX/\sqrt{X^2+Y^2}, dY/\sqrt{X^2+Y^2})$ in the real camera image to the pixel (x, y) the buffer. If the coordinates are not integers, we appropriately interpolate th image value from those of the surrounding pixels. Repeating this process for a pixels of the buffer, we obtain in the end an image as if taken by a perspecti camera of focal length f.

(3) If the optical axis of the assumed hypothetical perspective camera is not directe in the z-axis, we rotate the celestial sphere so that that direction moves to th z-axis. Let R be that rotation. This process goes as follows:

(i) For each pixel (x, y) of the buffer to store the perspective image to be create compute the point (X, Y, Z) on the celestial sphere as in the above (1).

(ii) Let (X', Y', Z') be the point obtained by rotating (X, Y, Z) around the len center by the inverse rotation R^{-1}.

(iii) Compute the incidence ange θ' of that point and the distance $d' = d(\theta')$ fro the principal point.

(iv) Copy the image value of the point in the real camera image with coordinat $(d'X'/\sqrt{X'^2+Y'^2}, d'Y'/\sqrt{X'^2+Y'^2})$ to the pixel (x, y) of the buffer. D appropriate interpolation if necessary.

9.5. (1) Rewrite Eq. (9.16) in the form

$$|FP|^2 = (2a - |PF'|)^2.$$

Since $|FP| = \sqrt{(x-f)^2+y^2}$ and $|PF'| = \sqrt{(x+f)^2+y^2}$, as shown i Fig. 9.18(a), the above equation is written as

$$\begin{aligned}(x-f)^2+y^2 &= (2a - \sqrt{(x+f)^2+y^2})^2 \\ &= 4a^2 - 4a\sqrt{(x+f)^2+y^2} + (x+f)^2+y^2,\end{aligned}$$

Squaring both sides, we obtain

$$a^2((x+f)^2+y^2) = f^2x^2 + 2fa^2 + a^4,$$

which is rearranged in the form

$$(a^2 - f^2)x^2 + a^2y^2 = a^2(a^2 - f^2).$$

Since $a^2 - f^2 = b^2$ from Eq. (9.16), we obtain Eq. (9.14) after some manipulations.

(2) Rewrite Eq. (9.20) in the form

$$|FP|^2 = (|PF'| \pm 2a)^2.$$

Since $|FP| = \sqrt{(x-f)^2+y^2}$ and $|PF'| = \sqrt{(x+f)^2+y^2}$, as shown in Fig. 9.19(a), the above equation is written as

$$\begin{aligned}(x-f)^2+y^2 &= (\sqrt{(x+f)^2+y^2} \pm 2a)^2 \\ &= 4a^2 \pm 4a\sqrt{(x+f)^2+y^2} + (x+f)^2+y^2,\end{aligned}$$

from which we obtain

$$\pm a\sqrt{(x+f)^2+y^2} = -fx - a^2.$$

Squaring both sides, we obtain

$$a^2((x+f)^2+y^2) = f^2x^2 + 2fa^2 + a^4,$$

which is rearranged in the form

$$(a^2 - f^2)x^2 + a^2y^2 = a^2(a^2 - f^2).$$

Since $a^2 - f^2 = -b^2$ from Eq. (9.19), we obtain Eq. (9.18) after some manipulations.

(3) Since $|HP| = x + f$ and $|PF'| = \sqrt{(x-f)^2+y^2}$, as shown in Fig. 9.19(b), we can rewrite Eq. (9.24) in the form

$$x + f = \sqrt{(x-f)^2+y^2}.$$

Squaring both sides, we obtain

$$x^2 + 2fx + f^2 = x^2 - 2fx + f^2 + y^2,$$

from which we obtain Eq. (9.23) after some manipulations.

Bibliography

[1] R. Benosman and S. Kang (Eds.), *Panoramic Vision*, Springer, New York, 2000.

[2] E. Bayro-Corrochano, *Geometric Computing for Perception Action Systems: Concepts, Algorithms and Scientific Applications*, Springer, London, 2001.

[3] E. Bayro-Corrochano, *Geometric Computing: For Wavelet Transforms, Robot Vision, Learning, Control and Action*, Springer, London, 2010.

[4] C. Doran and A. Lasenby, *Geometric Algebra for Physicists*, Cambridge University Press, Cambridge, UK, 2003.

[5] L. Dorst, D. Fortijne, and S. Mann, *Geometric Algebra for Computer Science: An Object-Oriented Approach to Geometry*, Morgan Kaufmann, Burlington, MA, 2007.

[6] L. Dorst, D. Fortijne, and S. Mann, `GAViewer`, 2007. http://www.geometricalgebra.net

[7] O. Faugeras and Q.-T. Luong, *The Geometry of Multiple Images*, MIT Press, Cambridge, MA, 2001.

[8] H. Flanders, *Differential Forms with Applications to the Physical Sciences*, Academic Press, New York, 1963.

[9] C. Geyer and K. Daniilidis, Catadioptric projective geometry, *International Journal of Computer Vision*, Vol. 45. No. 3, December 2001, pp. 223–243.

[10] R. Hartley and A. Zisserman, *Multiple View Geometry in Computer Vision*, 2nd Ed., Cambridge University Press, Cambridge, UK, 2003.

[11] D. Hestenes, A. Rockwood, and H. Li, *System for Encoding and Manipulating Models of Objects*, U.S. Patent No. 6,853,964, 2005.

[12] D. Hestenes and G. Sobczyk, *Clifford Algebra to Geometric Calculus*, D. Reidel, Dordrecht, The Netherlands, 1984.

[13] K. Kanatani, *Group-Theoretical Methods in Image Understanding*, Springer, Berlin, 1990.

[14] K. Kanatani, *Geometric Computation for Machine Vision*, Oxford University Press, Oxford, UK, 1993

[15] K. Kanatani, Calibration of ultra-wide fisheye lens cameras by eigenvalue minimization, *IEEE Transactions on Pattern Analysis and Machine Intelligence*, Vol. 35, No. 4, April 2013, pp. 813–822.

[16] C. Perwass, *Geometric Algebra with Applications in Engineering*, Springer, Berlin, 2009.

Boca Raton, FL, 1987.

[19] J. G. Semple and G. T. Kneebone, *Algebraic Projective Geometry*, Oxford Universi Press, Oxford, UK, 1952.

[20] J. G. Semple and L. Roth, *Introduction to Algebraic Geometry*, Oxford Universi Press, Oxford, UK, 1949.

[21] J. A. Schouten, *Tensor Analysis for Physicists*, 2nd ed., Oxford University Press, O ford, UK, 1954.

[22] J. A. Schouten, *Ricci-Calculus: An Introduction to Tensor Analysis and Its Geometric Applications*, 2nd ed., Springer, Berlin, 1954.

[23] J. Stolfi, *Oriented Projective Geometry: A Framework for Geometric Computation* Academic Press, San Diego, CA, 1991.

Printed and bound by CPI Group (UK) Ltd, Croydon, CR0 4YY
30/10/2024
01781017-0001